Performance-Based Specifications and Control of Concrete Durability

RILEM STATE-OF-THE-ART REPORTS

Volume 18

RILEM, The International Union of Laboratories and Experts in Construction Materials, Systems and Structures, founded in 1947, is a non-governmental scientific association whose goal is to contribute to progress in the construction sciences, techniques and industries, essentially by means of the communication it fosters between research and practice. RILEM's focus is on construction materials and their use in building and civil engineering structures, covering all phases of the building process from manufacture to use and recycling of materials. More information on RILEM and its previous publications can be found on www.RILEM.net.

The RILEM State-of-the-Art Reports (STAR) are produced by the Technical Committees. They represent one of the most important outputs that RILEM generates—high level scientific and engineering reports that provide cutting edge knowledge in a given field. The work of the TCs is one of RILEM's key functions.

Members of a TC are experts in their field and give their time freely to share their expertise. As a result, the broader scientific community benefits greatly from RILEM's activities.

RILEM's stated objective is to disseminate this information as widely as possible to the scientific community. RILEM therefore considers the STAR reports of its TCs as of highest importance, and encourages their publication whenever possible.

The information in this and similar reports are mostly pre-normative in the sense that it provides the underlying scientific fundamentals on which standards and codes of practice are based. Without such a solid scientific basis, construction practice will be less than efficient or economical.

It is RILEM's hope that this information will be of wide use to the scientific community.

More information about this series at http://www.springer.com/series/8780

Hans Beushausen · Luis Fernandez Luco
Editors

Performance-Based Specifications and Control of Concrete Durability

State-of-the-Art Report RILEM TC 230-PSC

Springer

Editors
Hans Beushausen
Department of Civil Engineering
University of Cape Town
Cape Town
South Africa

Luis Fernandez Luco
Facultad de Ingeniería
Universidad de Buenos Aires
Buenos Aires
Argentina

ISSN 2213-204X ISSN 2213-2031 (electronic)
RILEM State-of-the-Art Reports
ISBN 978-94-017-7308-9 ISBN 978-94-017-7309-6 (eBook)
DOI 10.1007/978-94-017-7309-6

Library of Congress Control Number: 2015945321

Springer Dordrecht Heidelberg New York London

Printed on acid-free paper

Springer Science+Business Media B.V. Dordrecht is part of Springer Science+Business Media
(www.springer.com)

Preface

The increasing amount of prematurely deteriorating concrete infrastructures worldwide, which is linked mainly to the effect of reinforcement corrosion, has resulted in significant efforts to develop technically sound methods for concrete durability design and specification. Performance-based approaches for concrete durability offer the advantage of providing relevant test parameters for the quantity and quality of the concrete cover, which are the main aspects to consider when designing concrete structures for prevention of reinforcement corrosion.

RILEM TC 230-PSC was established in 2008 with the main aim to provide useful guidance on suitable test methods and their application in performance-based specifications for concrete durability. The scope of the TC was limited to the following:

- In-situ durability assessment of concrete structures in view of reinforcement corrosion.
- Concrete penetrability properties such as permeability, conductivity, and sorptivity.
- Concrete cover thickness.

The committee came together for the first time in Varenna, Italy, in September 2008, with subsequent meetings held in Toulouse, Aachen, Leipzig, Amsterdam, Cape Town, Zurich, and Zagreb. In 2012, the TC organized and conducted Application Tests at BAS in Venlo, The Netherlands, during which several TC members used performance test methods to characterize the durability properties and make service life predictions for various concrete test panels. The final TC meeting in Zagreb, Croatia, in June 2014 was accompanied by the International Conference on Performance-Based Specifications and Control of Concrete Durability, which was attended by more than 150 delegates from around the world.

The main outcome of RILEM TC 230-PSC is this State-of-the-Art Report, which is divided into 12 chapters on various topics relating to performance-based specification and control of concrete durability. Each chapter had a coordinator, who was also the main author. All TC members who made contributions to the various chapters were made co-authors in alphabetical order. The final chapter layouts and contents were discussed and approved in meetings and via email correspondence.

The editors thank all TC members who have actively contributed to this report through meeting attendance, discussions, participation in the Application Testing, and direct input to the various chapters.

South Africa Hans Beushausen

Members of RILEM TC 230-PSC

A/Prof. Hans Beushausen, University of Cape Town, South Africa—*TC Chairman*
Prof. Luis Fernández Luco, University of Buenos Aires, Argentina—*TC Secretary*
Dr. Karim Ait-Mokhtar, University of La Rochelle, France
Prof. Mark Alexander, University of Cape Town, South Africa
Prof. Carmen Andrade, Eduardo Torroja Institute (IETcc-CSIC), Spain
Prof. Yunus Ballim, University of the Witwatersrand, South Africa
Dr. Veronique Baroghel-Bouny, Laboratoire Central des Ponts et Chaussées (LCPC), France
Prof. P.A. Muhammed Basheer, University of Leeds, UK
Dr. Siham Kamali Bernard, Institut National des Sciences Appliquées (INSA), France
Prof. Dubravka Bjegović, Civil Engineering Institute Zagreb, Croatia
Mr. David Corbett, Proceq SA, Switzerland
Dr. Renata d'Andréa, Eduardo Torroja Institute (IETcc-CSIC), Spain
Prof. Rui Miguel Ferreira, University of Minho, Portugal
Prof. Christoph Gehlen, University of Munich, Germany
Prof. Arlindo Gonçalves, Laboratório Nacional de Engenharia Civil—LNEC, Portugal
Prof. Christian Grosse, University of Munich, Germany
Mr. Joost Gulikers, Rijkswaterstaat Bouwdienst, The Netherlands
Prof. R. Doug Hooton, University of Toronto, Canada
Prof. Kei-ichi Imamoto, Tokyo University of Science, Japan
Dr. Mohamed A. Ismail, Universiti Teknologi Malaysia (UTM), Malaysia
Dr. Frank Jacobs, TFB, Switzerland
Prof. Toshiharu Kishi, The University of Tokyo, Japan
Dr. Michael Khrapko, CBE Consultancy Ltd, New Zealand
Prof. Tang Luping, CBI Swedish Cement and Concrete Research Institute, Sweden
Prof. W. John McCarter, Heriot Watt University, Scotland
Prof. André Valente Monteiro, National Laboratory for Civil Engineering, Portugal
Dr. Fabrizio Moro, Holcim Technology Ltd, Switzerland
Dr. Sreejeth Nanukuttan, Queen's University Belfast, UK

Prof. Rui Duarte Neves, Escola Superior de Tecnologia do Barreiro—IPS, Portugal
Prof. Lars-Olof Nilsson, Lunds University, Sweden
Dr. Mike Otieno, University of the Witwatersrand, South Africa
Dr. Peter Paulini, University Innsbruck, Austria
Prof. Rob Polder, TNO Built Environment and Geosciences, TU Delft, The Netherlands
Prof. Manu Santhanam, IIT Madras, India
Dr. Marijana Serdar, University of Zagreb, Croatia
Prof. Irina Stipanović Oslaković, University of Twente, The Netherlands
Dr. Roberto Torrent, Materials Advanced Services, Argentina
Dr. Andrés Antonio Torres-Acosta, Instituto Mexicano del Transporte, Mexico

RILEM Publications

The following list is presenting the global offer of RILEM Publications, sorted by series. Each publication is available in printed version and/or in online version.

RILEM Proceedings (PRO)

PRO 1: Durability of High Performance Concrete (ISBN: 2-912143-03-9); Ed. H. Sommer

PRO 2: Chloride Penetration into Concrete (ISBN: 2-912143-00-04); Eds. L.-O. Nilsson and J.-P. Ollivier

PRO 3: Evaluation and Strengthening of Existing Masonry Structures (ISBN: 2-912143-02-0); Eds. L. Binda and C. Modena

PRO 4: Concrete: From Material to Structure (ISBN: 2-912143-04-7); Eds. J.-P. Bournazel and Y. Malier

PRO 5: The Role of Admixtures in High Performance Concrete (ISBN: 2-912143-05-5); Eds. J. G. Cabrera and R. Rivera-Villarreal

PRO 6: High Performance Fiber Reinforced Cement Composites—HPFRCC 3 (ISBN: 2-912143-06-3); Eds. H. W. Reinhardt and A. E. Naaman

PRO 7: 1st International RILEM Symposium on Self-Compacting Concrete (ISBN: 2-912143-09-8); Eds. Å. Skarendahl and Ö. Petersson

PRO 8: International RILEM Symposium on Timber Engineering (ISBN: 2-912143-10-1); Ed. L. Boström

PRO 9: 2nd International RILEM Symposium on Adhesion between Polymers and Concrete ISAP '99 (ISBN: 2-912143-11-X); Eds. Y. Ohama and M. Puterman

PRO 10: 3rd International RILEM Symposium on Durability of Building and Construction Seal-ants (ISBN: 2-912143-13-6); Eds. A. T. Wolf

PRO 11: 4th International RILEM Conference on Reflective Cracking in Pavements (ISBN: 2-912143-14-4); Eds. A. O. Abd El Halim, D. A. Taylor and El H. H. Mohamed

PRO 12: International RILEM Workshop on Historic Mortars: Characteristics and Tests (ISBN: 2-912143-15-2); Eds. P. Bartos, C. Groot and J. J. Hughes

PRO 13: 2nd International RILEM Symposium on Hydration and Setting (ISBN: 2-912143-16-0); Ed. A. Nonat
PRO 14: Integrated Life-Cycle Design of Materials and Structures—ILCDES 2000 (ISBN: 951-758-408-3); (ISSN: 0356-9403); Ed. S. Sarja
PRO 15: Fifth RILEM Symposium on Fibre-Reinforced Concretes (FRC)—BEFIB'2000 (ISBN: 2-912143-18-7); Eds. P. Rossi and G. Chanvillard
PRO 16: Life Prediction and Management of Concrete Structures (ISBN: 2-912143-19-5); Ed. D. Naus
PRO 17: Shrinkage of Concrete—Shrinkage 2000 (ISBN: 2-912143-20-9); Eds. V. Baroghel-Bouny and P.-C. Aïtcin
PRO 18: Measurement and Interpretation of the On-Site Corrosion Rate (ISBN: 2-912143-21-7); Eds. C. Andrade, C. Alonso, J. Fullea, J. Polimon and J. Rodriguez
PRO 19: Testing and Modelling the Chloride Ingress into Concrete (ISBN: 2-912143-22-5); Eds. C. Andrade and J. Kropp
PRO 20: 1st International RILEM Workshop on Microbial Impacts on Building Materials (CD 02) (e-ISBN 978-2-35158-013-4); Ed. M. Ribas Silva
PRO 21: International RILEM Symposium on Connections between Steel and Concrete (ISBN: 2-912143-25-X); Ed. R. Eligehausen
PRO 22: International RILEM Symposium on Joints in Timber Structures (ISBN: 2-912143-28-4); Eds. S. Aicher and H.-W. Reinhardt
PRO 23: International RILEM Conference on Early Age Cracking in Cementitious Systems (ISBN: 2-912143-29-2); Eds. K. Kovler and A. Bentur
PRO 24: 2nd International RILEM Workshop on Frost Resistance of Concrete (ISBN: 2-912143-30-6); Eds. M. J. Setzer, R. Auberg and H.-J. Keck
PRO 25: International RILEM Workshop on Frost Damage in Concrete (ISBN: 2-912143-31-4); Eds. D. J. Janssen, M. J. Setzer and M. B. Snyder
PRO 26: International RILEM Workshop on On-Site Control and Evaluation of Masonry Structures (ISBN: 2-912143-34-9); Eds. L. Binda and R. C. de Vekey
PRO 27: International RILEM Symposium on Building Joint Sealants (CD03); Ed. A. T. Wolf
PRO 28: 6th International RILEM Symposium on Performance Testing and Evaluation of Bituminous Materials—PTEBM'03 (ISBN: 2-912143-35-7; e-ISBN: 978-2-912143-77-8); Ed. M. N. Partl
PRO 29: 2nd International RILEM Workshop on Life Prediction and Ageing Management of Concrete Structures (ISBN: 2-912143-36-5); Ed. D. J. Naus
PRO 30: 4th International RILEM Workshop on High Performance Fiber Reinforced Cement Composites—HPFRCC 4 (ISBN: 2-912143-37-3); Eds. A. E. Naaman and H. W. Reinhardt
PRO 31: International RILEM Workshop on Test and Design Methods for Steel Fibre Reinforced Concrete: Background and Experiences (ISBN: 2-912143-38-1); Eds. B. Schnütgen and L. Vandewalle
PRO 32: International Conference on Advances in Concrete and Structures 2 vol. (ISBN (set): 2-912143-41-1); Eds. Ying-shu Yuan, Surendra P. Shah and Heng-lin Lü
PRO 33: 3rd International Symposium on Self-Compacting Concrete (ISBN: 2-912143-42-X); Eds. Ó. Wallevik and I. Níelsson

PRO 34: International RILEM Conference on Microbial Impact on Building Materials (ISBN: 2-912143-43-8); Ed. M. Ribas Silva
PRO 35: International RILEM TC 186-ISA on Internal Sulfate Attack and Delayed Ettringite Formation (ISBN: 2-912143-44-6); Eds. K. Scrivener and J. Skalny
PRO 36: International RILEM Symposium on Concrete Science and Engineering—A Tribute to Arnon Bentur (ISBN: 2-912143-46-2); Eds. K. Kovler, J. Marchand, S. Mindess and J. Weiss
PRO 37: 5th International RILEM Conference on Cracking in Pavements—Mitigation, Risk Assessment and Prevention (ISBN: 2-912143-47-0); Eds. C. Petit, I. Al-Qadi and A. Millien
PRO 38: 3rd International RILEM Workshop on Testing and Modelling the Chloride Ingress into Concrete (ISBN: 2-912143-48-9); Eds. C. Andrade and J. Kropp
PRO 39: 6th International RILEM Symposium on Fibre-Reinforced Concretes—BEFIB 2004 (ISBN: 2-912143-51-9); Eds. M. Di Prisco, R. Felicetti and G. A. Plizzari
PRO 40: International RILEM Conference on the Use of Recycled Materials in Buildings and Structures (ISBN: 2-912143-52-7); Eds. E. Vázquez, Ch. F. Hendriks and G. M. T. Janssen
PRO 41: RILEM International Symposium on Environment-Conscious Materials and Systems for Sustainable Development (ISBN: 2-912143-55-1); Eds. N. Kashino and Y. Ohama
PRO 42: SCC'2005—China: 1st International Symposium on Design, Performance and Use of Self-Consolidating Concrete (ISBN: 2-912143-61-6); Eds. Zhiwu Yu, Caijun Shi, Kamal Henri Khayat and Youjun Xie
PRO 43: International RILEM Workshop on Bonded Concrete Overlays (e-ISBN: 2-912143-83-7); Eds. J. L. Granju and J. Silfwerbrand
PRO 44: 2nd International RILEM Workshop on Microbial Impacts on Building Materials (CD11) (e-ISBN: 2-912143-84-5); Ed. M. Ribas Silva
PRO 45: 2nd International Symposium on Nanotechnology in Construction, Bilbao (ISBN: 2-912143-87-X); Eds. Peter J. M. Bartos, Yolanda de Miguel and Antonio Porro
PRO 46: ConcreteLife'06—International RILEM-JCI Seminar on Concrete Durability and Service Life Planning: Curing, Crack Control, Performance in Harsh Environments (ISBN: 2-912143-89-6); Ed. K. Kovler
PRO 47: International RILEM Workshop on Performance Based Evaluation and Indicators for Concrete Durability (ISBN: 978-2-912143-95-2); Eds. V. Baroghel-Bouny, C. Andrade, R. Torrent and K. Scrivener
PRO 48: 1st International RILEM Symposium on Advances in Concrete through Science and Engineering (e-ISBN: 2-912143-92-6); Eds. J. Weiss, K. Kovler, J. Marchand, and S. Mindess
PRO 49: International RILEM Workshop on High Performance Fiber Reinforced Cementitious Composites in Structural Applications (ISBN: 2-912143-93-4); Eds. G. Fischer and V.C. Li

PRO 50: 1st International RILEM Symposium on Textile Reinforced Concrete (ISBN: 2-912143-97-7); Eds. Josef Hegger, Wolfgang Brameshuber and Norbert Will
PRO 51: 2nd International Symposium on Advances in Concrete through Science and Engineering (ISBN: 2-35158-003-6; e-ISBN: 2-35158-002-8); Eds. J. Marchand, B. Bissonnette, R. Gagné, M. Jolin and F. Paradis
PRO 52: Volume Changes of Hardening Concrete: Testing and Mitigation (ISBN: 2-35158-004-4; e-ISBN: 2-35158-005-2); Eds. O. M. Jensen, P. Lura and K. Kovler
PRO 53: High Performance Fiber Reinforced Cement Composites—HPFRCC5 (ISBN: 978-2-35158-046-2); Eds. H. W. Reinhardt and A. E. Naaman
PRO 54: 5th International RILEM Symposium on Self-Compacting Concrete (ISBN: 978-2-35158-047-9); Eds. G. De Schutter and V. Boel
PRO 55: International RILEM Symposium Photocatalysis, Environment and Construction Materials (ISBN: 978-2-35158-056-1); Eds. P. Baglioni and L. Cassar
PRO 56: International RILEM Workshop on Integral Service Life Modelling of Concrete Structures (ISBN 978-2-35158-058-5); Eds. R. M. Ferreira, J. Gulikers and C. Andrade
PRO 57: RILEM Workshop on Performance of cement-based materials in aggressive aqueous environments (e-ISBN: 978-2-35158-059-2); Ed. N. De Belie
PRO 58: International RILEM Symposium on Concrete Modelling—CONMOD'08 (ISBN: 978-2-35158-060-8); Eds. E. Schlangen and G. De Schutter
PRO 59: International RILEM Conference on On Site Assessment of Concrete, Masonry and Timber Structures—SACoMaTiS 2008 (ISBN set: 978-2-35158-061-5); Eds. L. Binda, M. di Prisco and R. Felicetti
PRO 60: Seventh RILEM International Symposium on Fibre Reinforced Concrete: Design and Applications—BEFIB 2008 (ISBN: 978-2-35158-064-6); Ed. R. Gettu
PRO 61: 1st International Conference on Microstructure Related Durability of Cementitious Composites 2 vol. (ISBN: 978-2-35158-065-3); Eds. W. Sun, K. van Breugel, C. Miao, G. Ye and H. Chen
PRO 62: NSF/ RILEM Workshop: In-situ Evaluation of Historic Wood and Masonry Structures (e-ISBN: 978-2-35158-068-4); Eds. B. Kasal, R. Anthony and M. Drdácký
PRO 63: Concrete in Aggressive Aqueous Environments: Performance, Testing and Modelling, 2 vol. (ISBN: 978-2-35158-071-4); Eds. M. G. Alexander and A. Bertron
PRO 64: Long Term Performance of Cementitious Barriers and Reinforced Concrete in Nuclear Power Plants and Waste Management—NUCPERF 2009 (ISBN: 978-2-35158-072-1); Eds. V. L'Hostis, R. Gens, C. Gallé
PRO 65: Design Performance and Use of Self-consolidating Concrete—SCC'2009 (ISBN: 978-2-35158-073-8); Eds. C. Shi, Z. Yu, K. H. Khayat and P. Yan
PRO 66: 2nd International RILEM Workshop on Concrete Durability and Service Life Planning—ConcreteLife'09 (ISBN: 978-2-35158-074-5); Ed. K. Kovler
PRO 67: Repairs Mortars for Historic Masonry (e-ISBN: 978-2-35158-083-7); Ed. C. Groot

PRO 68: Proceedings of the 3rd International RILEM Symposium on 'Rheology of Cement Suspensions such as Fresh Concrete (ISBN 978-2-35158-091-2); Eds. O. H. Wallevik, S. Kubens and S. Oesterheld

PRO 69: 3rd International PhD Student Workshop on 'Modelling the Durability of Reinforced Concrete (ISBN: 978-2-35158-095-0); Eds. R. M. Ferreira, J. Gulikers and C. Andrade

PRO 70: 2nd International Conference on 'Service Life Design for Infrastructure' (ISBN set: 978-2-35158-096-7, e-ISBN: 978-2-35158-097-4); Ed. K. van Breugel, G. Ye and Y. Yuan

PRO 71: Advances in Civil Engineering Materials—The 50-year Teaching Anniversary of Prof. Sun Wei' (ISBN: 978-2-35158-098-1; e-ISBN: 978-2-35158-099-8); Eds. C. Miao, G. Ye, and H. Chen

PRO 72: First International Conference on 'Advances in Chemically-Activated Materials—CAM'2010' (2010), 264 pp., ISBN: 978-2-35158-101-8; e-ISBN: 978-2-35158-115-5, Eds. Caijun Shi and Xiaodong Shen

PRO 73: 2nd International Conference on 'Waste Engineering and Management—ICWEM 2010' (2010), 894 pp., ISBN: 978-2-35158-102-5; e-ISBN: 978-2-35158-103-2, Eds. J. Zh. Xiao, Y. Zhang, M. S. Cheung and R. Chu

PRO 74: International RILEM Conference on 'Use of Superabsorsorbent Polymers and Other New Addditives in Concrete' (2010) 374 pp., ISBN: 978-2-35158-104-9; e-ISBN: 978-2-35158-105-6; Eds. O.M. Jensen, M.T. Hasholt, and S. Laustsen

PRO 75: International Conference on 'Material Science—2nd ICTRC—Textile Reinforced Concrete—Theme 1' (2010) 436 pp., ISBN: 978-2-35158-106-3; e-ISBN: 978-2-35158-107-0; Ed. W. Brameshuber

PRO 76: International Conference on 'Material Science—HetMat—Modelling of Heterogeneous Materials—Theme 2' (2010) 255 pp., ISBN: 978-2-35158-108-7; e-ISBN: 978-2-35158-109-4; Ed. W. Brameshuber

PRO 77: International Conference on 'Material Science—AdIPoC—Additions Improving Properties of Concrete—Theme 3' (2010) 459 pp., ISBN: 978-2-35158-110-0; e-ISBN: 978-2-35158-111-7; Ed. W. Brameshuber

PRO 78: 2nd Historic Mortars Conference and RILEM TC 203-RHM Final Workshop—HMC2010 (2010) 1416 pp., e-ISBN: 978-2-35158-112-4; Eds J. Válek, C. Groot, and J. J. Hughes

PRO 79: International RILEM Conference on Advances in Construction Materials Through Science and Engineering (2011) 213 pp., e-ISBN: 978-2-35158-117-9; Eds Christopher Leung and K.T. Wan

PRO 80: 2nd International RILEM Conference on Concrete Spalling due to Fire Exposure (2011) 453 pp., ISBN: 978-2-35158-118-6, e-ISBN: 978-2-35158-119-3; Eds E.A.B. Koenders and F. Dehn

PRO 81: 2nd International RILEM Conference on Strain Hardening Cementitious Composites (SHCC2-Rio) (2011) 451 pp., ISBN: 978-2-35158-120-9, e-ISBN: 978-2-35158-121-6; Eds R.D. Toledo Filho, F.A. Silva, E.A.B. Koenders and E.M. R. Fairbairn

PRO 82: 2nd International RILEM Conference on Progress of Recycling in the Built Environment (2011) 507 pp., e-ISBN: 978-2-35158-122-3; Eds V.M. John, E. Vazquez, S.C. Angulo and C. Ulsen

PRO 83: 2nd International Conference on Microstructural-related Durability of Cementitious Composites (2012) 250 pp., ISBN: 978-2-35158-129-2; e-ISBN: 978-2-35158-123-0; Eds G. Ye, K. van Breugel, W. Sun and C. Miao

PRO 85: RILEM-JCI International Workshop on Crack Control of Mass Concrete and Related issues concerning Early-Age of Concrete Structures—ConCrack 3—Control of Cracking in Concrete Structures 3 (2012) 237 pp., ISBN: 978-2-35158-125-4; e-ISBN: 978-2-35158-126-1; Eds F. Toutlemonde and J.-M. Torrenti

PRO 86: International Symposium on Life Cycle Assessment and Construction (2012) 414 pp., ISBN: 978-2-35158-127-8, e-ISBN: 978-2-35158-128-5; Eds A. Ventura and C. de la Roche

PRO 87: UHPFRC 2013—RILEM-fib-AFGC International Symposium on Ultra-High Performance Fibre-Reinforced Concrete (2013), ISBN: 978-2-35158-130-8, e-ISBN: 978-2-35158-131-5; Eds F. Toutlemonde

PRO 88: 8th RILEM International Symposium on Fibre Reinforced Concrete (2012) 344 pp., ISBN: 978-2-35158-132-2, e-ISBN: 978-2-35158-133-9; Eds Joaquim A.O. Barros

PRO 89: RILEM International workshop on performance-based specification and control of concrete durability (2014) 678 pp., ISBN: 978-2-35158-135-3, e-ISBN: 978-2-35158-136-0; Eds. D. Bjegović, H. Beushausen and M. Serdar

PRO 90: 7th RILEM International Conference on Self-Compacting Concrete and of the 1st RILEM International Conference on Rheology and Processing of Construction Materials (2013) 396 pp., ISBN: 978-2-35158-137-7, e-ISBN: 978-2-35158-138-4, Eds. Nicolas Roussel and Hela Bessaies-Bey

PRO 91 draft: CONMOD 2014—RILEM International Symposium on Concrete Modelling (2014), ISBN: 978-2-35158-139-1; e-ISBN: 978-2-35158-140-7

PRO 92: CAM 2014—2nd International Conference on advances in chemically-activated materials (2014) 392 pp., ISBN: 978-2-35158-141-4; e-ISBN: 978-2-35158-142-1, Eds. Caijun Shi and Xiadong Shen

PRO 93: SCC 2014—3rd International Symposium on Design, Performance and Use of Self-Consolidating Concrete (2014) 438 pp., ISBN: 978-2-35158-143-8; e-ISBN: 978-2-35158-144-5, Eds. Caijun Shi, Zhihua Ou, Kamal H. Khayat

PRO 94 (online version): HPFRCC-7—7th RILEM conference on High performance fiber reinforced cement composites, e-ISBN: 978-2-35158-146-9, Eds. H.W. Reinhardt, G.J. Parra-Montesinos, H. Garrecht

PRO 95: International RILEM Conference on Application of superabsorbent polymers and other new admixtures in concrete construction, ISBN: 978-2-35158-147-6; e-ISBN: 978-2-35158-148-3, Eds. Viktor Mechtcherine, Christof Schroefl

PRO 96 (online version): XIII DBMC: XIII International Conference on Durability of Building Materials and Components, e-ISBN: 978-2-35158-149-0, Eds. M. Quattrone, V.M. John

PRO 97: SHCC3—3rd International RILEM Conference on Strain Hardening Cementitious Composites, ISBN: 978-2-35158-150-6; e-ISBN: 978-2-35158-151-3, Eds. E. Schlangen, M.G. Sierra Beltran, M. Lukovic, G. Ye
PRO 98: FERRO-11—11th International Symposium on Ferrocement and 3rd ICTRC—International Conference on Textile Reinforced Concrete, ISBN: 978-2-35158-152-0; e-ISBN: 978-2-35158-153-7, Ed. W. Brameshuber
PRO 99 (online version): ICBBM 2015—1st International Conference on Bio-Based Building Materials, e-ISBN: 978-2-35158-154-4, Eds. S. Amziane, M. Sonebi
PRO 100: SCC16—RILEM Self-Consolidating Concrete Conference, ISBN: 978-2-35158-156-8; e-ISBN: 978-2-35158-157-5
PRO 101 (online version): III Progress of Recycling in the Built Environment, e-ISBN: 978-2-35158-158-2, Eds. M. Quattrone, V.M. John

RILEM Reports (REP)
Report 19: Considerations for Use in Managing the Aging of Nuclear Power Plant Concrete Structures (ISBN: 2-912143-07-1); Ed. D. J. Naus
Report 20: Engineering and Transport Properties of the Interfacial Transition Zone in Cementitious Composites (ISBN: 2-912143-08-X); Eds. M. G. Alexander, G. Arliguie, G. Ballivy, A. Bentur and J. Marchand
Report 21: Durability of Building Sealants (ISBN: 2-912143-12-8); Ed. A. T. Wolf
Report 22: Sustainable Raw Materials—Construction and Demolition Waste (ISBN: 2-912143-17-9); Eds. C. F. Hendriks and H. S. Pietersen
Report 23: Self-Compacting Concrete state-of-the-art report (ISBN: 2-912143-23-3); Eds. Å. Skarendahl and Ö. Petersson
Report 24: Workability and Rheology of Fresh Concrete: Compendium of Tests (ISBN: 2-912143-32-2); Eds. P. J. M. Bartos, M. Sonebi and A. K. Tamimi
Report 25: Early Age Cracking in Cementitious Systems (ISBN: 2-912143-33-0); Ed. A. Bentur
Report 26: Towards Sustainable Roofing (Joint Committee CIB/RILEM) (CD 07) (e-ISBN 978-2-912143-65-5); Eds. Thomas W. Hutchinson and Keith Roberts
Report 27: Condition Assessment of Roofs (Joint Committee CIB/RILEM) (CD 08) (e-ISBN 978-2-912143-66-2); Ed. CIB W 83/RILEM TC166-RMS
Report 28: Final report of RILEM TC 167-COM 'Characterisation of Old Mortars with Respect to Their Repair (ISBN: 978-2-912143-56-3); Eds. C. Groot, G. Ashall and J. Hughes
Report 29: Pavement Performance Prediction and Evaluation (PPPE): Interlaboratory Tests(e-ISBN: 2-912143-68-3); Eds. M. Partl and H. Piber
Report 30: Final Report of RILEM TC 198-URM 'Use of Recycled Materials' (ISBN: 2-912143-82-9; e-ISBN: 2-912143-69-1); Eds. Ch. F. Hendriks, G. M. T. Janssen and E. Vázquez
Report 31: Final Report of RILEM TC 185-ATC 'Advanced testing of cement-based materials during setting and hardening' (ISBN: 2-912143-81-0; e-ISBN: 2-912143-70-5); Eds. H. W. Reinhardt and C. U. Grosse

Report 32: Probabilistic Assessment of Existing Structures. A JCSS publication (ISBN 2-912143-24-1); Ed. D. Diamantidis
Report 33: State-of-the-Art Report of RILEM Technical Committee TC 184-IFE 'Industrial Floors' (ISBN 2-35158-006-0); Ed. P. Seidler
Report 34: Report of RILEM Technical Committee TC 147-FMB 'Fracture mechanics applications to anchorage and bond' Tension of Reinforced Concrete Prisms—Round Robin Analysis and Tests on Bond (e-ISBN 2-912143-91-8); Eds. L. Elfgren and K. Noghabai
Report 35: Final Report of RILEM Technical Committee TC 188-CSC 'Casting of Self Compacting Concrete' (ISBN 2-35158-001-X; e-ISBN: 2-912143-98-5); Eds. Å. Skarendahl and P. Billberg
Report 36: State-of-the-Art Report of RILEM Technical Committee TC 201-TRC 'Textile Rein-forced Concrete' (ISBN 2-912143-99-3); Ed. W. Brameshuber
Report 37: State-of-the-Art Report of RILEM Technical Committee TC 192-ECM 'Environment-conscious construction materials and systems' (ISBN: 978-2-35158-053-0); Eds. N. Kashino, D. Van Gemert and K. Imamoto
Report 38: State-of-the-Art Report of RILEM Technical Committee TC 205-DSC 'Durability of Self-Compacting Concrete' (ISBN: 978-2-35158-048-6); Eds. G. De Schutter and K. Audenaert
Report 39: Final Report of RILEM Technical Committee TC 187-SOC 'Experimental determination of the stress-crack opening curve for concrete in tension' (ISBN 978-2-35158-049-3); Ed. J. Planas
Report 40: State-of-the-Art Report of RILEM Technical Committee TC 189-NEC 'Non-Destructive Evaluation of the Penetrability and Thickness of the Concrete Cover' (ISBN 978-2-35158-054-7); Eds. R. Torrent and L. Fernández Luco
Report 41: State-of-the-Art Report of RILEM Technical Committee TC 196-ICC 'Internal Cur-ing of Concrete' (ISBN 978-2-35158-009-7); Eds. K. Kovler and O. M. Jensen
Report 42: 'Acoustic Emission and Related Non-destructive Evaluation Techniques for Crack Detection and Damage Evaluation in Concrete'—Final Report of RILEM Technical Committee 212-ACD (e-ISBN: 978-2-35158-100-1); Ed. M. Ohtsu

Contents

Chapter 1
Introduction

H. Beushausen

1.1 Background to the Work of TC 230-PSC

For the design of concrete structures, durability and service life prediction have increasingly gained importance in recent years. This comes as a result of the inadequate durability performance of many reinforced concrete structures built in the past decades, which places enormous strain on construction budgets worldwide. The dominant cause of premature deterioration of concrete structures is reinforcement corrosion related to carbonation or chloride ingress. Traditional durability design approaches are based on prescribed limiting values for selected mix design parameters such as water/binder ratio, compressive strength and cement content. However, prescriptive mix design parameters fail to adequately characterize the concrete's resistance against carbonation or chloride ingress, because they ignore to a large extent the different performance of various binder types and of mineral components added to the cements or to the concrete itself, as well as the type of aggregate, and do not allow to take into account the influences of on-site practice during the construction process. Prescriptive approaches also cannot explicitly account for a rational service life requirement.

Performance approaches, in contrast, are based on the measurement of material properties that can be linked to deterioration mechanisms under the prevalent exposure conditions. The measurement of actual concrete material properties of the as-built structure allows accounting for the combined influences of material composition, construction procedures, and environmental influences and therefore forms a rational basis for durability prediction and service life design. Performance approaches can be applied in different stages and for different purposes, including design, specification, pre-qualification and conformity assessment of the as built

H. Beushausen (✉)
University of Cape Town, Cape Town, South Africa

H. Beushausen and L. Fernandez Luco (eds.), *Performance-Based Specifications and Control of Concrete Durability*,
RILEM State-of-the-Art Reports 18, DOI 10.1007/978-94-017-7309-6_1

structure. Most test methods for the assessment of the structure's resistance against reinforcement corrosion are based on the quantity and quality of the cover concrete.

Transport properties of cementitious materials are key performance parameters for predicting the quality of the cover zone, since deterioration mechanisms such as chloride ingress or carbonation relate to the ease with which a fluid or ion can move through the concrete microstructure. The passage of potentially aggressive species (ions or molecules in the form of liquids and gases) is primarily influenced by the penetrability of the concrete. Penetrability is broadly defined as the degree to which the concrete permits gases, liquids, or ionic species to move through its pore structure. It embraces the concepts of permeation, sorption, diffusion and migration and is quantified in terms of the transport parameters. Various methods for testing concrete penetrability properties of as-built concrete structures have been developed worldwide, some of which have for many years already been successfully used not only for research, but also for durability specifications and quality control.

An important driver for producing performance-based approaches is the increasing development and use of innovative and new concrete types and constituent materials. Prescriptive requirements often fail to resemble the durability characteristics of modern concrete types and hinder innovation and economic design and construction. Further, based on the often premature deterioration of concrete infrastructure built in the past decades, owners of structures are increasingly reluctant to accept black-box construction solutions and are beginning to ask for technical proof that their structure can meet service life requirements. In this respect, one of the advantages of performance-based design specifications is that the quality of the as-built structure can be evaluated and actions can be taken in case of non-conformity, i.e. in case the as-built structure does not meet the specified limiting values for durability characteristics.

Performance-based approaches for concrete structures are not limited to durability characteristics and have for many decades already been successfully applied, for example for mechanical properties. The most widely accepted performance approach for concrete is that for compressive strength, which was developed some time in the early part of the 20th century. Prior to that time, strength was controlled through the specification of limiting w/c ratios, which is similar to the traditional and still widely applied prescriptive design approaches for concrete durability. The implementation of compressive strength as a clearly defined performance criterion enabled not only economic design but also created a very efficient quality control tool for construction quality. The same can be expected from the implementation of performance approaches for concrete durability.

The principle of performance-based design and quality control for concrete durability has been subject to significant worldwide research efforts for more than 25 years and the literature reports on many examples of successful implementation. The work of this TC 230-PSC is largely a follow-up from the work done by RILEM TC 189-NEC (Non-destructive evaluation of the penetrability and thickness of concrete cover), chaired by Roberto Torrent. TC 189-NEC published a State-of-the-Art Report in 2007 [1], concluding that several suitable test methods exist to characterize the penetrability, and hence the durability, of in situ concrete in a statistically significant manner.

In engineering practice, performance approaches are often still used in combination with prescriptive requirements. This is largely because, for most durability test methods, sufficient practical experience still has to be gained before engineers and owners are prepared to fully rely on them. In this respect, the exchange of relevant knowledge and experience between researchers and practitioners worldwide will help to successfully build the foundation for the full implementation of performance-based approaches. This State-of-the-Art Report, produced by RILEM TC 230-PSC (Performance-Based Specifications and Control of Concrete Durability), is intended to assist in such efforts.

Important aspects to consider for development and implementation of performance-based design approaches for durability include service life prediction models, deterioration mechanisms, performance test methods and their application, interpretation and limitations, responsibilities of owners, engineers and contractors, and appropriate actions in case of non-conformance to design specifications. This report addresses these issues and presents practical guidance for the selection and application of suitable test methods, statistical analysis, and interpretation of data.

1.2 Terminology

The authors of the various chapters adopted a standard terminology, as outlined in the following paragraphs. The suggested terminology relates to the specific case of durability of concrete with respect to the resistance against reinforcement corrosion. The mentioned terms may have different/additional meanings for other aspects of material technology. Definitions given in EN-206-1:2000 [2] have been added where relevant.

1.2.1 Compliance Assessment

Compliance assessment refers to the quality control of the as-built structure, with the aim to establish if specified performance criteria have been met. This involves experimental investigations on the structure, or on samples removed from the structure, or on laboratory-cured specimens made from the same concrete batch as the one used in the structure.

1.2.2 Designed Concrete

Concrete for which the required properties and additional characteristics are specified to the producer who is responsible for providing a concrete conforming to the required properties and additional characteristics (EN 206:2013, [2]).

1.2.3 Deterioration Model

A deterioration model allows predicting concrete deterioration over time. In the scope of this publication this commonly links to the analytical or numerical modelling of chloride ingress or carbonation.

1.2.4 Durability Indicators

Durability indicators are measurable material properties that can be used to predict the concrete's resistance against deterioration. Most commonly, these include transport properties or results from performance simulation tests.

1.2.5 Durability Potential

The durability potential of a certain concrete mix composition is established in the laboratory through experimental investigations of durability indicators, which is commonly done under near-ideal conditions for production processes and curing conditions. The as-built concrete structure may not achieve the full durability potential of the concrete, due to the influences of on-site workmanship and environmental conditions.

1.2.6 Initial Test

> Test or series of tests to check before the production starts how a new concrete or concrete family shall be composed in order to meet all the specified requirements in the fresh and hardened states (EN 206:2013, [2]).

1.2.7 Non-destructive Test

A test to quantify a specific concrete material property on an in situ structure without affecting the serviceability of the structure.

1.2.8 Non-invasive Test

Once the testing has been completed, a non-invasive test does not leave any evidence of the testing on the structure (such as holes, surface damage, surface contamination, or surface discolouration).

1.2.9 Performance-Based Design for Durability (General)

Performance-based design for durability involves the assessment of relevant material properties of a specific concrete through experiments, analytical modelling, numerical modelling, or experience in order to predict the concrete's resistance against deterioration for a certain period under certain environmental exposure conditions.

1.2.10 Performance-Based Design for Durability (Specific to This Publication)

In the scope of this publication, performance-based design for durability involves the assessment of relevant concrete properties through experimental investigations in the laboratory as well as on-site.

1.2.11 Performance Criteria

Performance criteria are limiting material parameters or properties that are established in the design process, usually linked to the concrete's resistance against chloride ingress or carbonation. Typical performance criteria in the scope of this publication refer to durability indicators.

1.2.12 Performance Simulation Tests

In the scope of this publication, performance simulation tests encompass the direct measurement of the concrete's resistance against the ingress of chlorides or the progress of carbonation, typically under accelerated conditions, i.e. under the influence of an artificial environment with chloride or carbon dioxide concentrations higher than those usually existent in real exposure conditions.

1.2.13 Prescribed Concrete

Concrete for which the composition of the concrete and the constituent materials to be used are specified to the producer who is responsible for providing a concrete with the specified composition (EN 206:2013, [2]).

1.2.14 Prescriptive Design for Durability

Prescriptive design for durability involves the specification of limiting values for constituent materials and mix design parameters, typically covering binder type, compressive strength, water/binder ratio, and binder content in relation to the environmental exposure class, cover depth, and required service life. The concrete is assumed to be durable for the specified service life when these prescriptive specifications are met.

1.2.15 Pre-qualification

Pre-qualification refers to the assessment of relevant concrete properties in the design process (prior to construction) in order to establish suitable concrete types and mix compositions for a given environmental exposure and required service life.

1.2.16 Producer

Person or body producing fresh concrete (EN 206:2013, [2]).

1.2.17 Semi-invasive Test Method

A semi-invasive test method leaves evidence of the testing on the structure (such as core or drill holes, minor surface damage, uncritical surface contamination, or surface discolouration).

1.2.18 Service Life

The period of time during which the performance of the concrete in the structure will be kept at a level compatible with the fulfilment of the performance requirements of the structure, provided it is properly maintained (EN 206:2013, [2]).

1.2.19 Service Life Model

A model for the prediction of the service life duration of concrete structures, based on deterioration models and limit state criteria such as corrosion initiation or propagation, damage indicators, etc. Service life models may have numerous input parameters such as material properties and mix proportions, durability indicators, environmental conditions, protective measures such as stainless reinforcing steel and concrete surface coatings, corrosion inhibitors, etc.

1.2.20 Specification

Final compilation of documented technical requirements given to the producer in terms of performance or composition (EN 206:2013, [2]).

1.2.21 Transport Properties

Concrete transport properties relevant to the scope of this publication include permeability, absorption, electrical resistivity, and conductivity, and are mostly used to predict/model the concrete's resistance against the ingress of harmful substances, such as chlorides or carbon dioxide.

1.2.22 User

Person or body using fresh concrete in the execution of a construction or a component (EN 206:2013, [2]).

1.2.23 Verification

Confirmation by examination of objective evidence that specified requirements have been fulfilled (EN 206:2013, [2]).

References

1. Torrent R, Fernandez Luco L (editors). Non-destructive evaluation of the penetrability and thickness of the concrete cover. France: State-of-the-Art-Report, RILEM TC 189-NEC, RILEM Publications S.A.R.L.; 2007. 223 pp.
2. EN-206:2013. Concrete: specification, performance, production and conformity. European Standard, CEN; 2013.

Chapter 2
Durability of Reinforced Concrete Structures and Penetrability

L.-O. Nilsson, S. Kamali-Bernard and M. Santhanam

2.1 Introduction

In this chapter a brief overview is given on the mechanisms causing reinforcement corrosion, on the concrete properties relating to the ingress of aggressive agents (penetrability and transport properties) and on the principles for service life design and deterioration models.

2.2 Mechanisms Causing Reinforcement Corrosion

Reinforcement steel in concrete is passivated because of the alkaline environment at the steel surface. The steel cannot corrode as long as this passivation is prevailing. This passivation can be broken in two ways:

- carbonation causing a drop in pH in the carbonated part of the concrete;
- chloride ingress causing a chloride content at the steel surface above a certain critical chloride content, the "threshold level".

After depassivation, and corrosion initiation, the rate of corrosion depends on the concrete properties, the thickness of the cover and the temperature and humidity conditions at the steel surfaces and in the cover. This "propagation process" is not dealt with here, since in most applications the service-life is defined to end at the

L.-O. Nilsson (✉)
Lund University, Lund, Sweden

S. Kamali-Bernard
Institut National des Sciences Appliquées (INSA), Rennes, France

M. Santhanam
IIT Madras, Chennai, India

H. Beushausen and L. Fernandez Luco (eds.), *Performance-Based Specifications and Control of Concrete Durability*,
RILEM State-of-the-Art Reports 18, DOI 10.1007/978-94-017-7309-6_2

start of the propagation period. Additionally, penetrability is not very relevant in the propagation period.

2.2.1 Carbonation

Carbonation is a combined process of diffusion of CO_2 through the open pores of concrete into a carbonation "front" where a chemical reaction occurs where CO_2 reacts with the cement paste hydrates. The carbonation of portlandite ($Ca(OH)_2$) and calcium silicate hydrate (*C-S-H*) leads to the formation of calcium carbonates and silica. The carbonation reaction of portlandite is shown in Eq. (2.1).

$$CO_2 + CaO \rightarrow CaCO_3 \tag{2.1}$$

This chemical reaction must be "supplied" by CO_2 to be able to continue. The two decisive parameters are the diffusion resistance of the carbonated concrete and the amount of CaO that can be carbonated.

2.2.2 Chloride Ingress

Chloride ingress is a combined process of three parts:

a. diffusion of chloride ions in the pore liquid,
b. convection of chloride ions in the pore liquid by liquid transport and
c. binding of chloride to the cement gel.

Part 'a' and 'b' are "penetrability and transport properties" that are retarded by the chloride binding process 'c'.

2.3 Concrete Properties Relating to the Ingress of Aggressive Agents

From Sect. 2.2 it is obvious that the concrete properties that are related to initiation of reinforcement corrosion are several:

a. resistance against diffusion of CO_2,
b. amount and availability of CaO as a reactant in the carbonation reaction,
c. moisture fixation properties of carbonated concrete, which influence the diffusion resistance,

d. moisture fixation and moisture transport properties of carbonated and uncarbonated concrete, which affect the moisture conditions in the concrete cover and the convection of chloride ions,
e. resistance against diffusion and convection of chloride ions,
f. chloride binding properties and
g. chloride threshold level.

Among these concrete properties only properties 'a', 'd' and 'e' are "penetrability and transport properties", depending on the definition of "penetrability". These are further described in the next sections. Properties that are binding and fixation properties are not covered.

2.3.1 Resistance Against Diffusion of CO_2

The flux of diffusing CO_2 in carbonated concrete is based on Eq. (2.2),

$$J_{CO_2} = -D_{CO_2}(RH, \alpha)\frac{dc_{CO_2}}{dx} = \frac{c}{R_{CO_2}} \tag{2.2}$$

where the CO_2-diffusion coefficient D_{CO_2} depends on the humidity RH and degree of hydration α. The concentration difference over the carbonated layer with thickness X_{CO_3} is $\Delta c = c\text{-}0 = c$.

The resistance to diffusion of CO_2 of the carbonated layer is R_{CO_2}, which is given by,

$$R_{CO_2} = \int_{x=0}^{x=X_{CO_3}} \frac{dx}{D_{CO_2}(RH(x), \alpha(x))} \tag{2.3}$$

The CO_2-diffusion coefficient varies with depth for two reasons; a humidity profile and a "curing profile" in the concrete cover. The moisture dependency is shown in principle in Fig. 2.1. Figure 2.2 shows examples of profiles of the degree of hydration and the corresponding CO_2-diffusion coefficient.

2.3.2 Moisture Transport Properties

Moisture is present in concrete as adsorbed in the gel at pore surfaces, physically bound by menisci in large gel pores and small capillary pores and as vapour in the "empty" pores. This moisture will be transported in the pore system due to differences in the state of water, usually described with the relative humidity, or pore humidity, RH or φ. Since the different types of water, adsorbed, capillary and

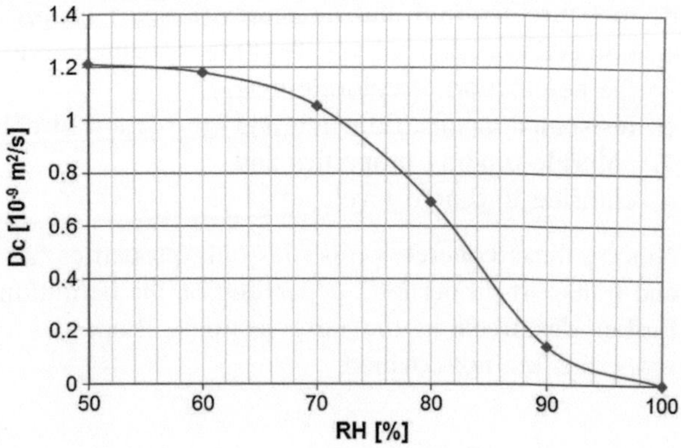

Fig. 2.1 The moisture dependency of the CO_2-diffusion coefficient; an example for a concrete with w/c = 0.4 [1]

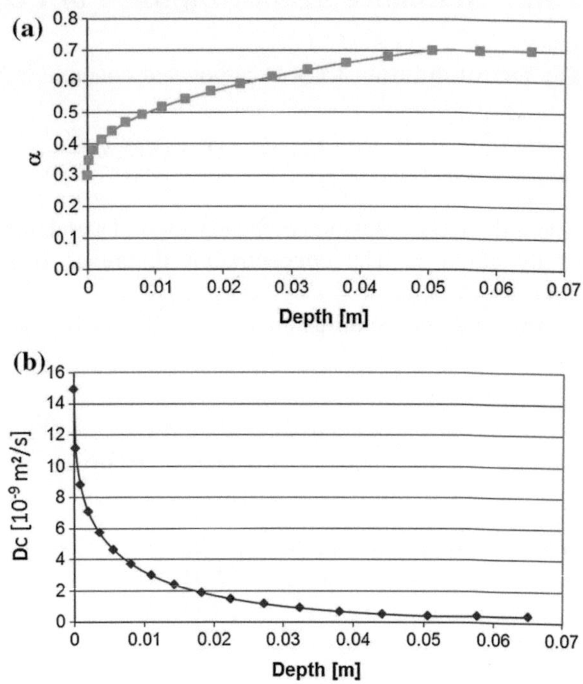

Fig. 2.2 An example of a profile of **a** degree of hydration (α) after bad curing, and **b** the corresponding CO_2-diffusion coefficient [2]

vapour, cannot be differentiated in a measurement, moisture transport properties are given as a total moisture transport coefficient δ_{RH}, e.g. expressed as in Eq. (2.4).

$$J_w = -\delta_{RH}(RH, \alpha)\frac{d\varphi}{dx} \tag{2.4}$$

where J_w is the total flux of moisture, α is the degree of hydration, φ is the moisture transport potential, and RH is the relative humidity.

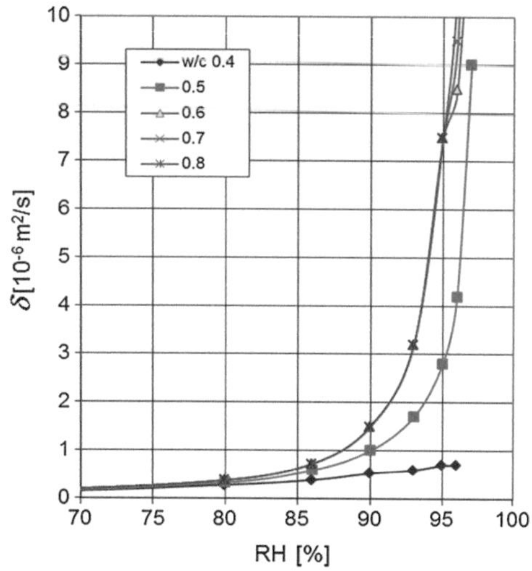

Fig. 2.3 The total moisture transport coefficient for various concretes [3]

The total moisture transport coefficient δ_{RH} is moisture dependent, since the flux due to the different moisture transport mechanisms depends on the state of moisture and moisture content in the pore system. An example of this moisture dependency is shown in Fig. 2.3 for concretes with different water-binder ratios.

2.3.3 *Resistance Against Chloride Diffusion and Convection*

Transport of chloride ions due to convection, by ions being transported with the pore liquid transport, is given by that part of the total moisture transport that can "carry" ions. What portion of the total moisture flux can carry ions is not yet known; this is still a topic of research.

Transport of chloride ions as diffusion in the pore liquid is usually described with Fick's 1st law or the Nernst-Planck equation. The latter considers the activity of various ions and the electrical field that is created by all the ions being present in the pore liquid. Fick's 1st law is shown in Eq. (2.5).

$$J_{Cl} = -D_{F1}(RH, \alpha)\frac{dc}{dx} \tag{2.5}$$

where J_{Cl} is the flux of ions, D_{F1} is the chloride diffusion coefficient, and c is the concentration of chloride ions in the pore liquid.

The chloride diffusion coefficient depends on the degree of hydration and the pore humidity RH or the degree of saturation S of the pore system, since ions can only move in the liquid part of the total moisture content. The moisture dependency is not well known; only a few measurements are known. One example is shown in

Fig. 2.4. This example is for the chloride diffusivity in Fick's 2nd law, however, see Eq. (2.10) (Fig. 2.4).

2.4 Service Life and Deterioration Models (Principles)

The traditional service-life and deterioration model for reinforcement corrosion is the one proposed by Tuutti [1], see Fig. 2.5.

The most common definition of service-life being used by owners of buildings and infrastructures is the one that marks the end of the service-life as the end of the corrosion initiation period. Then the service-life and deterioration models are "simple"; they are models that give the length of the initiation period.

The principles of models for carbonation-initiated corrosion are simple; corrosion is first initiated when the carbonation "front" reaches the steel bar, see Eq. (2.6).

$$x_{CO_3}(t_{SL}) = d \tag{2.6}$$

where x_{CO_3} is the depth of carbonation, t_{SL} is the service-life, and d is the thickness of the concrete cover.

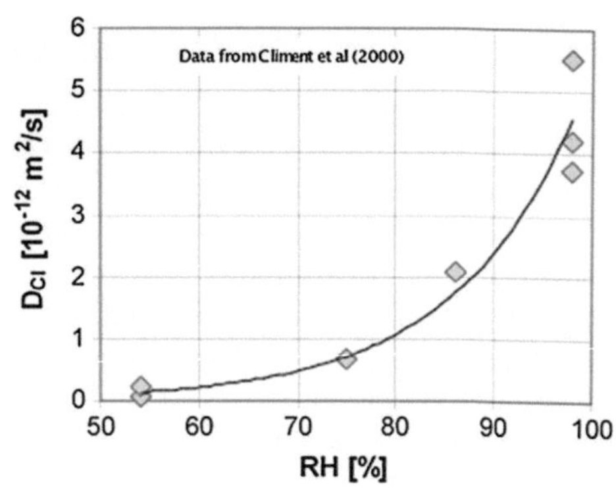

Fig. 2.4 An example of the moisture dependency of the chloride diffusivity D_{F2} [4]

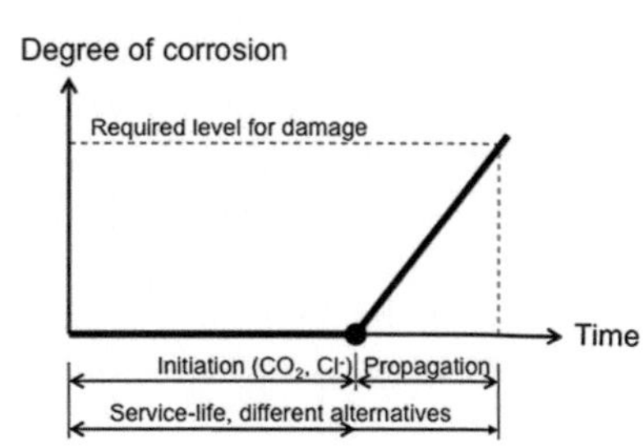

Fig. 2.5 The service-life model for reinforcement corrosion [1]

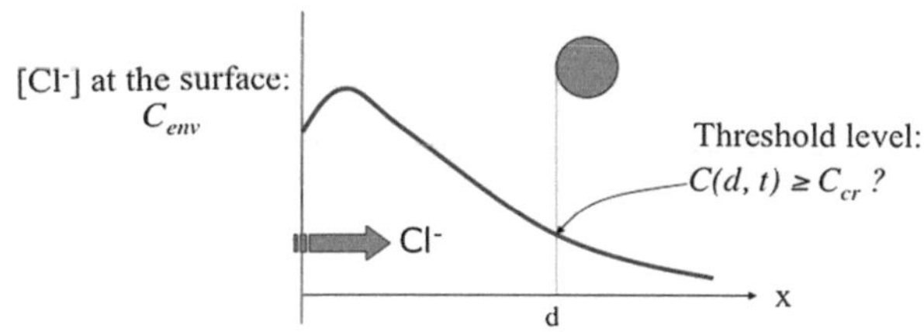

Fig. 2.6 The principle of most common service-life models for chloride initiated reinforcement corrosion

The principles of models for chloride initiated corrosion are somewhat more complicated since chloride ingress must give a certain chloride level $C(x = d, t_{SL})$ at the steel surface, above the chloride threshold level, C_{cr}, i.e. the principle of the service-life models is shown in Eq. (2.7) and Fig. 2.6.

$$C(x = d, t_{SL}) = C_{cr} \tag{2.7}$$

2.4.1 Carbonation Models, in Principle

Several carbonation models are proposed in the literature. A simple carbonation model that includes a "penetrability and transport property" is the one in Eq. (2.8) [5], see Fig. 2.7.

$$x_{CO_3}(t) = \sqrt{\frac{2D_{CO_2}c_{CO_2}}{a}} \cdot \sqrt{t} \tag{2.8}$$

where x_{CO_3} is the depth of carbonation, D_{CO_2} is the diffusion coefficient for carbon dioxide, c_{CO_2} is the concentration of carbon dioxide at the surface of the concrete, a is the amount of carbon dioxide required to carbonate a unit volume of concrete, and t is the time.

Fig. 2.7 A simple model for carbonation, with a diffusion and reaction process [5]

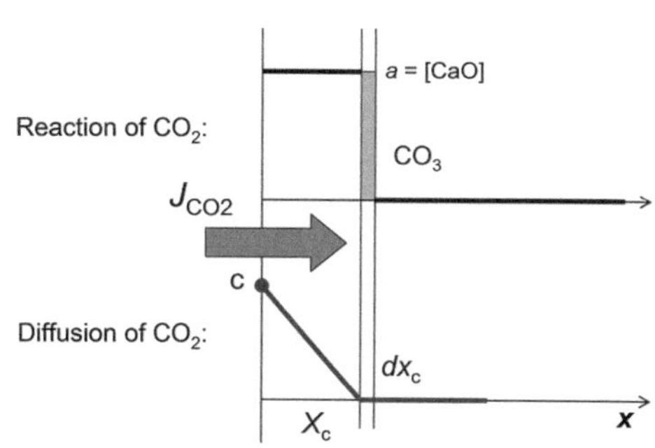

In the cover of a concrete structure exposed to real environmental actions, the moisture conditions are varying with time. This will affect the diffusion coefficient in such a way that it will vary with time and depth. To consider such effects, the carbonation process must be modelled with a "resistance" to diffusion of CO_2, i.e. Equation (2.8) must be differentiated and integrated over the carbonated part of the cover. This can be seen in [5].

2.4.2 Chloride Ingress Models, in Principle

A large number of models are proposed in literature [6]. All models could be said to be solutions to the mass balance equation for chloride, Eq. (2.9).

$$\frac{\partial C}{\partial t} = \frac{\partial c_b}{\partial t} + \frac{\partial c}{\partial t} = -\frac{\partial J_{Cl}}{\partial x} = \frac{\partial}{\partial x} D_{F1} \frac{\partial c}{\partial x} \tag{2.9}$$

where C is the total chloride content, c_b is the bound chloride content, c is the content of free chlorides, J_{Cl} is the flux of chlorides, D_{F1} is the diffusion coefficient in Fick's 1st law, x is the depth, and t is the time. Here the flux is simplified in the right-hand part of the equation to Fick's 1st law.

In its most simple form, the mass-balance Eq. (2.9) can be simplified to Fick's 2nd law, Eq. (2.10).

$$\frac{\partial C}{\partial t} = \frac{\partial}{\partial x} D_{F1} \frac{\partial c}{\partial x} \Leftrightarrow \frac{\partial C}{\partial t} = \frac{\partial}{\partial x} \frac{D_{F1}}{\frac{\partial C}{\partial c}} \frac{\partial C}{\partial x} \Leftrightarrow \frac{\partial C}{\partial t} = \frac{\partial}{\partial x} D_{F2} \frac{\partial C}{\partial x} \tag{2.10}$$

Note that the diffusivity D_{F2} is not only a "penetrability and transport property". The chloride binding capacity dC/dc is included in that parameter. Consequently, the penetrability of chloride ions is not directly proportional to the chloride ingress.

In most service-life models one of the two parameters D_{F1} or D_{F2} is used.

2.4.3 Discussion on the Influence of Cracks

Most transport and penetrability test methods are used to characterize un-cracked concrete. Also, service life models are usually used without taking into account the eventual presence of cracks, and only un-cracked concrete properties are considered. However, in situ, the presence of cracks in concrete is not rare, whatever is the cause of these cracks: early-age, thermal loading, shrinkage, or a simple mechanical overloading. Cracked concrete allows the corrosion process to initiate much faster than un-cracked concrete. Cracks may adversely affect concrete durability by providing easy access to aggressive agents and especially to chloride ions. Data in the literature show the importance of this parameter since it can lead to a significant increase of diffusivity. This increase depends on different parameters, among them the crack-width.

The quantification of this increase is not an easy task, however some data exists in the literature and one can expect that experimental data [7] and numerical simulations [8] on cracked concretes could be used to establish "correction factors" according to the type of the crack (self-healing or dynamic) and/or its geometrical properties (crack-width) and/or its density. These "correction factors" could be taken into account in the evaluation of the resistance against diffusion of CO_2 or chloride ions.

References

1. Tuutti K. Corrosion of steel in concrete. CBI Research Report 4.82, Swedish Cement and Concrete Research Institute, Stockholm, 1982.
2. Nilsson L-O. CO_2-cycle in cement and concrete. In: CO_2-uptake by concrete through carbonation. Part 7: Models. A new model for CO_2-absorption of concrete structures. Report TVBM-3158, Laboratory of building materials. Sweden: Lund University; 2011.
3. Hedenblad G. Moisture permeability of mature concrete, cement mortar and cement paste, Report TVBM-1014, Division of Building Materials. Sweden: Lund University; 1993.
4. Climent, M. A., de Vera, G., López, J., Garcia, D. and Andrade, C., Transport of chlorides through non-saturated concrete after an initial limited chloride supply. In: 2nd International RILEM Workshop on Testing and Modelling the Chloride Ingress into Concrete, 11–12 September 2000, Paris, France, 2000.
5. CEB. New Approach to durability design, CEB Bulletin No 238, Lausanne, 1997.
6. Nilsson L-O. Modelling of chloride ingress, WP4 Report of EU-Project ChlorTest, G6RD-CT-2002-00855, 112 pp, 2006.
7. Djerbi A, Bonnet S, Khelidj A, Baroghel-Bouny V. Influence of traversing crack on chloride diffusion into concrete. Cem Concr Res. 2008;38(6):877–83.
8. Kamali-Bernard S, Bernard F. Effect of tensile cracking on diffusivity of mortar: 3D numerical modelling. Comput Mater Sci. 2009;47(1):178–85.

References

Chapter 3
Prescriptive Durability Specifications

R. Torrent, R. d'Andréa, A. Gonçalves, F. Jacobs, K. Imamoto, A. Kanellopoulos, M. Khrapko, A.V. Monteiro and S.V. Nanukuttan

3.1 Introduction

It is important to start with two basic definitions [1] (compare also 1.2):

Prescriptive Specification

A specification in which the composition of materials and methods of installation are defined. A prescriptive specification for concrete focuses on the characteristics of raw materials, mix proportions, batching, mixing, and transport of fresh concrete and a range of construction operations from placing to curing. Prescriptive specifications rely on observed or implied relationships between the details specified and the desired final, in-place, or end-product concrete performance. Under a

R. Torrent (✉)
Materials Advanced Services, Buenos Aires, Argentina

R. d'Andréa
Eduardo Torroja Institute (IETcc-CSIC), Madrid, Spain

A. Gonçalves
Laboratório Nacional de Engenharia Civil - LNEC, Lisbon, Portugal

F. Jacobs
TFB, Wildegg, Switzerland

K. Imamoto
Tokyo University of Science, Katsushika, Japan

A. Kanellopoulos
University of Cambridge, Cambridge, UK

M. Khrapko
CBE Consultancy Ltd, Auckland, New Zealand

A.V. Monteiro
Laboratório Nacional de Engenharia Civil, Lisbon, Portugal

S.V. Nanukuttan
Queen's University Belfast, Belfast, UK

H. Beushausen and L. Fernandez Luco (eds.), *Performance-Based Specifications and Control of Concrete Durability*,
RILEM State-of-the-Art Reports 18, DOI 10.1007/978-94-017-7309-6_3

prescriptive specification, the desired end-product performance may or may not be described.

Performance Specification

A specification that defines required results, criteria by which performance will be judged, and methods of evaluation, without requirements for how the results are to be obtained.

Since the times of the famous Roman architect Marco Vitruvio Pollione, who practiced in the last 50 years of the 1st century BC, most codes of concrete constructions applied prescriptive requirements to ensure the durability of the structures. In his treaty *De Architectura,* in order to achieve durable roman concrete constructions especially in contact with water, Vitruvio specified the type of binder (lime—including the correct process to produce it—and pozzolan as well as the proportion pozzolan/lime), the quality of the sand, the proportion lime/sand, as well as indications on how the elements had to be built. About 100 years later, Plinio the Elder in his *Natural History*, recommended to use as little water as possible to produce the roman concrete and to thoroughly compact it [2].

In terms of specifications for concrete strength, a steady change from prescriptive to performance took place very early, possibly due to the fast development of suitable and widely accepted testing methods. Reportedly, the first systematic strength tests of concrete were conducted in Germany in 1836 [3] and the origin of ASTM C39 "Standard Test Method for Compressive Strength of Cylindrical Concrete Specimens", dates from 1921.

On the other hand, the development of test methods, suitable to measure in reasonably short-term the potential durability of a concrete, happened much later. For instance [4]:

- the water sorptivity test was developed by G. Fagerlund in the late 1970s; a site method (ISAT) was proposed in 1969 by M. Levitt
- the Rilem-Cembureau method to measure O_2 permeability was developed by C. D. Lawrence around 1980; J.W. Figg proposed a site test for air and water permeability in 1973
- the chlorides migration test method, today covered by ASTM C1202, was developed by D. Whiting also around 1980; the test method to measure the diffusivity to chlorides, today covered by NT Build 492, was developed by L. Tang and L.-O. Nilsson about 10 years later.

The lack of suitable and practical test methods to measure durability-related properties is possibly the main reason why prescriptive specifications for durability have lasted so long and are still the basis of the most widely used Codes for Structural Concrete (ACI and EN Standards).

Nevertheless, some national standards have already started to include performance requirements on top of the prescriptive ones that are still present. Canadian Standard A23.1-04/A23.2-04 includes requirements of maximum values of Coulombs (ASTM C1202) for certain exposure and service life conditions. The Swiss Standards SIA 262 and 262/1, possibly the most advanced in the world in terms of performance requirements, specify maximum values of water sorptivity or chlorides diffusivity (depending on the exposure classes) on cast specimens and of air-permeability measured on site.

Given the interest of many associations (RILEM, ACI, NRMCA, *fib*, etc.) in promoting the development of performance specification for concrete it is foreseen that, in the coming 10 years, we will see more standards following the P2P (Prescription to Performance) trend.

In this Chapter, a survey of relevant Prescriptive Codes and Standards (see Table 3.1), dealing with the durability of concrete structures, has been performed, comparing them and highlighting coincidences and discrepancies. The analysis is restricted, in line with the scope of TC230-PSC, to the cases of damage due to steel corrosion (Tables 3.2 and 3.3).

3.2 Exposure Classes

Table 3.4 presents a summary of the way the main Codes and Standards define Exposure Classes. An attempt has been made to include the exposure classes of different Codes and Standards in "equivalent" categories. However, a univocal equivalence between Exposure Classes for different standards is impossible, due to the different criteria used to establish them. When just one class exists per type of damage, it has been attributed to the "Severe" category.

ACI 318:
Defines 4 exposure categories for: Freezing and thawing (F), Sulfate (S), Low Permeability (P) and Corrosion protection of the reinforcement (C). Within each category, there may be more than one class, indicated with a digit (0 to 3) that rates the degree of severity of the exposure. Digit 0 indicates that the specific category is not applicable to the structure or element (e.g. Exposure Class F0 indicates that there is no risk of freezing and thawing damage).

It is the responsibility of the licensed design professional to assign exposure classes based on the severity of the anticipated exposure of structural concrete members for each exposure category.

In total, ACI 318 defines 4 exposure categories and 13 exposure classes, 3 of which refer to corrosion of steel (see Table 3.4).

EN 206:
Defines 6 exposure categories for: No risk of corrosion or attack (X0), Corrosion induced by carbonation (XC), Corrosion induced by chlorides other than from sea water (XD), Corrosion induced by chlorides from sea water (XS), Freeze/thaw attack with or without de-icing agents (XF) and Chemical attack (XA). Except for the first category, there are more than one class, indicated by a digit (1 to 3 or 4) that rates the degree of severity of the exposure.

In total, EN 206 defines 6 exposure categories and 18 exposure classes, out of which 15 apply within the scope of this report (see Table 3.4).

Table 3.1 List of standards covered or referred to in the survey: General Concrete Construction Codes/Standards

Country	Standard designation	Title or Brief Description
USA	ACI-318-08 ACI 201.2R	Structural concrete building code Guide to durable concrete
CEN European Committee for Standardization	EN 1992-1-1 EN206 EN 13670	Eurocode 2: Design of concrete structures—Part 1: General rules and rules for buildings Concrete—Part 1: Specification, performance, production and conformity Execution of concrete structures
Australia	AS 3600-2001	Australian Standards on Concrete Structures
Germany	DIN 1045-2	Application Rules for EN 206
México	NMX C403 pNMX C155	Mexican standard on structural use of concrete—version 1999 Mexican standard on structural use of concrete—draft version 2010
Portugal	LNEC E 464	Concrete. Prescriptive methodology for a design working life of 50 and of 100 years under environmental exposure
Spain	EHE-08	Spanish Instructions on Structural Concrete
Switzerland	SIA 262:2003 SIA 262/1:2013 Annex A Annex B Annex E Annex H SN EN 206	Swiss concrete construction code Concrete construction—complementary specifications: Capillary suction Chloride resistance Air-permeability on site Water content of fresh concrete National Annex of EN 206
UK	BS 8500-1:2006 BS 8500-2:2006	Concrete—complementary British Standard to BS EN 206 Part 1: Method of specifying and guidance for the specifier. Part 2: Specification for constituent materials and concrete

Main standards and codes are shown in bold characters

3.2.1 Complementary Local Standards

AS 3600:

Defines 5 exposure categories for: Surface of members in contact with the ground (1), Surfaces of members in interior environments (2), Surfaces of members in above-ground exterior environments (3), Surfaces of members in water (4), Surfaces of marine structures in sea water (5), and Surfaces of members in other environments (6). Within each category there are subcategories (e.g. in category 4,

Table 3.2 Standards for Concrete Constituents

Country	Standard designation	Title or brief description
USA	ASTM C 33	Standard specification for concrete aggregates
	ASTM C 150	Standard specification for Portland cement
	ASTM C 595	Standard specification for blended hydraulic cements
	ASTM C 618	Standard specification for coal fly ash and raw or calcined natural pozzolan for use in concrete
	ASTM C 845	Standard specification for expansive hydraulic cement
	ASTM C 989	Standard specification for slag cement for use in concrete and mortars
	ASTM C 1157	Standard performance specification for hydraulic cement
	ASTM C 1218	Standard test method for water-soluble chloride in mortar and concrete
	ASTM C 1240	Standard specification for silica fume used in cementitious mixtures
	ASTM C 1602	Standard specification for mixing water used in the production of hydraulic cement concrete
CEN	EN 12620	Aggregates for concrete
	EN 1097-6	Tests for mechanical and physical properties of aggregates—Part 6: determination of particle density and water absorption
	EN 197-1	Cement—Part 1: composition, specifications and conformity criteria for common cements
	EN 450-1	Fly ash for concrete—Part 1: definition, specifications and conformity criteria
	EN 13263-1	Silica fume for concrete—Part 1: Definitions, requirements and conformity criteria
	EN 15167-1	Ground granulated blast furnace slag for use in concrete, mortar and grout—Part 1: definitions, specifications and conformity criteria
	EN 1008	Mixing water for concrete

a differentiation is made on whether the member is in fresh or soft/running water). For each particular case, an exposure class is attributed, namely A1, A2, B1, B2, C1, C2, with the requirements becoming more stringent in that order. A further class U (undefined) is also attributed to cases of exposure to other environments, but predominantly referred to chemical attack.

For exposure category 3, the following classes apply:

- inland (>50 km from the coastline) depending on the geographical location, with the country divided into four regions: Non-industrial aid, Non-industrial Temperate, Non-industrial Tropical, and Industrial.
- Near-coastal (1 to 50 km from coastline), and
- Coastal

Table 3.3 Standards for Reinforcing Steel

USA	ASTM A615	Standard specification for deformed and plain Carbon-steel bars for concrete reinforcement
	ASTM A706	Standard specification for low-alloy steel deformed and plain bars for concrete reinforcement
	ASTM A767	Standard specification for zinc-coated (Galvanized) Steel bars for concrete reinforcement
	ASTM A775	Standard specification for epoxy-coated steel Reinforcing bars
	ASTM A934	Standard specification for epoxy-coated prefabricated steel reinforcing bars
	ASTM A955	Standard specification for deformed and plain stainless-steel bars for concrete reinforcement
	ASTM A996	Standard specification for rail-steel and axle-steel deformed bars for concrete reinforcement
	ASTM A1035	Standard specification for deformed and plain, low-carbon, chromium, steel bars for concrete reinforcement
CEN	EN 10080	Steel for the reinforcement of concrete
	EN 10088-1	Stainless steels—Part 1: list of stainless steels
	EN 10088-2	Stainless steels—Part 2: technical delivery conditions for sheet/plate and strip of corrosion resisting steels for general purposes
	EN 10088-3	Stainless steels—Part 3: technical delivery conditions for semi-finished products, bars, rods, wire, sections and bright products of corrosion resisting steels for general and construction purposes
	EN 10348	Steel for the reinforcement of concrete—Galvanized reinforcing steel
Sweden	SS 14 23 40	Stainless steel reinforcement—steel 2340
UK	BS 6744	Stainless steel bars for the reinforcement of and use in concrete. Requirements and test methods

The severity of the exposure rises in that order.

In total, AS 3600 defines 7 exposure categories and 6 exposure classes, out of which 6 apply within the scope of this report (see Table 3.4).

EHE-08:
Defines 7 exposure categories for: Non aggressive (I), Normal (II), Marine (III), Chlorides other than from sea (IV), Chemical attack (Q), Frost (H/F) and Erosion (E). For some categories there are more than one class, indicated by a letter (a to c) that rates the degree of severity of the exposure, in that order.

In total, EHE-08 defines 7 exposure categories and 13 exposure classes, out of which 9 apply within the scope of this report (see Table 3.4).

NMX C403 and pNMX C155:
Presently the valid standard in México is NMX C403-1999, which defines 5 exposure categories for: Dry (1), Moist or submerged in water (2a), Moist with

Table 3.4 Classification of exposure conditions according to various standards

Exposure type		Standard/code					
Damage	Severity	ACI 318	EN 206	AS 3600	EHE-08	NMX C403	pNMX C155
No risk		C0 /F0	X0	–	I	1	1
Corrosion induced by carbonation	Mild	–	XC1	–	–	–	–
	Moderate	–	XC2, XC3	–	IIa	–	–
	Severe	C1	XC4	–	IIb	2a	2a
Corrosion induced by Seaborne chlorides	Mild	–	XS1	B1	IIIa	–	4a, 4b
	Moderate	–	XS2	B2	IIIb	–	4c
	Severe	C2	XS3	C1, C2	IIIc	4	4d
Corrosion induced by de-icing chlorides	Mild	–	XD1	–	–	–	–
	Moderate	–	XD2	–	–	–	–
	Severe	C2	XD3	–	IV	3	3
Frost-thaw with/without chlorides	Mild		XF1	–	–	–	–
	Moderate	F1	XF2	–	H	–	–
	Severe	F2	XF3	–	F	2b	3
	Very severe	F3	XF4	–*	–	–	–

– Class not defined in the standard
–* AS3600 does not provide a classification to freeze/thaw exposure as such, but provides requirements for specific conditions

freezing and thawing (2b), Moist with freezing and thawing and deicing salts (3) and Marine (4).

In total, NMX C403 defines 5 exposure categories and 9 exposure classes, out of which 5 fall within the scope of this report (see Table 3.4).

Under discussion is the draft of the new standard pNMX C155-2010. It is of interest to compare the criteria of both standards, because the draft introduces significant changes in the prescriptive requirements, as well as performance requirements not existing in the current standard NMC C403. The draft pNMX C155 defines 6 exposure categories for: Dry (1), Moist or submerged in water (2), Moist with freezing and thawing (3), Marine (4), Chemical (5), Erosion and cavitation (6). For some categories there are more than one class, indicated by a letter (a to d) that rates the degree of severity of the exposure, in that order.

In total, pNMX C155 defines 6 exposure categories and 16 exposure classes, out of which 7 fall within the scope of this report (see Table 3.2). The main difference between this draft and the current standard, in terms of classification of exposure, is the more detailed definition of the marine environment, that has 4 classes instead of just one in the current standard.

3.3 Materials

3.3.1 General

EN 206:
Constituent materials shall not contain harmful ingredients in such quantities as may be detrimental to the durability of the concrete or cause corrosion of the reinforcement and shall be suitable for the intended use in concrete.

3.3.2 Cements

ACI 318:
All types of cements complying with ASTM C 150, 595, 845 and 1157, except IS Cement (slag content ≥ 70 %) can be used for exposure classes C1, C2, F1, F2 and F3.

EN 206:
All types of cements complying with EN 197-1, applicable to Italy, France, Belgium and Netherlands, can be used indifferently. On the other hand, some countries (e.g. Austria, Croatia, Germany, Portugal, Romania, Switzerland) limit the types of cement that can be used for particular Exposure Classes. The German standard DIN 1045-2 is one of the most restrictive in terms of limitations to the cement types useable (see Table 3.5 showing the restrictions for steel corrosion and frost attack exposure classes).

3.3.3 Supplementary Cementitious Materials (SCM)

ACI 318:
Fly-ash and pozzolans (ASTM C618), Ground-granulated blast-furnace slag (ASTM C989) and Silica fume (C1240) can be freely used for exposure classes C1, C2, F1 and F2.

For class F3, a maximum percentage of addition is established for single blends (25 % for fly-ash and pozzolans, 50 % for slag and 10 % for silica fume) as well as for ternary blends (see Table 3.4 of ACI 318).

EN 206:
Fly Ash conforming to EN 450-1 and Silica Fume conforming to EN 13263-1 can be used as type II (pozzolanic and latent hydraulic) additions. No explicit reference is made to the use of Ground Granulated Blast Furnace Slag (GGBS), covered by European Standard EN 15167-1 since 2006, although it will be included in next revision.

Table 3.5 Cement types allowed and not allowed after DIN 1045-2

			Carbonation-induced corrosion				Chloride-induced corrosion						Freezing and Thawing			
							Non-marine chlorides			Marine chlorides						
			XC1	XC2	XC3	XC4	XD1	XD2	XD3	XS1	XS2	XS3	XF1	XF2	XF3	XF4
CEM I			+	+	+	+	+	+	+	+	+	+	+	+	+	+
CEM II	A/B	S	+	+	+	+	+	+	+	+	+	+	+	+	+	+
	A	D	+	+	+	+	+	+	+	+	+	+	+	+	+	+
	A/B	P/Q	+	+	+	+	+	+	+	+	+	+	+	–	+	–
	A	V	+	+	+	+	+	+	+	+	+	+	+	–	+	–
	B		+	+	+	+	+	+	+	+	+	+	+	–	–	–
	A	W	+	+	–	–	–	–	–	–	–	–	–	–	–	–
	B		–	+	–	–	–	–	–	–	–	–	–	–	–	–
	A/B	T	+	+	+	+	+	+	+	+	+	+	+	+	+	+
	A	LL	+	+	+	+	+	+	+	+	+	+	+	+	+	+
	B		+	+	–	–	–	–	–	–	–	–	–	–	–	–
	A	L	+	+	+	+	+	+	+	+	+	+	–	–	–	–
	B		+	+	–	–	–	–	–	–	–	–	–	–	–	–
	A	M	+	+	–	–	–	–	–	–	–	–	–	–	–	–
	B		–	+	–	–	–	–	–	–	–	–	–	–	–	–
CEM III	A		+	+	+	+	+	+	+	+	+	+	+	+	+	xx
	B		+	+	+	+	+	+	+	+	+	+	+	+	+	xx
	C		–	+	–	–	–	–	–	–	–	–	–	–	–	–
CEM IV	A		–	+	–	–	–	–	–	–	–	–	–	–	–	–
	B		–	+	–	–	–	–	–	–	–	–	–	–	–	–
CEM V	A		–	+	–	–	–	–	–	–	–	–	–	–	–	–
	B		–	+	–	–	–	–	–	–	–	–	–	–	–	–

+ Applicable
– Non-applicable
xx Applicable with restrictions

There are some limits to the maximum amount of Fly Ash and Silica Fume that can be considered as contributing a "cementitious" value to the compliance with w/c_{max} and $cement_{min}$ requirements for durability. If a greater amount of fly ash than 33 % or Silica Fume than 11 % of the content of CEM I (OPC) is used, the excess shall not be taken into account for the calculation of the water/cementitious ratio and the minimum cement content.

In some national annexes different regulations were made.

EHE-08:
Use restricted to fly-ash and silica fume; GGBS not allowed to be used in concrete.

3.3.4 Aggregates

ASTM C33:
No specific provisions for aggregates regarding steel corrosion.

EN 12620:2000:
When required, the water-soluble chloride ion content of aggregates for concrete shall be determined and shall, on request, be declared by the producer. If the water-soluble chloride ion content of the combined aggregate is known to be 0.01 % or lower this value can be used in the calculation of the chloride content of concrete.

3.3.5 Admixtures

ACI 318:
Calcium chloride or admixtures containing chloride from sources other than impurities in admixture ingredients shall not be used in prestressed concrete, in concrete containing embedded aluminum, or in concrete cast against stay-in-place galvanized steel forms.

EN 206:
Calcium chloride and chloride based admixtures shall not be added to concrete containing steel reinforcement, prestressing steel reinforcement or other embedded metal.

3.3.6 Water

ACI 318:
Refers to ASTM C1602, where an optional limit of chloride content (as Cl^-) in mixing water of 500 ppm for prestressed concrete and bridge decks and of 1000 ppm for reinforced concrete is established.

EN 206:
Refers to EN 1008, where a limit of chloride content (as Cl^{-}) in mixing water of 500 mg/l for prestressed concrete or grout and of 1000 mg/l for reinforced concrete is established. For plain concrete without embedded metals the limit is 4500 mg/l. However, these limits can be exceeded if it is proven that the chloride content of concrete does not exceed the limits described in the next section.

3.3.7 Chloride Content of Concrete

ACI 318:
For corrosion protection of reinforcement in concrete, the maximum water soluble chloride ion concentrations in hardened concrete at ages from 28 to 42 days contributed from the ingredients including water, aggregates, cementitious materials, and admixtures shall not exceed the limits of Table 3.6 (4.4.1 in the standard). When testing is performed to determine the water soluble chloride ion content in concrete, test procedures shall conform to ASTM C 1218.

An initial evaluation may be performed by testing individual concrete ingredients for total chloride ion content. If the total chloride ion content, calculated on the basis of concrete proportions, exceeds the values permitted in Table 3.6, it may be necessary to test samples of the hardened concrete for water-soluble chloride ion content described in the ACI 201.2R guide. Some of the total chloride ions present in the ingredients will either be insoluble or will react with the cement during hydration and become insoluble under the test procedures described in ASTM C 1218.

When epoxy or zinc-coated bars are used, the limits in Table 3.6 may be more restrictive than necessary.

EN 206:
For reinforced concrete not prestressed, there are two classes Cl 0,20 and Cl 0,40 that accept a maximum chloride ion content by mass of cement of 0.20 and 0.40 %, respectively. For the determination of the chloride content of the concrete, the sum of the contributions from the constituent materials shall be determined based on the

Table 3.6 Maximum admissible content of water-soluble chloride of concrete

Type of member	Maximum water soluble chloride ion (Cl^{-}) in concrete, percent by weight of cement
Prestressed concrete	0.06
Reinforced concrete exposed to chloride in service	0.15
Reinforced concrete that will be dry or protected from moisture in service	1.00
Other reinforced concrete construction	0.30

maximum chloride content of the constituent either permitted in the standard for the constituent or declared by the producer of each constituent material.

For prestressed reinforced concrete, there are two classes Cl 0,10 and Cl 0,20 that accept a maximum chloride ion content by mass of cement of 0.10 and 0.20 %, respectively.

The applicable class depends upon the national regulations valid in the place of use of the concrete.

3.3.8 *Reinforcing Steel*

ACI 318 (3.5.3.1):
Deformed reinforcing bars shall conform to the requirements for deformed bars in one of the following specifications:

(a) Carbon steel: ASTM A615 (marked as S)
(b) Low-alloy steel: ASTM A706 (marked as W for "enhanced weldability");
(c) Stainless steel: ASTM A955;
(d) Rail steel and axle steel: ASTM A996. Bars from rail steel shall be Type R.

ACI 318 (3.5.3.8):
Galvanized reinforcing bars shall conform to ASTM A767. Epoxy-coated reinforcing bars shall comply with ASTM A775 or with ASTM A934. Bars to be galvanized or epoxy-coated shall conform to one of the specifications listed in Sect. 3.5.3.1 of ACI 318.

EN 1992-1-1:
According to CEN expectations, weldable reinforcing steel (bars, de-coiled rods, wire fabrics and lattice girders) have to comply with EN 10080, which establishes the test methods, verification procedures and limits for the chemical composition of steel. For stainless reinforcing steel, EN 10088-1, EN 10088-2 and EN 10088-3 apply instead. Concerning galvanized reinforcing steel EN 10348 applies.

Since none of the above standards are harmonized, the majority of the European countries uses their own procedures for certifying reinforcing steel by referring in the National Annex of EN 1992-1-1 to the relevant National standards. Examples of these National standards for stainless steel are: BS 6744 and SS 14 23 40.

The requirements for the properties of the reinforcement are established in EN 1992-1-1, which defines three ductility classes: Class A, B and C, in increasing order of ductility.

3.4 Service Life

Table 3.7 presents the service life that is expected for structures built according to the requirements of the respective codes and standards.

Table 3.7 Expected service life corresponding to different standards

Code/Standard	Expected service life	Comments
ACI 318-08	Not disclosed	
Eurocode 2—EN 1992-1-1 (Design of concrete structures—Part 1: general rules and rules for buildings)	50 years for structural class S4	Increase minimum cover, e.g. by 10 mm, for 100 years
EHE-08	50 years	Increase minimum cover, e.g. by 10 mm for 100 years
AS 3600	40–60 years	

3.5 Concrete Strength Grades

Table 3.8 shows the requirements for minimum strength grade as function of Exposure Class, according to the respective codes and standards. Strengths indicated correspond to cylindrical specimens. The Strength Grade or Specified Strength corresponds to the 5 or 9 % lower fractile in EN 206 or ACI 318, respectively.

Table 3.8 Minimum strength grades required for different exposure classes of Table 3.4

Exposure type		Minimum strength grade—Cylinder (MPa)					
Damage	Severity	ACI 318	EN 206^{x}	AS 3600	EHE-08	NMX C403	pNMX C155
No risk		17.5	12.0	–	25.0	20.0	–
Corrosion induced by carbonation	Mild	–	20.0	–	–	–	–
	Moderate	17.5	25.0, 30.0	–	o	–	–
	Severe	–	30.0	–	o	27.0	o
Corrosion induced by seaborne Cl$^-$	Mild	–	30.0	32.0	o	–	o
	Moderate	–	35.0	40.0	o	–	o
	Severe	35.0	35.0	50.0	o	30.0	o
Corrosion induced by de-icing Cl$^-$	Mild	–	30.0	–	–	–	–
	Moderate	–	30.0	–	–	–	–
	Severe	35.0	35.0	–	o	25.0	o
Frost-thaw with/without chlorides	Mild	–	30.0	32.0	–	–	–
	Moderate	31.5*	25.0*	40.0	o	25.0	–
	Severe	31.5*	30.0*	–	o*	25.0	o
	Very severe	31.5*	30.0*	–	–	–	–

*Minimum entrained air specified
$^{-}$Class not defined in the standard
^{o}Class defined without specified requirement
^{x}According to an informative Annex F

Extreme cases are the Spanish EHE-08 and the draft Mexican pNMX C155 that have no specifications for minimum strength grade and the Australian AS 3600, for which the only requirement for the different durability exposure classes is the compressive strength (5 % fractile). Note the very demanding strength requirements in general, but especially for severe marine exposure, of AS 3600.

3.6 Mix Composition Prescriptions

Table 3.9 presents the constraints to the mix design prescribed by different standards, as a function of the exposure classes presented in Table 3.4. They refer to maximum water/cement ratio and minimum cement content.

The following can be noticed:

- All standards excluding AS 3600 specify a maximum w/c ratio
- All standards excluding ACI 318 and AS 3600 specify a minimum cement content
- The Mexican standard, revised in 2009, reduces the maximum w/c ratio allowed for sea water exposure from a general value of 0.55 to 0.40–0.45 (depending on the severity of the exposure)

It has to be mentioned that the individual countries that adopted the EN 206 have modified the limits shown in Table 3.8 to a very large extent. This can be seen in Figs. 3.1 and 3.2, where the lower and upper limits for the composition constraints are indicated (for more details see [5]).

Figures 3.1 and 3.2 show the very large differences of criteria that exist within the CEN member countries regarding prescriptive requirements. The numbers below the exposure classes indicate the number of countries where no requirement is specified for the particular constraint.

Typically a range of 0.15 in w/c ratio exists (maximum range of 0.30 for XC1). The range in cement content is enormous, to a large extent due to the low values specified in Denmark and Sweden (150 kg/m^3). The countries with a tendency to specify high limits to the minimum cement content are Ireland, Italy and Portugal (also the UK in some specific cases).

3.7 Contribution of Supplementary Cementitious Materials

The constraints shown in Table 3.9 to the composition of the mixes involve the cement content (directly or indirectly through the w/c ratio).

One issue of high technical, economical and environmental impact is how the standards treat the "cementitious" contribution of supplementary cementitious materials, batched at the concrete plant as a separate ingredient.

Table 3.9 Requirements for maximum w/c or w/c_m and minimum cement (cementitious) content as function of exposure class

Exposure type		Maximum w/c ratio						Minimum cement content (kg/m^3)					
Damage	Severity	ACI 318	EN 206[x]	AS 3600	EHE-08	NMX C403	pNMX C155	ACI 318	EN 206[x]	AS 3600	EHE-2007	NMX C403	pNMX C155
No risk		o	–	–	0.65	0.65	0.65	o	–	–	–	270	270
	Mild	–	0.65	–	–	–	–	–	260	–	–	–	–
	Moderate	Damage	Severity	–	0.60	–	–	o	280, 280	–	–	–	–
	Severe	No risk	0.50	–	0.55	0.60	0.60	–	300	–	–	300	300
Corrosion induced by Seaborne Cl^-	Mild	Corrosion induced by carbonation	Mild	–	0.50	–	0.45	–	300	–	300	–	300
	Moderate		Moderate	–	0.50	–	0.45	–	320	–	325	–	300
	Severe	0.40	0.45	–	0.45	0.55	0.40	o	340	–	350	300	300
Corrosion induced by de-icing Cl^-	Mild	–	0.55	–	–	–	–	–	300	–	–	–	–
	Moderate	–	0.55	–	–	–	–	–	300	–	–	–	–
	Severe	0.40	0.45	–	0.45	0.55	0.55	o	320	–	325	300	300
Frost-thaw with/without chlorides	Mild	–	0.55	–	–	–	–	–	300	–	–	–	–
	Moderate	0.45	0.55	–	0.55	0.55	–	o	300	–	–	300	–
	Severe	0.45	0.50	–	0.50	0.55	0.55	o	320	–	325	300	300
	Very severe	0.45	0.45	–	–	–	–	o	340	–	–	–	–

[–]Class not defined in the standard
[o]Class defined without specified requirement
[x]According to normative Annex

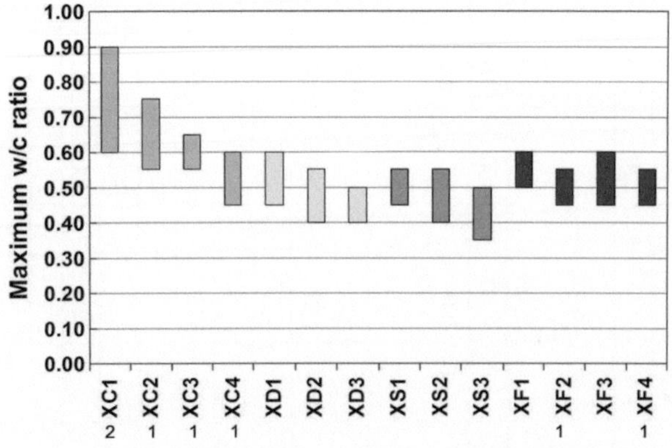

Fig. 3.1 Upper and lower limits of maximum w/c ratios specified in CEN member countries for several exposure classes (the numbers below the exposure classes indicate the number of countries where no requirement is specified for the particular constraint)

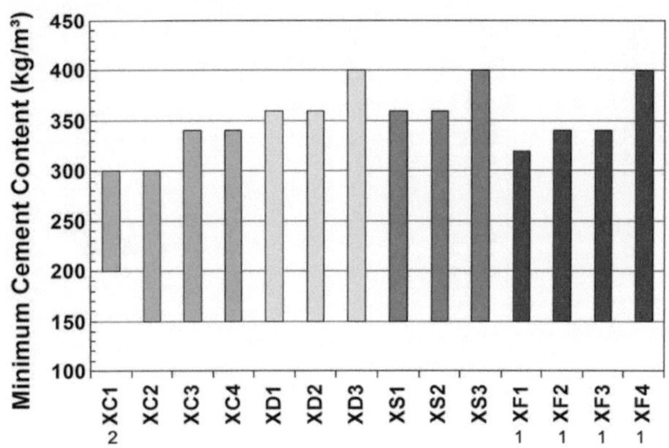

Fig. 3.2 Upper and lower limits of minimum cement contents specified in CEN member countries for several exposure classes (the numbers below the exposure classes indicate the number of countries where no requirement is specified for the particular constraint)

This is done through the "cementitious contribution" factor k, such that:

$$c_m = c + ka \tag{3.1}$$

where c_m is the cementitious content, c is the cement content and a is the supplementary cementitious content. The composition constraints are now expressed as minimum c_m and maximum w/c_m ratio.

Table 3.10 shows how different standards assign predefined values to the factor k.

Table 3.10 Maximum *k* values according to different standards

Standard	Predefined value of k for:				Comments
	GGBFS	PFA	Pozzolan	Silica Fume	
ACI 318-08	1.0	1.0	1.0	1.0	
EN 206	–	0.4[a] a/c ≤ 0.33	–	2.0[b] a/c ≤ 0.11	[a]Depending on CEM I class; [b]Depending on w/c_m and exposure class
LNEC E 464	–	–	–	–	The additions count fully to the cementitious content if there is a cement covered by EN 197-1 with the same composition of the combination of cement and addition
EHE-08		0.3–0.5[a] a/c ≤ 0.35		1.0–2.0[b] a/c ≤ 0.10	[a]When supported by tests [b]Depending on w/c_m and exposure class
NMX C403					Not allowed

[a, b]There are further restrictions to the additions and maximum contents

EN 206 presents an alternative approach to the "*k*" value by which a mix can be accepted if it is proven that the concrete has an equivalent performance, especially with respect to its reaction to environmental actions and to its durability, when compared with a reference concrete in accordance with the requirements for the relevant exposure class. Annex J of EN 206-1 gives some principles for the assessment of the equivalent concrete performance concept, which does not seem to be widely applied. In Holland this concept is implemented through CUR-Recommendation 48.

LNEC has developed a quite comprehensive procedure, based on performance indicators (oxygen permeability, accelerated carbonation, water sorptivity and chloride migration) to establish whether a candidate concrete has a durability performance equivalent to a reference concrete that complies with the prescriptive indicators (see Sect. 8.12).

BS 8500 presents a special approach by which blends of CEM I and PFA or GGBFS, manufactured in the concrete mixer, are given Broad Designations (similar to the cement designations given in EN 197-1). Additions content may be fully taken into account regarding the cement content for compliance with durability constraints if the suitability is established. A similar procedure is being used in Portugal (LNEC E 464) and in Ireland (I.S. EN 206). The general principles of all these approaches will be considered by the next revision of EN 206, under the common designation of Equivalent Performance of Combinations Concept (EPCC).

3.8 Cover Depth Prescriptions

ACI 318:
Tolerances on cover (in minus): 10 or 13 mm depending on the effective depth of the element.

Minimum Cover depth (mm):

- Concrete cast against and permanently exposed to earth: 75 mm for ø ≥ 19 mm
- Concrete exposed to earth or weather: 51 mm for ø ≤ 16 mm
- Concrete not exposed to weather and not in contact with ground:
  - Slabs, walls, beams: 19 mm for ø ≤ 36 mm
  - Beams, columns: Longitudinal reinforcement, ties, stirrups, spirals: 38 mm
  - Shells, folded plate members: 19 mm for ø ≥ 19 mm; 12.5 mm for ø ≤ 16 mm
- Concrete exposed to chlorides, recommended ≥51 mm for walls and slabs and 63 mm for other members

For precast concrete elements, some 25 % reduction in cover thickness is accepted.

EN 1992-1-1 (Eurocode 2):
Nominal cover depth c_{nom} should be specified in the design drawings and corresponds to the minimum cover c_{min} plus the absolute value of the accepted negative deviation Δc_{dev}:

$$c_{nom} = c_{min} + \Delta c_{dev} \tag{3.2}$$

The recommended acceptable deviation is 10 mm. However, allowance is given for the reduction of this value in cases where concrete cover depths are monitored under an appropriate quality system.

The minimum cover c_{min} is established for durability, fire resistance and reinforcement bond purposes. Only the first one is addressed in this document.

Minimum cover depth due to the environmental conditions for reinforced structures with normal weight concrete can be found in Table 3.11. The choice of the structural class is up to each country to decide. For a design working life of 50 years the recommended structural class is S4 (highlighted in the table in bold font).

The following modifications to the values of Table 3.11 are also recommended, provided that in any case the minimum allowed structural class is S1:

- Design working life of 100 years: increase by 2 structural classes;
- Excluding XC3, if concrete has two strength classes (or one strength class, in cases where more than 4 % air is entrained) above those required in Table 3.8: reduce by 1 structural class;

Table 3.11 Minimum cover regarding the durability of reinforced concrete structures according to EN 1992-1-1

Structural class	Minimum Cover c_{min} (mm)						
	Exposure Class						
	X0	XC1	XC2/XC3	XC4	XD1/XS1	XD2/XS2	XD3/XS3
S1	10	10	10	15	20	25	30
S2	10	10	15	20	25	30	35
S3	10	10	20	25	30	35	40
S4	**10**	**15**	**25**	**30**	**35**	**40**	**45**
S5	15	20	30	35	40	45	50
S6	20	25	35	40	45	50	55

- For XC3, if concrete has one strength above that required in Table 3.8 or if more than 4 % air is entrained: reduce by 1 structural class;
- Structural members with slab geometry: reduce by 1 structural class;
- Special quality control of concrete production: reduce by 1 structural class.

Allowance is also given for the reduction of minimum cover in structures where stainless steel or additional protection (e.g. coating) is used, but no values for these reductions are established in the standard (refers to the national annexes). In LNEC E 464, for instance, the reduction with the use of stainless steel is of 15 mm and, when a coating is applied, of 5 mm. In any case, the minimum cover shall not be less than the one corresponding to structural class S2 or S4 (S1 and S3 when using stainless steel) for working lives of 50 and 100 years, respectively.

For structures where freeze/thaw action on concrete is expected, the above limits for minimum cover are considered sufficient.

For concrete with uneven surfaces the following minimum values must prevail:

- Concrete cast against prepared ground: ≥40 mm;
- Concrete cast directly against soil ≥75 mm;
- Concrete with exposed aggregate: the minimum cover depth should be increased by at least 5 mm.

BS 8500-1: 2006 Approach

This document allows to design the system by linking the nominal cover for a given exposure class to the quality of the concrete (characterized by its compressive strength, maximum w/c ratio, cement type and minimum content), as shown in Table 3.12 for a service life ≥50 years. A similar table exists for a service life ≥100 years. The principle behind this approach is that a better quality concrete cover will have a lower "penetrability", thus allowing a reduction of the cover depth.

AS 3600:

This standard follows a similar approach to BS 8500 in that the required cover for a given Exposure Class (see Sect. 3.2) is a function of the Strength Class chosen (Characteristic Strength), as shown in Table 3.13 for standard formwork and

Table 3.12 Durability recommendations for reinforced and prestressed elements with an intended working life of at least 50 years

Nominal cover (mm)	Compressive Strength Class where recommended, maximum w/c ratio and minimum cement								
	15 + Δc	20 + Δc	25 + Δc	30 + Δc	35 + Δc	40 + Δc	45 + Δc	50 + Δc	Cement type
Corrosion induced by carbonation (XC exposure classes)									
XC1	C20/25 0.70; 240	C20/25 0.70; 240	C20/25 0.70; 240	C20/25 0.70; 240	C20/25 0.70; 240	C20/25 0.70; 240	C20/25 0.70; 240	C20/25 0.70; 240	All
XC2	–	–	C25/30 0.65; 260	C25/30 0.65; 260	C25/30 0.65; 260	C25/30 0.65; 260	C25/30 0.65; 260	C25/30 0.65; 260	All
XC3/4	–	C40/50 0.45; 340	C30/37 0.55; 300	C28/35 0.60; 280	C25/30 0.65; 260	C25/30 0.65; 260	C25/30 0.65; 260	C25/30 0.65; 260	All but IVB-V
	–	–	C40/50 0.45; 340	C30/37 0.55; 300	C28/35 0.60; 280	C25/30 0.65; 260	C25/30 0.65; 260	C25/30 0.65; 260	IVB-V
Corrosion induced by chlorides (XS from sea water, XD other than sea water) *Also adequate for any associated carbonation induced corrosion (XC)*									
XD1	–	–	C40/50 0.45; 360	C32/40 0.55; 320	C28/35 0.60; 300	C28/35 0.60; 300	C28/35 0.60; 300	C28/35 0.60; 300	All
XS1	–	–	–	C45/55 0.35; 380	C35/45 0.45; 360	C32/40 0.50; 340	C32/40 0.50; 340	C32/40 0.50; 340	CEM I, IIA, IIB-S, SRPC
	–	–	–	C40/50 0.35; 380	C32/40 0.45; 360	C28/35 0.50; 340	C25/30 0.55; 320	C25/30 0.55; 320	IIB-V, IIIA
	–	–	–	C32/40 0.40; 380	C25/30 0.50; 340	C25/30 0.50; 340	C25/30 0.55; 320	C25/30 0.55; 320	IIIB
	–	–	–	C32/40 0.40; 380	C28/35 0.50; 340	C25/30 0.50; 340	C25/30 0.55; 320	C25/30 0.55; 320	IVB-V

(continued)

Table 3.12 (continued)

Nominal cover (mm)	Compressive Strength Class where recommended, maximum w/c ratio and minimum cement								
	15 + Δc	20 + Δc	25 + Δc	30 + Δc	35 + Δc	40 + Δc	45 + Δc	50 + Δc	Cement type
XD2 or XS2	–	–	–	C40/50 0.40; 380	C32/40 0.50; 340	C28/35 0.55; 320	C28/35 0.55; 320	C28/35 0.55; 320	CEM I, IIA, IIB-S, SRPC
	–	–	–	C35/45 0.40; 380	C28/35 0.50; 340	C25/30 0.55; 320	C25/30 0.55; 320	C25/30 0.55; 320	IIB-V, IIIA
	–	–	–	C32/40 0.40; 380	C25/30 0.50; 340	C20/25 0.55; 320	C20/25 0.55; 320	C20/25 0.55; 320	IIIB, IVB-V
XD3	–	–	–	–	–	C45/55 0.35; 380	C40/50 0.40; 380	C35/45 0.45; 360	CEM I, IIA, IIB-S, SRPC
	–	–	–	–	–	C35/45 0.40; 380	C32/40 0.45; 360	C28/35 0.50; 340	IIB-V, IIIA
	–	–	–	–	–	C32/40 0.40; 380	C28/35 0.45; 360	C25/30 0.50; 340	IIIB, IVB-V
XS3	–	–	–	–	–	–	C45/55 0.35; 380	C40/50 0.40; 380	CEM I, IIA, IIB-S, SRPC
	–	–	–	–	–	C35/45 0.40; 380	C32/40 0.45; 360	C28/35 0.50; 340	IIB-V, IIIA
	–	–	–	–	–	C32/40 0.40; 380	C28/35 0.45; 360	C25/30 0.50; 340	IIIB, IVB-V

Note For certain exposure classes there are some restrictions on the usable cement types

Table 3.13 Required cover where standard formwork and compaction are used

Exposure class	Required cover (mm)				
	Characteristic strength f'$_c$ (MPa)				
	20	25	32	40	≥50
A1	20	20	20	20	20
A2	(50)	30	25	20	20
B1		(60)	40	30	25
B2			(65)	45	35
C				(70)	50

Note Bracketed figures are the appropriate covers when only one surface of the element is exposed

compaction. The required cover can be reduced if intensive compaction or self-compacting concrete is applied to rigid steel forms for precast elements.

3.9 Concrete Practices

ACI 318:
No special recommendations for durable concrete, just general descriptive recommendations on how to perform the tasks of conveying, depositing (including compaction) and curing. Regarding curing, concrete shall be maintained above 10 °C and in a moist condition for at least the first 7 days after placement.

EN 13670:
EN 13670 gives rules for pre-concreting operations, delivery, reception and site transport of fresh concrete and for its placing, curing and protection.

The requirements for most of the above operations are generic rules of good practice, highlighting some relevant aspects in the case of placing special concretes (e.g. self-compacting concrete and underwater concreting).

Concerning curing, four Curing Classes are defined (Class 1–4).

For Curing Class 1 the curing period shall be at least 12 h (provided that the initial setting time is not greater than 5 h and the temperature of the concrete surface is not less than 5 °C) and for Curing Classes 2, 3 and 4 the curing period shall last until surface concrete achieves, at least, 35, 50 and 70 % of its characteristic compressive strength, respectively.

Informative values for the minimum curing periods as a function of the concrete surface temperature and the concrete strength development (ratio between compressive strengths at 2 days and 28 days) are also given for Curing Classes 2, 3 and 4. For Curing Class 3, used in most common structures, these minimum values are presented in Table 3.14.

Requirements for the temperature of concrete or of surfaces in contact with it are also established. Thereby, concrete shall not be placed over frozen grounds (unless special procedures are followed) or over concrete surfaces under 0 °C at the time of concreting.

Table 3.14 Informative values for the minimum curing periods—curing class 3

Minimum curing period, days[a]			
Concrete strength development[c, d]; $r = (f_{cm2}/f_{cm28})$			
Surface concrete temperature (t), °C	Rapid $r \geq 0.50$	Medium $0.50 > r \geq 0.30$	Slow $0.30 > r \geq 0.15$
$t \geq 25$	1.5	2.5	3.5
$25 > t \geq 15$	2.0	4	7
$15 > t \geq 10$	2.5	7	12
$10 > t \geq 5$[b]	3.5	9	18

[a]Plus any period of set exceeding 5 h
[b]For temperatures below 5 °C, the duration should be extended for a period equal to the time below 5 °C
[c]The concrete strength development is the ratio of the mean compressive strength after 2 days to the mean compressive strength after 28 days determined from initial tests or based on known performance of concrete of comparable composition (see EN 206)
[d]For very slow concrete strength development, special requirements should be given in the execution specification

The temperature of concrete shall not fall below 0 °C until the surface concrete achieves a minimum compressive strength of 5 MPa.

The peak temperature of early age concrete shall not exceed 70 °C unless proven that no significant adverse effects may arise.

Concerning concrete cover, no guidelines or standards for the placement of spacers and chairs are recommended.

3.10 Compliance/Conformity Control

ACI 318:
Regarding strength, compliance control is based on requirements for individual test results and for the overlapping (moving) average of 3 successive test results of concrete cylinders cast at the discharge point of the mixer (plant or jobsite).

No provisions for test methods or conformity control of the specified maximum w/c_m ratio.

No provisions for control of the finished structure, either for concrete quality or for compliance with concrete cover and tolerances.

EN 206:
Conformity with strength requirements consists of a requirement for individual test results and another for the non-overlapping averages of n results, with n = 3 or ≥15 for initial or continuous production conditions, respectively.

Regarding conformity with w/c ratio and cement content, there is a tolerance of 0.02 above the maximum w/c specified, for single results. For cement content, that tolerance is of 10 kg/m^3 below the minimum value specified.

Regarding the determination of the cement, water, or addition content, the values shall be taken either as recorded on the print-out of the batch recorder or where recording equipment is not used, from the production record in connection with the batching instruction.

There are some provisions for the experimental determination of the w/c ratio, although no procedure is described or quoted. Where the water/cement ratio of concrete is to be determined, it shall be calculated on the basis of the determined cement content and effective water content. The water absorption of the aggregate after one hour of immersion shall be determined in accordance with EN 1097-6 and deducted from the measured water content.

No provisions for control of the finished structure, either for concrete quality or for compliance with concrete cover and tolerances.

SIA 262 and SIA 262/1:
The Swiss standards have two peculiarities, as follows. First, the Swiss Code for Concrete Construction SIA 262 states:

1. With regard to durability, the quality of the cover concrete is of particular importance
2. The impermeability of the cover concrete shall be checked by means of permeability tests (e.g. air permeability measurements) on the structure or on core samples taken from the structure

Hence, it recognizes that the durability of a concrete structure is determined by the "in situ" quality of the surface layers and not by cast specimens. Therefore, it is controlling not just the concrete as supplied by the producer, but the end-product including the care and dedication placed on all concrete processing practices applied on site. Swiss Standard 262/1 includes as Annex E a test method to measure the coefficient of air-permeability of concrete on site *kT*, specifying maximum characteristics values as function of exposure classes and a compliance criterion.

The second peculiarity is that SIA 262/1 has an Annex H ("Water content of fresh concrete"), where a standard procedure to determine the total water content of fresh concrete (i.e. the effective water plus that absorbed by the aggregates) is given. Basically it consists in carefully stirring a drying 10 kg sample (for D_{max} 32 mm) of fresh concrete placed on a pan subjected to strong heating, until reaching a constant weight (usually about 15–20 min).

3.11 Discussion and Conclusions

3.11.1 Exposure Classes

The EN 206 presents a large number of Exposure Classes, loosely defined through literary variables ("very low", "low" or "moderate" humidity, "wet, rarely dry", "near the coast", etc.). There is little or no quantification of these variables, nor of particular aspects such as dominant wind direction. In essence, it would be unrealistic to expect that the complexities of macro and micro-environments can ever be fully described qualitatively or quantitatively. These factors may make its application in real life rather difficult for the architect or engineer.

Different opinions exist on this topic, even among the authors of this Chapter. Some [6] suggest that there are too many Exposure Classes, presenting Fig. 3.3 as an example showing that a simple concrete structure for a small house (not exposed to frost, marine or chemical attack) involves 5 different exposure classes.

On the contrary, in Switzerland, Exposure Class XD2 has been subdivided into XD2a and XD2b in function of the chloride content of the solution in contact with the concrete.

ACI 318 follows exactly the opposite approach. For "Corrosion protection of reinforcement", there are just 2 Classes (besides C0), very clearly delimited, without any ambiguity; similarly for "Freezing and thawing". Another good aspect of ACI 318 is that a rating has to be given for each exposure category, e.g. F0, S0, P0, C1 that means no Frost, no Sulphate, no low Permeability, but risk of Corrosion due to carbonation).

Australian Standard AS 3600 is an interesting example, where at least what is "Inland", "Near-Coast" or "Coastal" is defined as function of the distance of the building to the seashore. Similarly, "Industrial" environment refers to areas that are within 3 km of industries that discharge atmospheric pollutants. The "Coastal" delimitation takes into account the direction of the prevailing winds. The "Inland non-industrial" environment can be "arid", "temperate" or "tropical", depending on the geographical location of the site (see Fig. 3.4).

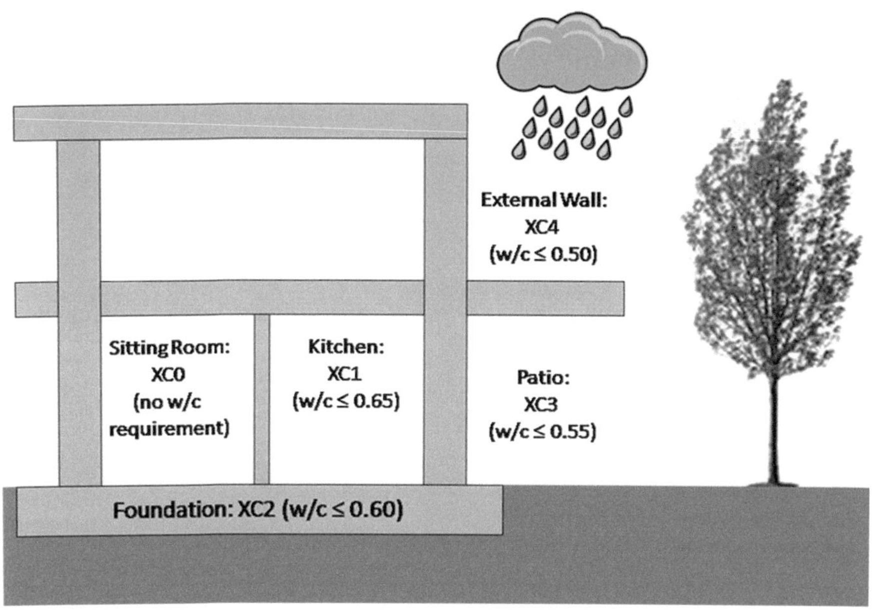

Fig. 3.3 Different exposure classes involved in a small house project in mild climate [6]

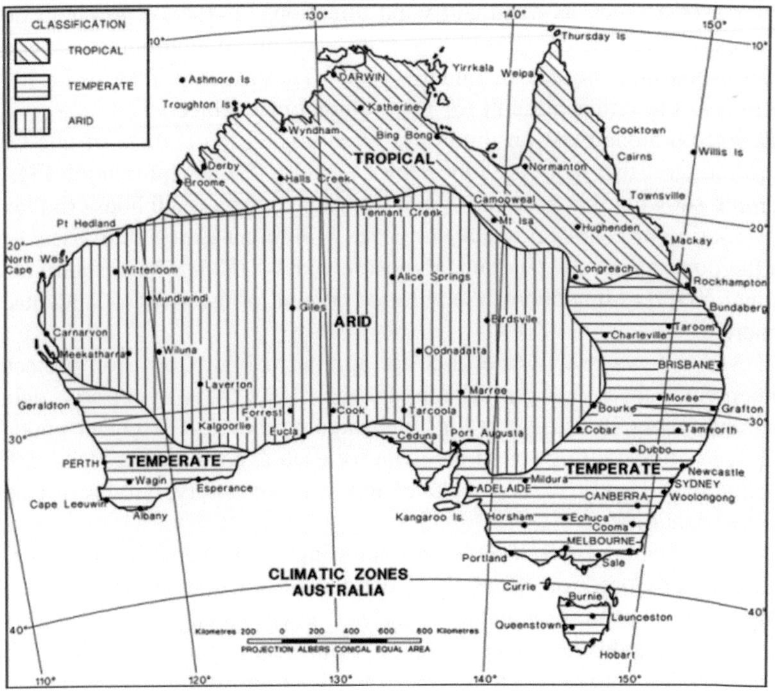

Fig. 3.4 Australia's climatic zones

3.11.2 *Durability Indicators*

3.11.2.1 Concrete Strength

The most used durability indicator is, besides mix composition constraints, the Concrete Strength Grade (Class). It is applied in all surveyed Standards and Codes, with the sole exception of Spain's EHE-08 and is the main durability indicator in AS3600. Often, it is used as an indirect way of controlling the w/c ratio as shown by the following commentaries:

ACI 318 (R4.1.1.):
"Because it is difficult to accurately determine the w/cm of concrete, the *f′c* specified should be reasonably consistent with the w/cm required for durability. Selection of an fc′ that is consistent with the maximum permitted w/cm for durability will help ensure that the maximum w/cm is not exceeded in the field. For example, a maximum w/cm of 0.45 and *f′c* of 3000 psi (about 20 MPa) should not be specified for the same concrete mixture. Because the usual emphasis during inspection is on concrete compressive strength, test results substantially higher than the specified compressive strength may lead to a lack of concern for quality and could result in production and delivery of concrete that exceeds the maximum w/cm".

BS 8500-1:2006 (A.4.2):
"Resistance to chloride ingress is mainly dependent upon the cement or combination type and the w/c ratio, with aggregate quality being a secondary factor. Compressive strength is included as an indirect control on these parameters".

In the Portuguese prescriptive requirements, established in LNEC E 464, the compressive strength has also been set as the controlling parameter of the concrete mix design for each particular exposure class, due to the difficulties of controlling the minimum C and maximum w/c on the construction site.

Both ACI and EN standards specify rather similar minimum strengths for equivalent exposure classes. The most demanding Standard regarding specification of minimum compressive strength is Australian Standard AS3600 (particularly for severe marine exposure) and the more lax is the Mexican Standard NMX C403 (currently under revision).

The procedures for testing and evaluating conformity of compressive strength are robust and are well established worldwide, just with minor variations from country to country. However, the validity of compressive strength as durability indicator is being increasingly questioned [7], even by former advocates of that concept [8]. The compressive strength of concrete is related to the capillary porosity which, in turn, can be related to the permeability of the material, as discussed in [9], where some relations between both properties are proposed.

The *fib* Model Code 2010 proposes some equations relating 'penetrability' properties with the compressive strength of concrete [10] (coefficient of permeability to water and gas, coefficient of diffusion of water, gases and chlorides and coefficient of water absorption).

However, an obvious simple example will show the intrinsic weakness of associating strength with durability. An air-entrained concrete requires, to achieve the same strength, a much lower w/c ratio than a normal concrete, and will present a much lower "penetrability" (on top of its higher frost resistance) than a non-air-entrained concrete of the same strength.

It is important to stress, when dealing with the "penetrability" of concrete on site, that we are concerned with the quality of the first centimetres of concrete, not the bulk. Strength tests on drilled cores will measure the contribution of the whole volume of concrete involved in the specimen and not just the 'covercrete', which will give a distorted information on the potential durability.

3.11.2.2 Water/Cement Ratio

The w/c ratio is extensively used as a durability indicator. It is applied in all reviewed Standards and Codes, with the sole exception of Australia's AS 3600.

However, the discrepancies between the limiting values stipulated by different standards are large. In particular, the maximum w/c ratios established by the different European countries (that adopt EN 206) differ widely, with differences typically of 0.15 for the same exposure class. ACI 318 tends to be more conservative with limiting values typically 0.05 below those stipulated by EN 206 for equivalent Exposure Classes (although the way both standards deal with the contribution of mineral additions makes a true comparison difficult).

The water/cement ratio is a measure of the degree of dispersion of cement particles in the "effective water", i.e. the distance between neighbouring cement particles that has to be bridged by hydration products. The w/c durability indicator is based on the assumption that all cements (or at least those allowed for a particular Exposure Class) perform identically after hydration, which is certainly not true.

This is reflected in Fig. 3.5, showing the Chloride Migration coefficient (after SIA 262/1-B, similar to NT Build 492) of concretes made with different binders, as a function of w/c or w/c_{eq} ratio.

Due to ecological constraints, the composition of cements has been evolving in the past and will continue to change. In Switzerland, up to the end of the 20th century, nearly all cements were of type CEM I (OPC), while today less than 20 % of the total cement consumption belongs to that type. Having in mind that the variety and content of Type II additions and the use of non-CEM I cements will increase in the future, the requirements based solely on prescriptive composition constraints are likely to disappear.

The consideration of the contribution of additions (when batched separately by the concrete producer), through the "*k*" values, to the "cement" denominator of the w/c ratio is a matter of controversy. The equivalent performance concept of EN 206 is promising, but is not clearly defined in the standard and thus not widely applied, although in essence contains implicitly the same limitations, because the reference concrete is defined based on prescriptive requirements and may end being a bad reference.

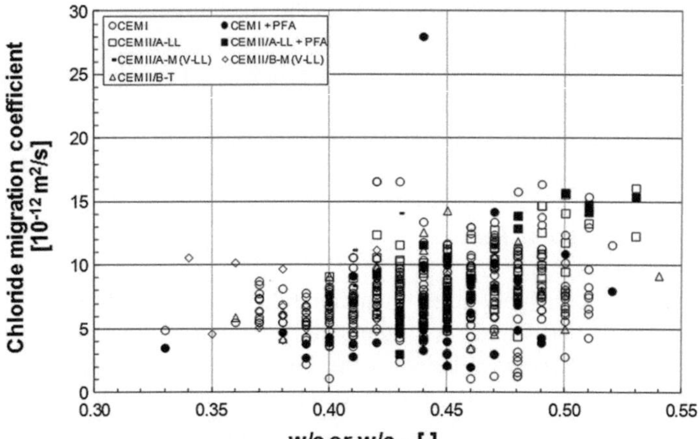

Fig. 3.5 Chloride Migration coefficient versus w/c ratio for different binders [11]

Since practical tests to control the w/c ratio of fresh concrete are not well established, the specification of a maximum value is largely irrelevant from the point of view of the consumer, who basically has to trust the producer for its compliance. For the producer it is basically a parameter for the mix design, which in most countries is seldom if ever checked during production control. The batching record is a weak proof of conformity, due to uncertainties in the effective water content, arising from:

- aggregates moisture: reported in batch record ≠ real
- "slumping" water at the plant/jobsite often not recorded
- eventual washing water left inside drum before batching not accounted for
- need to know accurately aggregates' water absorption

In some countries (Switzerland, Germany, Thailand) the w/c ratio is often determined semi- experimentally, by measuring the total water by drying a sample of fresh concrete, deducting the water absorbed by the aggregates and dividing by the cement and addition content declared by the producer.

3.11.2.3 Minimum Cement Content

The cement content is used as a durability indicator in all reviewed Standards and Codes, with the exceptions of ACI 318 and Australia's AS 3600.

Here, the two main global codes (ACI and EN) follow completely different approaches. Even within EN 206, the criterion applied by the various European countries is widely different, with a typical difference of 200 kg/m^3 in minimum cement content specified for the same Exposure Class.

The inclusion of the Cement Content as durability indicator can only be justified on its role to slow down carbonation and chlorides penetration rates, due to chemical binding of the aggressive species. However, a higher cement content at the same w/c ratio means a proportionally higher paste content in the concrete and, hence, more pores allowing a proportionally higher flow of penetrating aggressive species. Therefore, the higher cement content/m^3 is balanced by a larger quantity of CO_2 or chloride ions per m^3, besides the higher risk of thermal and shrinkage cracking, [12]. This fact has been confirmed experimentally [13]. The same considerations regarding the influence of the cement types and the "k-value" approach for additions, formulated in the previous section, apply here as well.

3.11.3 Final Remarks

Exposure Classes, that are a need for both prescriptive and performance specifications, should be sufficiently clear and unambiguous that, reasonably, two different specifiers (architects, engineers) are able to attribute the same Exposure Class to the elements of a given projected structure. This may not be the case with EN 206, which seems to be rather complex to apply; ACI 318 may be too simple but it is certainly practical and AS 3600 constitutes a good balance between the other two.

The durability indicators considered in most codes, i.e. compressive strength, maximum water/cement ratio and minimum cement content may be inadequate to provide sufficient protection of the concrete structures against most aggressive species. Therefore, prescriptive-based Codes and Standards may fail in achieving potentially durable concrete designs.

In particular, these durability indicators are loosely linked to the rate at which aggressive species penetrate the concrete and develop their actions. Moreover, the prescriptive constraints to the water/cement ratio are difficult to control by the consumer.

No provision is made in most codes to control both the penetrability and the thickness of the concrete cover on site, which to a large extent condition the service life of concrete elements in environments prone to develop steel corrosion. And this, despite the fact that suitable (semi-invasive and non-invasive) site methods to assess them are available [4]. In Switzerland, and probably in other countries as well, the Inspector/Engineer is responsible to check the cover before concreting; he does not always do it and, if a low cover is detected, he would be responsible too.

An important aspect that is missing in almost all Standards (Switzerland and South-Africa are exceptions) is the on-site quality control (i.e. conformity assessment of the structure). Approaches and methods need to be developed which allow assessing the goodness of the concrete practices applied (e.g. curing) through the evaluation of the real quality achieved on site.

References

1. ACI ITG-8R-10. Report on Performance-Based Requirements for Concrete. 2010. p. 50.
2. Collepardi M. Dal calcestruzzo antico a quello moderno, parte II—Calcestruzzo Romano. Enco J. 2008; 43, Treviso, Italy. http://www.enco-ournal.com/journal/ej43/collepardi.html. Accessed 2013.
3. Cemex. History of Concrete and Cement. http://www.cemexusa.com/ProductsServices/ReadyMixConcreteHistoryFacts.aspx. Accessed 2013.
4. RILEM TC 189-NEC. Non-destructive evaluation of the penetrability and thickness of the concrete cover. In: Torrent R, Fernández Luco L, editors. RILEM Report 40. 2007. p. 223.
5. CEN/TC 104/SC 1. Survey of national provisions for EN 206-1, N 485. 30 Jan 2007. p. 148.
6. Troli R, Collepardi M. Technical contradictions in the European Norm EN 206 for concrete durability. In: Proceedings International Symposium. Dedicated to Prof. S. Shah, Dundee, Scotland; 3–4 Sept. 2003. p. 665–674.
7. Swamy RN. Concept of strength through durability, sustainable concrete for the 21st century, Japanese Society of Civil Engineers, Concrete Committee Newsletter 13, 2008.
8. Neville, A., Durability, Chapter 3 of *Neville on Concrete*, American Concrete Institute, Michigan, 2003.
9. Hilsdorf HK. Concrete compressive strength, transport characteristics and durability. In: Kropp J, Hilsdorf HK, editors. RILEM TC 116-PCD Report on *Performance Criteria for Concrete Durability*. London: E & FN Spon; 1995.
10. *fib*, Model Code 2010, First complete draft - Volume 1, fib Bull. 2010;55:157–163.

11. Jacobs F, Leemann A. Betoneigenschaften nach SN EN 206-1, ASTRA Report VSS Nr. 615. 2007. p. 27.
12. Torrent R. From Theorecrete to Labcrete to Realcrete: the Swiss P2P Road. 2nd P2P Technical Seminar, Saudi Arabia. 25–29 May 2013. http://p2psaudiarabia.com/2013pdfs/dr_robertotorrent.pdf.
13. Wasserman R, Katz A, Bentur A. Minimum cement content requirements: a must or a myth? Mater Struct. 2009;42:973–82.

Chapter 4
Test Methods for Concrete Durability Indicators

D. Bjegović, M. Serdar, I.S. Oslaković, F. Jacobs, H. Beushausen, C. Andrade, A.V. Monteiro, P. Paulini and S. Nanukuttan

4.1 Introduction

Durability of concrete structures is primarily dependent on the environmental influences, i.e. the penetration of aggressive substances in the structural element from the environment. This is why the penetrability of fluids and aggressive substances through the porous structure of hardened cement paste is the main parameter one should be familiar with to predict the potential durability of a reinforced concrete structure. The transport of substances within concrete directly depends on the very cause of transport that can take place due to hydraulic gradient, concentration gradient, or moisture movement. Depending on the driving force of

D. Bjegović (✉) · M. Serdar
University of Zagreb, Faculty of Civil Engineering, Zagreb, Croatia

D. Bjegović
Institute IGH, Zagreb, Croatia

I.S. Oslaković
University of Twente, Enschede, The Netherlands

F. Jacobs
TFB, Zurich, Switzerland

H. Beushausen
University of Cape Town, Cape Town, South Africa

C. Andrade
Eduardo Torroja Institute (IETcc-CSIC), Madrid, Spain

A.V. Monteiro
National Laboratory for Civil Engineering, Lisbon, Portugal

P. Paulini
University Innsbruck, Innsbruck, Austria

S. Nanukuttan
Queen's University Belfast, Belfast, UK

H. Beushausen and L. Fernandez Luco (eds.), *Performance-Based Specifications and Control of Concrete Durability*,
RILEM State-of-the-Art Reports 18, DOI 10.1007/978-94-017-7309-6_4

the process and the nature of the transported matter, different transport processes for deleterious substances through concrete are distinguished. They can be categorised as follows:

1. absorption—movement of fluid due to the capillary forces created inside the capillary pores
2. permeation—movement of fluid due to the action of pressure
3. diffusion—movement of fluid due to a concentration gradient.

Penetrability is an important durability indicator of concrete and by specifying different classes of penetrability of concrete it should be possible to design a structure with the required resistance to environmental loads. Nowadays, many testing procedures for testing penetrability properties of concrete are standardized or have already been used for long periods, and have proven to have satisfactory precision. But for a certain property to be used as a durability indicator in the performance based design procedure it needs to be quantifiable by laboratory and on-site tests in a reproducible manner and with clearly defined test procedures. Furthermore, limiting values of the property required for a specific environmental class and required service life of a structure need to be established. Only then can such a durability indicator of concrete be prescribed during the design of concrete structures, obtained during prequalification testing, used in service life models, tested during construction as part of quality control on site, or tested during the service life to assess the condition of the structure.

This chapter covers descriptions of the available and commonly applied in situ and laboratory, non-invasive and semi-invasive test methods for evaluating concrete penetrability properties. The intention of this chapter is to give an overview of the methods that are most frequently used in engineering practice and research, and with which a significant experience is available. Both methods for laboratory and on-site testing are described. An overview of all described methods and examples for recommended limiting values for different concrete quality categories are given at the end of each group of methods. The application of those methods that have prescribed limiting values, which can be used during design, compliance testing, and quality control of "as-built" concrete are described in further detail in Chap. 8.

4.2 Gas Permeability

Gas permeability of concrete is defined as a property characterizing the ease by which gas under pressure passes through the concrete. The gas permeability depends on the properties of concrete (e.g. water/cement ratio, porosity, narrowness and tortuosity of the pores and the cracks, friction at the pore and crack walls, etc.) as well as the environmental influences (e.g. moisture, temperature, viscosity of the gas, applied pressure gradient) [1]. The permeability to gas is generally accepted as a durability related property of concrete. Being theoretically related to gas diffusivity [2, 3], this property is of particular interest for assessing concrete performance against CO_2 and oxygen penetration, responsible for the depassivation and corrosion of the reinforcement. It is also used in comparative tests for quality assessment; especially to support the choice of concrete mix design parameters for structural elements exposed to environments where carbonation-induced corrosion is the main deterioration mechanism (e.g. environmental class XC). In some countries the permeability to oxygen is also used in performance-based methodologies in order to estimate the corrosion initiation period of concrete structures subjected to carbonation, based on carbonation models such as the one developed by the CEN TC104 [4].

4.2.1 *Principle and Mechanism*

The flow in the capillary pores in saturated concrete can be described as a laminar flow of Newtonian fluids through a porous medium [5]:

$$\frac{dq}{dt}\frac{1}{A} = \frac{K}{\eta L}\frac{\left(p^2 - p_a^2\right)}{2p_0} \tag{4.1}$$

where dq/dt is the rate of gas flow in m^3/s, A is the cross-sectional area in m^2, K is the intrinsic permeability coefficient in m^2, p is the inlet pressure in N/m^2, p_a is the outlet pressure (usually atmospheric pressure) in N/m^2, p_O is the pressure at which the rate of flow is measured in N/m^2, L is the thickness of the specimen in m, and η is the dynamic viscosity of the fluid in Ns/m^2.

4.2.2 *Test Methods*

Gas permeability tests can be performed in the laboratory and on-site. Tests can be performed under steady state conditions when a constant pressure over the specimen is maintained and under non-steady state conditions of flow [6]. Tests performed on site are usually those with non-steady state condition of flow, since it is difficult to maintain a constant gas pressure over the concrete.

For the on-site measurement of air permeability of concrete cover, based on the way of provoking the air flow in concrete, testing methods can be divided into two groups:

1. non-invasive or surface methods and
2. invasive methods.

Non-invasive or surface methods involve the creation of an air pressure gradient between the surface and the pore network below the surface of the concrete by using a vacuum chamber. Methods of this group are completely non-invasive.

A so-called single-chamber method was developed by several researchers, e.g. Schönlin [7]. A so-called double chamber method was developed by e.g. Torrent [8].

In the invasive testing method, a hole is introduced into the concrete surface and air is introduced into or withdrawn from this hole until a certain air pressure is reached. By measuring afterwards the pressure change with time in the hole, due to the outward or inward movement of air through the pore network of concrete, an air permeability index of cover concrete is determined. An example of invasive methods is a method developed by Figg [9].

The air permeability test on site is a relatively simple and easy method for evaluating concrete penetrability properties, since it can be used on horizontal and vertical structural elements, which is often not the case with water permeability and absorption tests. It is, however, significantly influenced by the environmental conditions during the test, mainly by the humidity of the tested concrete and the temperature. These conditions have to be measured and noted during on site testing of air permeability, since they are valuable during the evaluation of results.

Measuring the gas permeability in the laboratory usually consists of placing the specimen in an air tight cell and allowing gas under pressure to go through the specimen. The flow of the gas passing through the specimen is recorded during the test, and the permeability coefficient calculated using Eq. (4.1).

4.2.2.1 Figg's Method

Figg's method [9] consists of creating a negative relative pressure (55 kPa below atmospheric pressure) inside a small hole drilled in the concrete. This is done by connecting a vacuum pump to a needle inserted inside the drilled hole through a rubber plug, Fig. 4.1. The rubber plug is used to make the hole airtight, allowing only air from the concrete to enter the hole and increase the pressure inside it. The time required for the absolute pressure to rise to 5 kPa is recorded as the air permeability index.

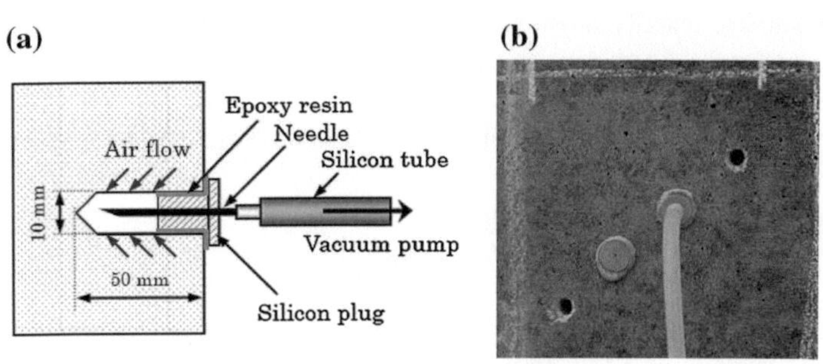

Fig. 4.1 **a** Schematic of Figg's permeability test, and **b** performing Figg's test on site [11]

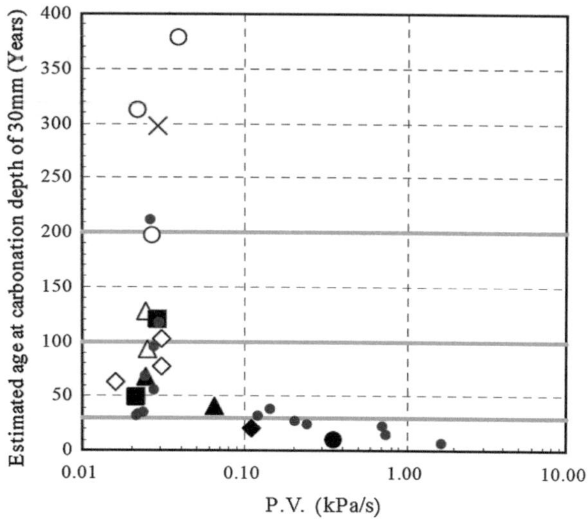

Fig. 4.2 Relationship between carbonation progress and air permeability [11]

Figg's invasive method has been modified by [10] and used by [11]. In this method the permeability velocity (PV) through concrete is calculated as the time required for the pressure in the hole to change from 21.3 to 25.3 kPa. According to this method a permeability velocity greater than 0.10 (kPa/s) would correspond to concrete durability classified as "poor"'.

Figure 4.2 shows the relationship between the permeability velocity and the carbonation progress, showing that those concrete mixtures that have lower permeability coefficients are more resistant to carbonation, since less CO_2 is able to penetrate through the concrete and react with cement hydration products.

4.2.2.2 Schönlin and Hilsdorf

This method was developed as an alternative to Figg's method, with the attempt of creating a purely non-invasive method [7, 12]. The method consists of a vacuum chamber mounted on the surface of the concrete, Fig. 4.3, in which pressure is decreased to less than 99 kPa below atmospheric pressure. The valve is then closed and the air is let to penetrate from concrete into the chamber, resulting in the increase of the pressure inside the chamber. Knowing the time required for the pressure inside the chamber to reach a predefined level (e.g. −70 kPa) and the volume of the chamber, the air permeability index in m^2/s can be calculated.

A simplified version of this method can be prepared as an in-house method, using a syringe instead of the vacuum pump to create high and lower pressure values. In this method the pressure between concrete and the chamber is set at −70 kPa with the syringe, Fig. 4.4a [13]. The change of the pressure in the chamber is monitored during testing and afterwards the rise of pressure over time is plotted. The slope of the linear regression on the natural logarithm of pressure versus time

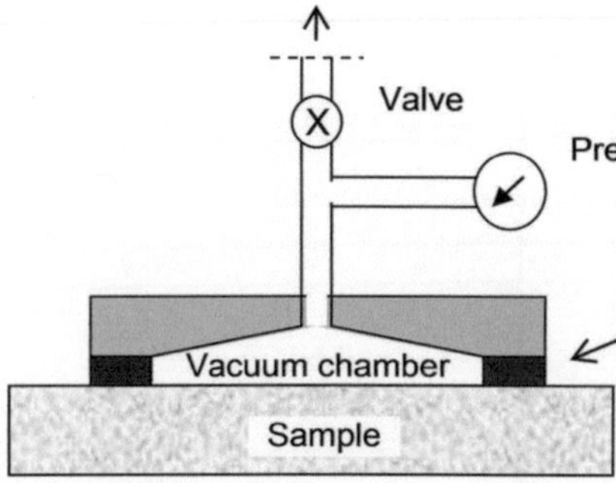

Fig. 4.3 Shematic of Schönlin and Hilsdorf permeability test

(a)

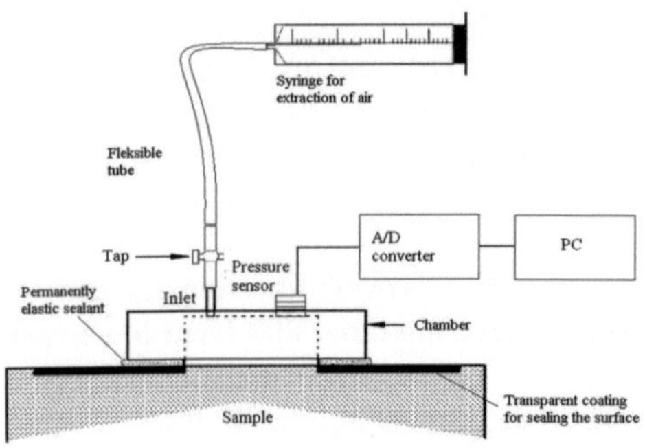

(b)

Fig. 4.4 **a** Schematic of in-house method for air permeability testing based on Schönlin and Hilsdorf method, and **b** application on site [13]

curve presents the air permeability index, in ln(bar)/min. This method is convenient for on-site testing because the entire system is simple, lightweight and no additional power source is needed, Fig. 4.4b.

One of the main problems during air permeability testing using this or similar methods is ensuring tightness between air chamber and concrete surface. This problem is solved with elastic sealants and additional fastening bolts. If there is no sufficient tightness between chamber and concrete, this area will be permeable and test results will not be reliable. To ensure that the test is working properly and that results indicate concrete permeability, it is recommended to repeat the test at the same location after some time. It is also recommendable to restrain the testing area and avoid dissipation of air through the surrounding concrete. To do so, the surface around the testing area can be sealed with impermeable coating a few minutes before performing the air permeability test.

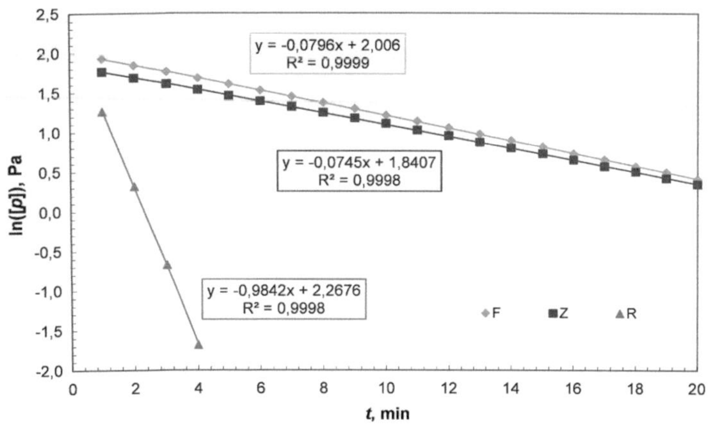

Fig. 4.5 Schönlin and Hilsdorf method used on concrete prepared in regular formwork and ones prepared in controlled permeability formwork [14]

Figure 4.5 shows the example of using the Schönlin & Hilsdorf method for comparison of air permeability properties of concrete prepared in controlled permeability formwork (specimens Z and F) and concrete prepared in regular wooden formwork (specimen R) [14]. The difference between gas permeability properties of three types of concrete tested with the Schönlin & Hilsdorf method can be easily seen. The coefficient of gas permeability is expressed as the slope of the lines correlating the decline of pressure inside the vacuum chamber during time.

4.2.2.3 Autoclam Method

The Autoclam Permeability System can be used to measure the air permeability, sorptivity and water permeability of concrete [6]. In order to carry out an Autoclam permeation test, an area of 50 mm diameter is isolated on the test surface with a metal ring, Fig. 4.6. The ring can be either bonded to the test surface with an adhesive or clamped with a rubber ring to provide an airtight seal. The air permeability test is carried out by increasing the air pressure on the test surface to 50 kPa and noting the decay of pressure with time. The decay of the pressure is monitored every minute for 15 min or until the pressure has diminished to zero. The plot of natural logarithm of pressure against time is linear and the slope of the linear regression fit of data between the 5th and the 15th minute for tests lasting for 15 min is reported as an air permeability index, in ln(pressure)/min. When the pressure becomes zero before the test duration of 15 min, the data from the start is used to determine the slope.

Fig. 4.6 Autoclam systems for testing air permeability

4.2.2.4 Torrent Method

The Torrent method is based on creating a vacuum on the surface of the concrete and monitoring the rate at which the pressure is rising in the test chamber after the vacuum pump has been disconnected, Fig. 4.7 [8]. The distinctive features of the method are a double-chamber cell and a pressure regulator that balances the pressure in both chambers during the test. The special features of the apparatus create a controlled, unidirectional flow of air from the pores of the concrete into the inner chamber, while the outer chamber acts as a guard-ring. Under these conditions it is possible to calculate the coefficient of permeability to air, the so-called *kT* of the concrete. The Torrent method is standardized in the Swiss Standard SIA 262/1:2013 on "Concrete Structures – Complementary Specifications" [15].

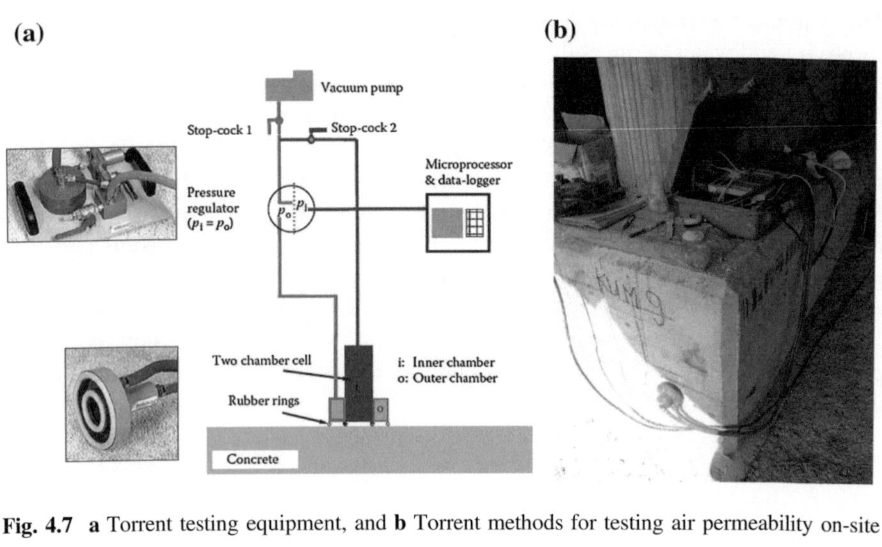

Fig. 4.7 **a** Torrent testing equipment, and **b** Torrent methods for testing air permeability on-site

In order to obtain reproducible and accurate values of the coefficient of air permeability, several recommendations on the performance of the measurements and the evaluation of the data are given [16, 17]:

- The temperature of the concrete element has to be higher than 5–10 °C
- specific electrical resistivity (Wenner probe) >10–20 kΩ cm or moisture content <5.5 wt%, measured by an impedance based instrument
- 6 measurements have to be made on the selected area under investigation
- if no more than 1 out the 6 measurements exceeds the limiting value, the area under investigation fulfils the requirements
- if more than two out of six measurements exceed the limiting values, the area fails to meet the requirements
- if only two out the 6 measurements exceed the limiting value, a new set of six measurements on the same area under investigation are allowed to be made. From the new six measurements a maximum of one measurement may exceed the limiting value; if more measurements exceed the limiting value, the area fails to meet the requirements.

4.2.2.5 Permeability Exponent

The transport law for gases usually used for calculation of the coefficient of permeability, Eq. (4.1), describes a nonlinear relation between pressure and velocity of gas. Fine porous concrete has a strong nonlinear behaviour while concrete with high w/c ratio and coarser pores shows a linear relation between pressure and velocity [18]. Therefore, a power law for air/gas transport in concrete with two parameters was proposed [19], where the reference velocity and the permeability exponent define the air flow behaviour and the pore structure of concrete:

$$v = v_1 \left(\frac{ph_1}{hp_1} \right)^n = v_1 \left(\frac{p}{h} \right)^n \tag{4.2}$$

$$ln \left(\frac{v}{v_1} \right) = nln \left(\frac{p}{h} \right) \tag{4.3}$$

where v is velocity in m/s, v_1 is the reference velocity at pressure gradient 1 MPa/m in m/s, p is pressure in Pa/m, p_1 is the reference pressure (1 MPa/m), h is the thickness of the specimen in m, h_1 is the reference thickness (1 m) and n is the permeability exponent.

The test set-up for laboratory and on-site evaluation of the permeability exponent were also developed, Fig. 4.8. The test set-up consists of a flow cell, an air vessel, an amplifier and a computer for data storage. For laboratory testing, specimens need to be oven dried at 105 °C, mounted between 2 steel tubes, inserted in the cell and

Fig. 4.8 Test setup for evaluation of permeability exponent [19]

sealed by a rubber hose. For on-site tests the flow cell is replaced with a packer and a surface sealing plate. A 3 cm bore hole is used to fix the packer, so that a 1 cm deep air slit remains below the plate. The humidity of concrete is determined by drying the bore dust.

In both cases the test is performed by applying stepwise decreasing air pressure levels on the specimen. The applied pressure is controlled with a regulation valve between the vessel and the pressure line. Two pressure gauges—one in the vessel and the other in the pressure line—are connected to an amplifier and the data is stored into a file. After reaching steady state conditions, the air flow is maintained at least for the time needed to allow a drop of 5 kPa to occur in the vessel [19].

4.2.2.6 Cembureau Method

The Cembureau method is based on the Hagen-Poiseuille equation for laminar flow of compressive fluids [20]. The method consists of measuring the volume flow rate of the gas that passes through a specimen against which oxygen or nitrogen is pressurized. A standard concrete disc (100 or 150 mm diameter, 50 mm high) is placed inside a cell and sealed laterally with a tight fitting rubber collar pressed against its curved surface, Fig. 4.9a. After the cell is tightened, gas at different pressures (usually between 0.5 and 3.0×105 N/m^2 above the atmospheric pressure) is applied on the bottom surface of the specimen. The volume flow rate of the gas that passes across the specimen is then measured with a soap bubble flow meter connected with the top surface of the specimen, Fig. 4.9b. The gas permeability coefficient is then calculated for each applied gas pressure through Eq. (4.1) and the resulting average value is considered as the test result.

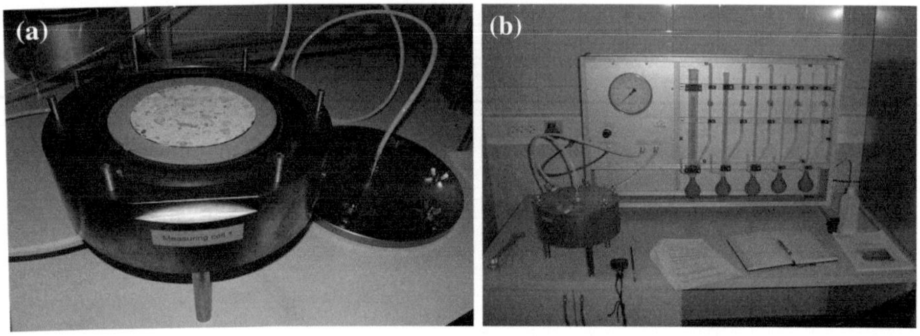

Fig. 4.9 **a** Standard concrete disc placed inside the Cembureau cell, and **b** Cembureau cell setup

A very important aspect to obtain reproducible results is the preconditioning procedure of the specimens prior to testing. There is still no consensus in the scientific community on this procedure and thus the results obtained by different laboratories are often not comparable. An attempt for the harmonization of a reproducible preconditioning procedure and for the improvement of this test (allowing the use of N_2 as the permeating medium) was done by RILEM TC 116—PCD [21]. Although this procedure allows to achieve a predefined hygrometric condition in the concrete specimen (equilibrium moisture concentration with 75 ± 2 % RH at 20 ± 1 °C), it has the disadvantage of being too laborious and time consuming. For this reason most countries continue to use their own procedures.

Since the assumption of laminar flow is not strictly true, the flow rate Q often shows a non-linear behaviour with $p^2 - p_a^2$, which makes K a function of the applied pressure p [20, 22], as shown in Fig. 4.10.

This behaviour compromises the "intrinsic nature" of the measured permeability coefficient. Some authors attribute this dependence of the permeability with the applied gas pressure to the "Knudsen flow" and suggest a method to calculate an "intrinsic gas permeability coefficient" by estimating (by extrapolation) the gas permeability for a hypothetical infinite gas pressure applied [22].

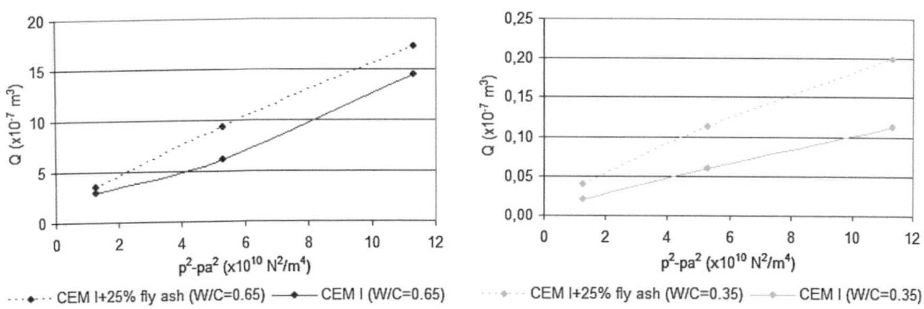

Fig. 4.10 Example of Cembureau test results

Regarding the precision of this test and its sensibility to assess concrete quality, RILEM TC 116-PCD stated that [21]: "The CEMBUREAU method is very reliable, easy to handle and exhibits very good repeatability. This method is recommended therefore as a standard test method for gas permeability measurements."

It was also chosen by RILEM TC 189-NEC as a reference test in order to assess the suitability of non-invasive tests to measure gas permeability of concrete [1]. One other limitation of this test is the need of drilling cores from the structures in cases of assessing the quality of in situ concrete.

4.2.2.7 Oxygen Permeability Index Test (South Africa)

The South African Oxygen Permeability Index (OPI) test method consists of measuring the pressure decay of oxygen passed through an oven dried, 30 mm thick slice (representing the cover concrete) of a (typically) 70 mm diameter core placed in a falling head permeameter, Fig. 4.11. The OPI is defined as the negative log of the coefficient of permeability. Common OPI values for South African concrete range from 8.5 to 10.5, a higher value indicating a lower permeability and thus a concrete of potentially higher quality. Note that oxygen permeability index is measured on a log scale; therefore the difference between 8.5 and 10.5 is substantial, the former being 100 times more permeable than the latter. Details on the test equipment and test procedure can be found in the literature [23–25].

The oxygen permeability test assesses the overall micro- and macrostructure of the outer surface of cast concrete, and is particularly sensitive to macro-voids and cracks which act as short-circuits for the permeating gas. Thus the test is useful to

(a) (b)

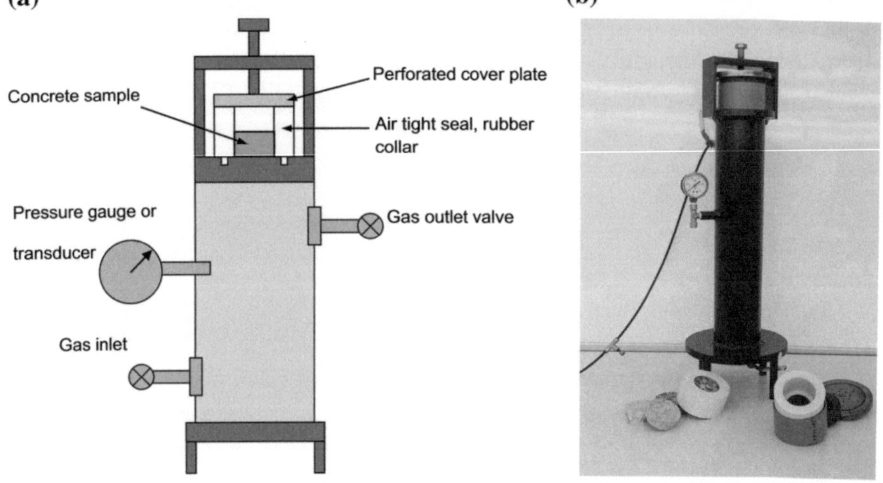

Fig. 4.11 **a** Schematic of an oxygen permeameter, and **b** photograph of an oxygen permeability test setup

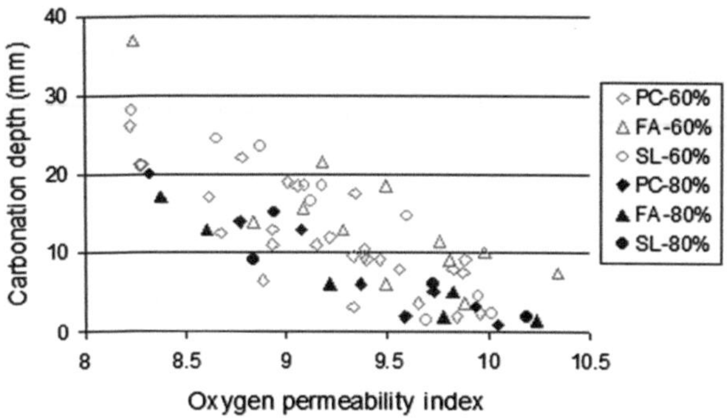

Fig. 4.12 Carbonation depth in various concretes (*PC* Portland cement, *FA* Fly ash, *SL* Slag) versus oxygen permeability index (measured at 28 days) for 4 years exposure at an average relative humidity of 60 or 80 % [24]

assess the state of compaction, presence of bleed voids and channels, and the degree of interconnectedness of the pore structure. Correlations between OPI values recorded at 28 days and carbonation depths after natural exposure have been found to be good, Fig. 4.12 [24].

In a comparative international study of various test methods for durability indicators, the OPI test was found to be able to detect differences in w/b ratio, binder type, and curing condition on a statistically highly significant level [1]. The same study revealed that results obtained with the OPI test equipment correlate well with other existing test methods for oxygen permeability, such as the Cembureau method and the Torrent Permeability Test.

The OPI test method is used for performance-based design and quality control of concrete structures exposed to environmental exposure class XC (carbonation) in South Africa. Limiting OPI values were developed based on scientific and empirical correlations between concrete permeability and carbonation of concrete for various environmental conditions and binder types (compare Sect. 8.3).

4.2.3 Overview and Criteria for Evaluation of Concrete Quality

From different methods of testing air permeability concrete quality can be evaluated using criteria listed in Table 4.1 [9, 15, 19, 23, 26, 27]. Table 4.2 presents an overview of different methods with advantages and limitations. Advantages and limitations are relative and based on the experience of the authors with the methods considered in this chapter.

Table 4.1 Criteria for concrete quality based on air permeability

Concrete permeability Figg	S&H/Autoclam, Air perm. index, ln(bar)/min	Torrent $kT \times 10^{-16} m^2$	Cembureau gas permeability coefficient, m^2	Permeability exponent n	Oxygen permeability index OPI	Concrete quality
–	≤0.10	<0.01	–	>2	>10	Very good
>200	0.10–0.50	0.01–0.1	$<10^{-18}$	1.5–2	9.5–10	Good
100–200	–	0.1–1.0	10^{-18}–10^{-16}			Normal
<100	0.50–0.90	1.0–10	$>10^{-16}$	1–1.5	9.0–9.5	Poor
–	>0.90	>10	–	<1	<9	Very poor

Criteria here are given solely as an overview, with the purpose of highlighting the importance of indicating the test method in the performance-based design approach to avoid misinterpretation due to different test methods

Table 4.2 Overview of different methods and their advantages and limitations

Advantages/limitations	Test method						
	Figg's method	Schönlin and Hilsdorf	Autoclam method	Torrent method	Permeab. exponent	Cembureau method	Oxygen Permeab. Test
Non-invasive	–	+	–	+	–	–	–
Applicable on-site	+	+	+	+	+	–	–
Applicable on drilled cores	–	–	–	–	+	+	+
Performance criteria correlated to exposure classes	–	–	–	+	–	+	+
Standardised	–	–	–	+	–	+	+
Used in service life models	–	–	–	+	–	+	+
Correction for moisture	–	–	–	+	–	–	–
Measures through thicker concrete layer	+	–	–	–	+	+	+
Same equipment can be used for other durability indicators	–	–	+	–	–	–	–

4.3 Water Permeability

Water permeability is measured as a flow of water under a constant or decreasing pressure gradient through concrete. This property is of great importance when performing an assessment of hydro structures, reservoirs or any other civil engineering structure that is in direct contact with water; it can further be used to characterize concrete for transport properties that link to durability.

4.3.1 Principle and Mechanism

Flow of water through capillary pores in saturated concrete follows D'Arcy's law for laminar flow through a porous medium [5]:

$$\frac{dq}{dt}\frac{1}{A} = \frac{K'\rho g}{\eta}\frac{\Delta h}{L} \tag{4.4}$$

where dq/dt is the rate of gas flow in m^3/s, A is the cross-sectional area in m^2, K is the intrinsic permeability in m^2, L is the thickness of the specimen in m, η is the dynamic viscosity of the fluid in Ns/m^2, ρ is the density of the fluid in kg/m^3, g is the acceleration due to gravity in m^2/s, and Δh is the drop in hydraulic head through the specimen (corresponds to the height of a water column) in m.

When the fluid in matter is water at room temperature, then the equation can be rewritten using the coefficient of water permeability K_w expressed in m/s:

$$\frac{dq}{dt}\frac{1}{A} = K_w\frac{\Delta h}{L} \tag{4.5}$$

4.3.2 Test Methods

The procedure for testing water permeability is similar to testing gas permeability, with the difference lying in the penetrating fluid. Water permeability can be tested on-site with non-invasive methods, or in the laboratory on specimens taken from the structure or prepared in the laboratory. Instruments for testing water permeability on-site usually consist of a reservoir filled with water, which is connected to the concrete surface, a syringe or vacuum pump for introducing pressure into the reservoir and a transducer for monitoring pressure. Most of the laboratory methods consist of placing the concrete specimen under a pressurised water flow and measuring the amount of water penetrated through the concrete.

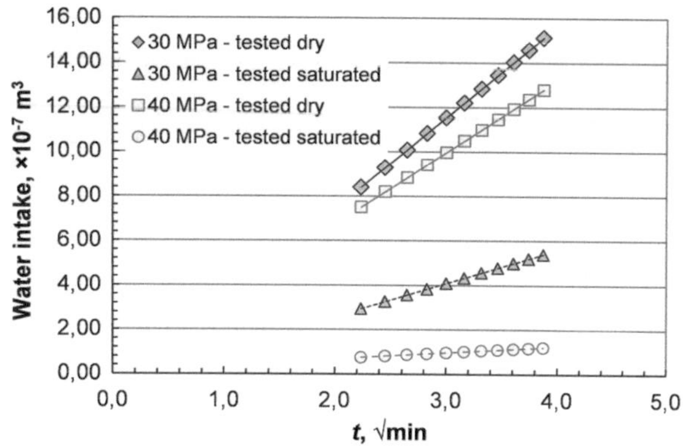

Fig. 4.13 Application of Autoclam water permeability test [13]

4.3.2.1 Autoclam Method

The procedure used for both the water permeability and capillary absorption with the Autoclam is similar; the main difference is in the test pressure. In the case of the water permeability test the pressure inside the chamber is maintained at 50 kPa. It is expected that the pressure of 50 kPa will cause the pressure induced flow dominating over capillary absorption [27]. The test lasts for 15 min and the cumulative water penetration into concrete is plotted against the square root of time, which is more or less linear, as shown in Fig. 4.13. The slope of the square root time plot between the 5th and the 15th minute is used to evaluate the water permeability index, expressed in $m^3/min^{0.5}$.

Figure 4.13 shows an example of using the Autoclam method to compare two concrete mixtures, one with a compressive strength of 30 MPa, and the other with a compressive strength of 40 MPa, both tested in saturated and in dry condition. It is evident that the mixture yielding higher compressive strength and higher quality is more resistant to water penetrability under pressure. Also, it is evident that saturated concrete absorbs less water under pressure, since the pores are already filled.

The water permeability test can be carried out at the location of the air permeability test, but with at least 1 h time difference between the two tests. If both the air permeability test and any of the water flow tests are to be performed at the same location, it is important to do the air permeability test first because the test area will get wet with the water flow test. Also, it is advised not to perform the water permeability test at the same test location where the absorption test was performed.

One of the main problems during water permeability testing using this or similar methods is ensuring tightness between the chamber and concrete surface, especially if the test is performed on non-horizontal surfaces. Some problems have been experienced in ensuring constant water pressure when similar methods were used on vertical concrete elements. It is also important to record the environmental conditions when the measuring is performed, as well as to evaluate the moisture

Fig. 4.14 Connection of the bottom part of the chamber to the concrete

content in the concrete, since it has a great influence on the water permeability results, Fig. 4.13. To ensure the connection between instrument and concrete element, the bottom of the chamber is connected to the concrete with a set of screws, as shown in Fig. 4.14. On completion of the test, small holes remain on the concrete surface.

4.3.2.2 Water Permeability Test

One of the non-invasive methods for evaluating water permeability on-site is the Water permeability Test (GWT), Fig. 4.15 [28]. During GWT testing, a sealed pressure chamber is attached to the concrete surface. Then, water is filled into the pressure chamber and a specified water pressure is applied to the surface. The pressure may be kept constant using a micro-meter gauge with an attached pin that reaches into the chamber. The testing can be performed on both vertical and horizontal faces. The result obtained in most cases represents a combination of the

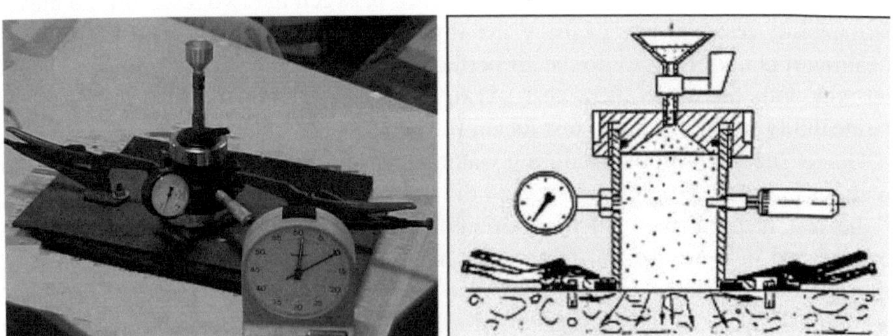

Fig. 4.15 Instrument for testing water permeability [28]

influence of three factors—surface porosity, water permeability and absorption. In each case, the test results are evaluated separately after planning of test conditions depending on the purpose of the testing.

The flux q may be calculated from Eq. (4.6):

$$q = \frac{B(g_1 - g_2)}{At} \quad (4.6)$$

where q is the flux in mm/s, B is the area of the micrometer pin pressed into the chamber water (78.6 mm^2 for a 10 mm pin diameter) in mm^2, A is the water pressure surface area (3018 mm^2 for a 62 mm diameter) in mm^2, g_1 and g_2 are the micrometer gauge readings in mm before and after the test has been performed, and t is the time the test is performed in s.

The surface permeability may be assessed by means of Darcy's law:

$$K_{cp} = \frac{q}{b\left(\frac{\Delta P}{L}\right)} \quad (4.7)$$

where K_{cp} is the concrete permeability coefficient in mm/s, b is the percentage of the concrete cement matrix (assuming the aggregates are impermeable) in %, ΔP is the pressure selected in Pa, and L is the length the pressure is applied over (15 mm, equal to the thickness of the pressure gasket) in mm.

During water permeability testing it is assumed that the water will flow parallel to the gasket, from the compression chamber to the outside. If the concrete is rather porous, this may not be valid. In such cases, the water will flow into the concrete building up a more and more stable pressure until the water flows below the pressure gasket as intended. Additionally, problems have been observed with obtaining constant water pressure on non-horizontal concrete elements [28].

4.3.2.3 Depth of Penetration of Water Under Pressure

The basic setup for measuring water permeability is similar to that for gas permeability. Generally, higher pressures are required to force the water through a saturated concrete, Fig. 4.16. At solely one pressure, for normal concrete between 4×10^5 and 8×10^5 N/mm^2, water permeability is determined on specimens stored for >24 h in water. As water permeability decreases with increasing measuring time, a constant measuring time (e.g. 24 h) should be selected [29, 30]. An appropriate control of the experiment is achieved if the quantity of the inflowing and outflowing water is determined.

After the specimen is saturated, the flow rate reading is taken using a burette by measuring the change of volume of water over time. The permeability is defined by Darcy's Law as follows:

Fig. 4.16 Water permeability testing in the laboratory

$$k = \frac{QL}{AH} \tag{4.8}$$

where k is the permeability coefficient in m/s, Q is the flow rate in m^3/s, A is the area in m^2, L is the depth of the specimen in m, and H is the head of water in m.

Another way to determine the water permeability is to determine the penetration depths of water [31]. In this method cylindrical specimens should be dried in the oven at 105 °C until reaching a constant mass. The specimens are then coated with epoxy on the circular side to prevent water penetration from the side during the test. A pressure of (500 ± 50) kPa should be applied to the specimens at a pressure head of 92.5 m. The pressure is maintained for 72 h, after which the specimens are split in half and the maximum depth of water penetration is measured.

4.3.3 *Overview and Criteria for Evaluation of Concrete Quality*

From different methods of testing water permeability, the concrete quality can be evaluated using criteria listed in Table 4.3 [3, 26–28].

Table 4.4 presents an overview of different methods with advantages and limitations. Advantages and limitations are relative and based on the experience of the authors with the methods considered in this chapter.

4.4 Capillary Absorption

In unsaturated concrete, the rate of ingress of water or other liquids is largely controlled by absorption due to capillary rise. The capillarity behaviour of concrete is one of the most important causes of chloride contamination of non-saturated

Table 4.3 Criteria for concrete quality based on water permeability

Autoclam water permeability index, $10^{-7} \times m^3/min^{0.5}$	Water permeability test, water permeability index	Coefficient of water permeability, m^2	Concrete quality
≤3.70	1.0×10^{-7}–1.0×10^{-9}		Very good
3.70–9.40	1.0×10^{-6}–1.0×10^{-7}	$<10^{-12}$	Good
	1.0×10^{-5}–1.0×10^{-6}	10^{-12}–10^{-10}	Normal
9.40–13.80	1.0×10^{-3}–1.0×10^{-4}	$>10^{-10}$	Poor
≥13.80	$\leq 1.0 \times 10^{-3}$		Very poor

Criteria here are given solely as an overview, with the purpose of highlighting the importance of indicating the test method in the performance-based design approach to avoid misinterpretation due to different test methods

Table 4.4 Overview of different methods and their advantages and limitations

Advantages/limitations	Test method		
	Autoclam method	Water permeability test (GWT)	Depth of penetration of water under pressure
Non-invasive	–	–	–
Applicable on-site	+	+	–
Applicable on drilled cores	–	–	+
Performance criteria correlated to exposure classes	–	–	+
Standardised	–	–	+
Used in service life models	+	+	+
Measures through thicker concrete layer	–	–	+
Same equipment can be used for other durability indicators	+	–	–

concrete [3]. For this reason, the absorption of water by capillarity is often used in quality comparative tests in order to support decisions regarding the choice of concrete to be placed to construct structural elements (e.g. columns of bridges and piles) exposed to fluids containing aggressive agents (commonly chlorides or sulphates) under wetting/drying cycles. Capillary absorption is also used as a method for evaluating concrete resistance to freezing and thawing. With this method the interconnected capillary pore structure of concrete can be assessed.

4.4.1 Principle and Mechanism

The penetration of liquids in concrete as a result of capillary forces is called absorption. For some types of construction materials, like clay bricks, the magnitude of capillary rise follows a linear relationship with the square root of time elapsed and the constant of proportionality is called sorptivity [32]. In general, measurements on concrete do not follow the previous mentioned linear relationship. In a realistic situation, during the testing of concrete in situ, achieving unidirectional penetration of water is difficult. Consequently, the absorption characteristics of concrete are usually measured indirectly [33].

Water absorption may be expressed by the following equation:

$$w = w_1 \left(\frac{t}{t_1} \right)^n = M_w t^n \quad (4.9)$$

where w is the water absorbed per unit area at time t in m^3/m^2, w_1 is the water absorbed at a given time t_1 in m^3/m^2, t is duration of water absorption in s, n is 0.5, and M_w is the coefficient of water absorption in $m/s^{0.5}$.

4.4.2 *Test Methods*

Capillary suction experiments have to be designed in such a way that the driving force of an absolute external pressure is excluded or minimized, and the only mechanism of water penetration is the action of capillary forces. The measurement of capillary absorption of concrete in the laboratory is straight-forward, and the prevailing step in the procedure is the preconditioning of the specimens. Nowadays there are several systems available on the market for automatic testing of absorption on-site. Similar to gas permeability testing on-site, methods for testing capillary absorption on-site can generally be divided into those that measure surface absorptivity (non-invasive methods) and drilled hole absorptivity (semi-invasive methods). However, non-invasive methods are nowadays used more frequently, compared to methods that rely on drilling holes into the concrete surface.

4.4.2.1 Initial Surface Absorption Test (ISAT)

The simplest method to test sorptivity on site is to use a reservoir filled with water and connect it to the concrete through a cap with known area [34, 35]. The water reservoir should be placed in a way that the level of water in the reservoir is 200 ± 5 mm above the concrete surface. The water level in the reservoir is monitored by the sensor placed at the side of the reservoir and the data is sent to the personal computer through an A/D (analog-to-digital) converter [13]. Before the start of the measurement a calibration should be made, which converts vertical movement of the water level in the reservoir to the volume of water outflow from the reservoir. The outflow from the reservoir is equal to the water inflow into the concrete. The initial surface absorption value is then calculated and expressed in the units $ml/(m^2 s)$. A schematic presentation of the system is shown in Fig. 4.17.

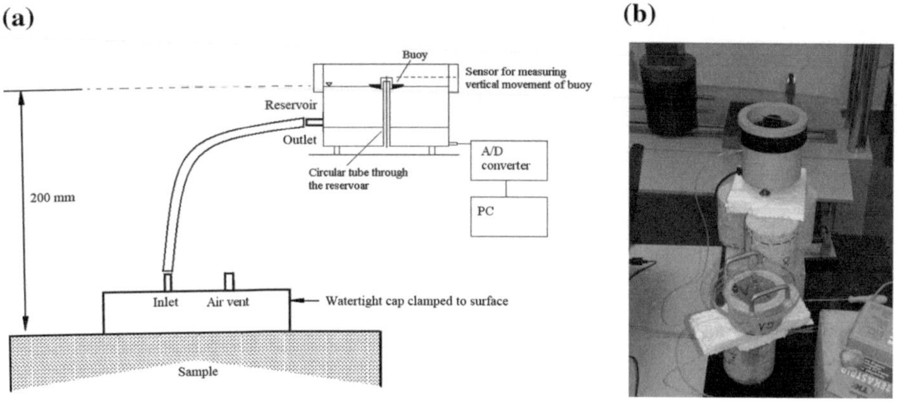

Fig. 4.17 Systems for testing absorption; **a** schematic setup, and **b** laboratory setup [13]

4.4.2.2 Autoclam Method

The Autoclam sorptivity test measures the cumulative inflow of water in the first 15 min from a water source of 50 mm diameter (i.e. from a base ring of internal diameter 50 mm) at an applied pressure of 2 kPa (approximately 200 mm water head) [6]. A plot of cumulative volume of water versus square root of time gives a linear relationship and the slope obtained from the graph is reported as a sorptivity index.

The moisture content of the concrete surface has been reported to influence the Autoclam sorptivity index and it has been proposed that the test should be carried out when the internal relative humidity of concrete at 10 mm depth is less than 80 % to eliminate this effect.

4.4.2.3 Water Absorption According to RILEM CPC11.2

A cross-section surface of a concrete specimen with the height at least twice as large as the edge or diameter, resting on stable supports inside a recipient under atmospheric pressure, is submerged in water with a constant level of about 5 mm above the bottom surface of the specimen [36]. The recipient and the specimen are covered with an enveloping vessel to avoid a rapid evaporation of the water from the specimen, Fig. 4.18.

The specimen is weighed at different periods of time t_i, initially and 3, 6, 24, 24 and 72 h after the first contact with water. Before each measurement the specimen should be allowed to drip the excess of water by resting it in a non-absorbent base for about 60 s. The absorption of water by capillary a_{cc} is then calculated by the following expression for each period of time t_i, [36]:

Fig. 4.18 Absorption of water by concrete by capillarity

$$a_{cc} = \frac{m_i - m_0}{A} \quad (4.10)$$

where m_0 is the initial mass of the specimen in g, m_i is the mass of the specimen after a specific time t_i in contact with the water in g, and A is the cross-sectional area of the specimen in mm^2.

Alternatively, the absorption of water can be expressed in terms of height of capillary rise measured over four vertical lines (mean value), equally spaced, along the lateral surface(s) of the specimen, after each period t_i.

The test results are frequently expressed as a function of $\sqrt{t}$ by analogy, for example, with the capillary action of water in an ideal cylindrical pore, which can be approximately expressed by the Washburn equation [37]:

$$m(t) = S_c\sqrt{t} \quad (4.11)$$

where *m(t)* is the mass increase of the specimen (or cumulative water absorption) after time t, and S_c is designated as the absorption coefficient or sorptivity. An example of results obtained in this test is presented in Fig. 4.19.

However, the linear behaviour of *m(t)* with $\sqrt{t}$ is often not observed during the tests. For this reason, it is usually preferred to express the capillary behaviour in terms of water adsorbed at a given time rather than in terms of absorption coefficient or sorptivity. Several reasons can be found in literature [38] to explain this non-linear behaviour.

As an addition to the recommendation by RILEM CPC11.2, the results of the test can also be expressed as the coefficient of water absorption R in h/m^2, calculated according to Eq. (4.12) [39]:

$$R = \frac{t_c}{x^2} \quad (4.12)$$

where R is the coefficient of absorption in h/m^2, t_c is the testing time in h, and x is the height of the absorbed water in m.

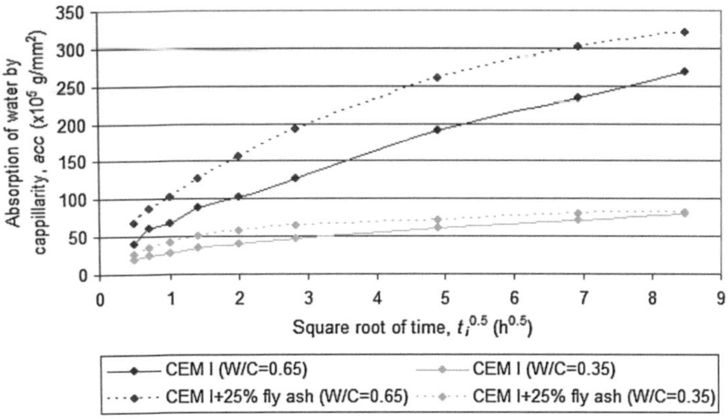

Fig. 4.19 Example of surface absorption test results

A very important aspect for obtaining reproducible results is the preconditioning procedure used in the specimens prior to testing. There is still no consensus in the scientific community on this procedure and thus the results obtained by different laboratories are often not comparable. An attempt for the harmonization of a reproducible preconditioning procedure and for the improvement of this test was done by RILEM TC 116—PCD: Permeability of Concrete as a Criterion of its Durability [21]. Although this procedure allows achieving a predefined hygrometric condition in the concrete specimen, it has the disadvantage of being too laborious and time consuming.

Other limitations of this test are the absence of established values for evaluating concrete quality (excluding national standards) and the need for drilling cores from the structures in cases of assessing the quality of in situ concrete.

4.4.2.4 Water Absorption According to ASTM C1585

The procedure of performing the water absorption test according to the ASTM standard is similar to that recommended by RILEM. In this standard, the preconditioning of the specimens is recognised as the prevailing step as well. Specimens need to be in the environmental chamber at a temperature of 50 ± 2 °C and a RH of 80 ± 3 % for 3 days. Subsequently, specimens should be placed inside a sealable container with free flow of air around the specimen. The container should be stored at 23 ± 2 °C for at least 15 days before the start of the absorption procedure. This preconditioning ensures the equilibration of the moisture distribution within the test specimens and has been found to provide an internal relative humidity of 50–70 %, which is similar to the relative humidity found near the surface in some field structures.

Similar to the procedure recommended in RILEM, the preconditioned specimens are left to absorb water for a given time. The absorption I is the change in mass divided by the product of the cross-sectional area of the test specimen and the density of water. For the purpose of this test, the temperature dependence of the density of water is neglected and a value of 0.001 g/mm^3 is used. The units of I are mm.

$$I = \frac{m_t}{a/d} \tag{4.13}$$

where I is the absorption in mm, m_t is the change in specimen mass in g at time t, a is the exposed area of the specimen in mm^2, and d is the density of the water in g/mm^3.

4.4.2.5 Water Sorptivity Test (South Africa)

The water sorptivity test [23] measures the rate of movement of a water front through the concrete under capillary suction, Fig. 4.20. It is particularly sensitive to the micro-structural properties of the near-surface zone of concrete and therefore reflects the nature and effectiveness of curing. Water sorptivity measures the rate of water uptake by a dry specimen, normalized by its porosity. The water sorptivity is measured in mm/h$^{0.5}$. The drying of the specimen is carried out by placing it in an oven at a constant temperature of 50 °C for a period of 7 days. The lower the water sorptivity index, the better is the potential durability of the concrete. The same specimens as those used to measure the Oxygen Permeability Index (compare Sect. 4.2.2.7) can be used for the test.

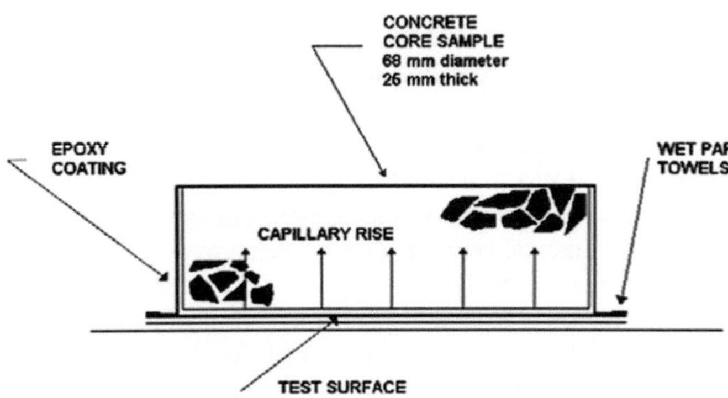

Fig. 4.20 Concrete specimen prepared for measuring the water intake according to water sorptivity test

Table 4.5 Criteria for concrete quality based on water absorption

ISAT absorption after 1 h, ml/m^2/s	Capillary absorption according to RILEM CPC11.2, %	Water sorptivity test, mm/h$^{0.5}$	Concrete quality
		<6	Very good
0.10	<3	6–10	Good
0.10–0.20	3–5	–	Normal
>0.20	>5	10–15	Poor
		>15	Very poor

Criteria here are given solely as an overview, with the purpose of highlighting the importance of indicating the test method in the performance-based design approach to avoid misinterpretation due to different test methods

4.4.3 *Overview and Criteria for Evaluation of Concrete Quality*

From the different methods of testing water absorption, the concrete can be evaluated using the criteria listed in Table 4.5 [23, 26, 36].

An overview of different test methods and their advantages and limitations for practice is presented in Table 4.6. Advantages and limitations are relative and based on the experience of the authors with the methods considered in this chapter.

Table 4.6 Overview of different methods and their advantages and limitations

Advantages/limitations	Test method				
	Initial surface absorption test	Autoclam method	Water sorptivity test	ASTM C 1585	RILEM CPC11.2
Non-invasive	+	–	–	–	–
Applicable on-site	+	+	–	–	–
Applicable on drilled cores	–	–	+	+	+
Performance criteria correlated to exposure classes	–	–	+	–	+
Standardised	+	–	+	+	+
Used in service life models	+	–	–	–	+
Easy to use on site	+	–/+	–	–	–
Same equipment can be used for other durability indicators	–	+	–	–	–

4.5 Chloride Penetration

For reinforced concrete structures subjected to chloride exposure, one of the most important issues is to determine the concrete's resistance to chloride penetration. The assessment of the resistance to chloride penetration can be used in the design and construction stage for quality control purposes, as well as for the assessment of concrete quality in existing structures. Since the penetration of chloride ions through the concrete is a slow process, it cannot be determined directly in a time frame that would be useful as a quality control and assessment procedure. Therefore, to assess chloride penetration indicators in a reasonable time, test methods that accelerate the process are usually used. In this document the most common methods are presented, which are based on the two most dominant physical processes of penetration, namely diffusion and migration.

4.5.1 *Principle and Mechanism*

Transport mechanisms of chlorides penetrating concrete are very complex phenomena, involving diffusion, capillary suction, hydrostatic pressure, convection, migration, etc., accompanied by physical and chemical binding.

The mechanisms of ionic transport in solution can be reliably described on the basis of an equation which in electrochemistry is known as the extended Nernst–Planck equation, describing unidirectional x flux of a particular ion J_i as a flow of mass due to the simultaneous action of a concentration gradient, an electrical field and a flow of the solvent, i.e. convection [40, 41]:

$$Flux = diffusion + migration + convection \qquad (4.14)$$

$$-J_i = D_i \frac{\partial C_i(x)}{\partial x} + \frac{z_i F}{RT} D_i C_i \frac{\partial E(x)}{\partial x} + C_i v_i(x) \qquad (4.15)$$

where J_i is the flux of the ionic species I in mol/(m^{2}s), D_i is the diffusion coefficient of the ionic species i as a function of location x in mols/m^3, z_i is the valences of ionic species I, F is Faraday's constant ($F = 9.648 \times 10^4$ J/(V × mol)) in J/(V × mol), R is the universal gas constant ($R = 8.314$ J/(mol × K)), T is the temperature in K, $E(x)$ is the applied electrical potential as a function of x in V, and v_i is the convection velocity of I in m/s.

This equation allows the calculation of D from the total ionic flux record. In the testing methods for the evaluation of concrete resistance to chloride penetration usually only one transport mechanism is present or is taken as a dominant one. Equation (4.15) is then simplified and the calculation of chloride diffusion or migration coefficients can be easily performed, which is explained in the following sections.

4.5.1.1 Diffusion

Diffusion of chlorides into concrete can be described by Fick's First Law, which, in the one-dimensional situation normally considered, states:

$$q = -D_{eff}\frac{\partial C}{\partial x} \tag{4.16}$$

where q is the flux of chloride ions in mol/(m^{2}s), D_{eff} is the effective diffusion coefficient in m^2/s, C is the concentration of chloride ions ins mol/m^3, and x is the position variable in m.

The general setup of a diffusion cell consists of a container which is separated by the test specimen of given thickness into two chambers in such a way that the chambers contain media of known concentrations c_1 and c_2 for the species under investigation, Fig. 4.21.

In practical terms, this equation for the diffusion process is only useful after steady-state conditions have been reached, i.e. when there is no change in concentration with time. It can be used, however, to derive the relevant equation for non-steady conditions (when concentrations are changing), often referred to as Fick's Second Law which includes the effect of changing concentration with time t:

$$\frac{\partial C}{\partial t} = D_{app}\frac{\partial^2 C}{\partial x^2} \tag{4.17}$$

where D_{app} is the apparent chloride diffusion coefficient in m^2/s. This has been solved using the boundary condition $C(x = 0, t > 0) = C_0$ (the surface concentration is constant at C_0), the initial condition $C(x > 0, t = 0) = 0$ (the initial concentration in the concrete is 0) and the infinite point condition $C(x = \infty, t > 0) = 0$ (far enough away from the surface, the concentration will always be 0). The solution is then as follows:

$$C(x,t) = C_0\left(1 - erf\frac{x}{\sqrt{4D_{app}t}}\right) = C_0 \cdot erfc\frac{x}{\sqrt{4D_{app}t}} \tag{4.18}$$

where $erf(y)$ is the error function, a mathematical construct found in math tables or as a function in common computer spreadsheets.

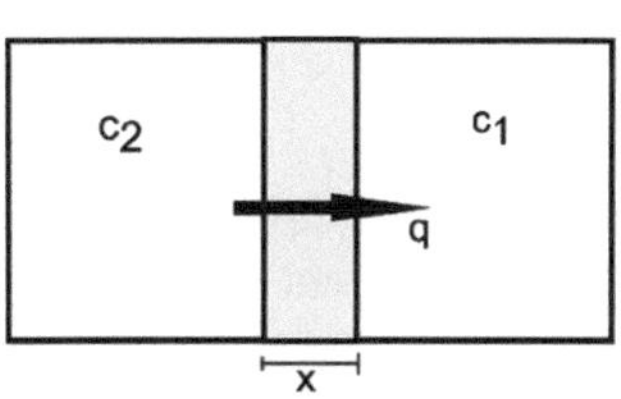

Fig. 4.21 Typical setup of a diffusion cell [42]

Frequently the *erf*-solution to Fick's 2nd law is fitted to measured chloride profiles which are obtained from specimens that were removed from an in situ structure, or from specimens used in a ponding test [42, 43].

4.5.1.2 Migration

Diffusion of ions in a liquid provokes an electrical field if the transport of cations is not balanced by a corresponding counter flow of anions. Considering that in natural conditions as well as in most experimental set-ups not only one type of ion is moving, and different ions have different mobility, the build-up of a voltage difference is likely to occur. This transport is best described with the Nernst-Planck flux equation (Eq. (4.15)). If the potential difference is the dominant transport driver, which is especially the case in accelerated migration tests with an applied electrical field, and by acceptance of several assumptions, Eq. (4.15) can be simplified in the following way [40]:

$$-J_i = \frac{z_i F}{RT} D_i C_i \frac{\partial E(x)}{\partial x} \tag{4.19}$$

For the application in concrete it can be used for the calculation of the migration coefficient:

$$D = \frac{JRTl}{zFC\Delta E} \tag{4.20}$$

where all parameters are known and flux J can be calculated from an experimental test in which the amount of chlorides is monitored over time [41].

Migration tests can be performed in steady and non-steady state conditions in order to obtain the corresponding "diffusion coefficient" D_s and D_{ns} which are also named "effective" and "apparent", respectively [42].

4.5.2 *Test Methods*

The test types used at present are all based on the contact of a *NaCl* solution with the concrete during times not exceeding 90 days. They can be grouped in several manners. In general they differ in the concentration of the *NaCl* solution used and in the test duration, and by:

a. those which enable chloride penetration by natural diffusion and aim at the calculation of the apparent diffusion coefficient D_{ns} as a parameter of reference; and
b. those which accelerate the penetration by applying an electrical field.

It can further be said that they differ in the parameters used to characterize the rate of penetration such as:

a. the D_s value (stationary regime)
b. the D_{ns} value (non-stationary regime)
c. the Coulombs recorded
d. the resistivity values.

They can be also grouped according to the place of testing, being either laboratory or on-site non-invasive testing methods. In the laboratory, measurements are performed on specimens from batch concrete or on core specimens removed from the structure. Laboratory tests should be used as a reference tests for the evaluation of non-invasive testing (NDT) on-site.

4.5.2.1 Non-steady State Chloride Diffusion Coefficient

This method is based on natural diffusion under a very high concentration gradient [44]. Specimens are firstly pre-conditioned, namely saturated with saturated lime-water, and then immerged in a solution of 165 g *NaCl* per litre for at least 35 days, Fig. 4.22. The upper surface is exposed to a *NaCl*-solution, while the other surfaces are isolated with epoxy coating.

The values of the apparent non-steady state chloride diffusion coefficient D_{nssd} and the apparent surface chloride content C_s are determined by curve-fitting the measured chloride profile to an error-function solution of Fick's 2nd law, according to the principle of least squares, as illustrated in Fig. 4.23:

$$C(x,t) = C_s - (C_s - C_i) \cdot erf \frac{x}{\sqrt{4D_{nssd}t}} \tag{4.21}$$

From the values of D_{ns} and C_s, the parameter K_{Cr}, called penetration parameter, can be derived:

$$K_{Cr} = 2\sqrt{D_{nssd}} \cdot erf^{-1}\left(\frac{C_S - C_r}{C_S - C_i}\right) \tag{4.22}$$

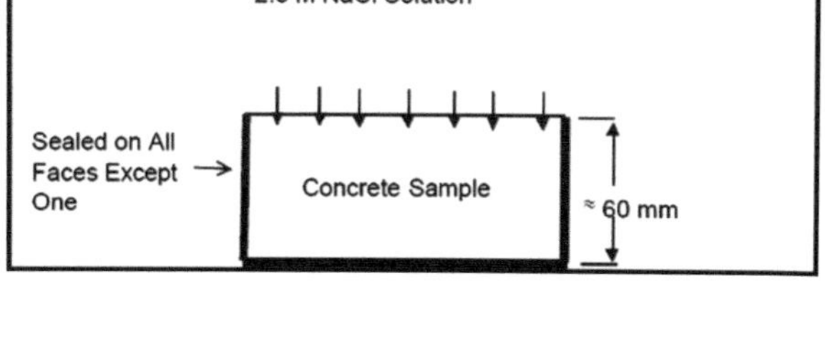

Fig. 4.22 NT BUILD 443 setup

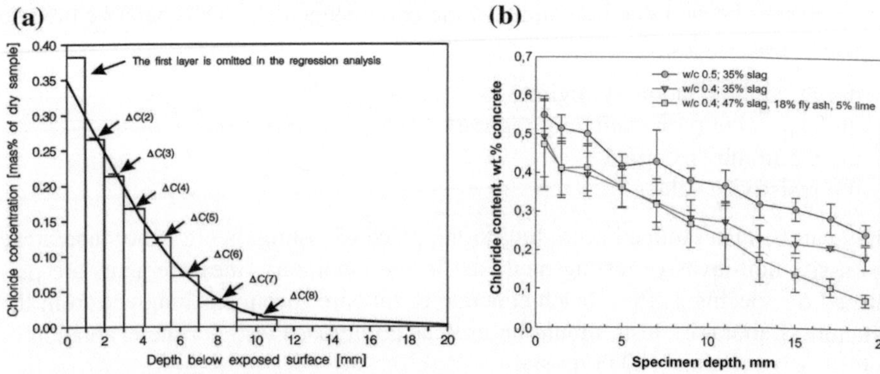

Fig. 4.23 **a** Example of regression analysis for curve fitting [43], and **b** example of chloride profile after 3 months exposure according to NT BUILD 443 [46]

where erf^{-1} is the inverse of the error function. K_{Cr} can be determined when $C_S > C_r > C_i$. The assumed value of C_r is 0.05 mass% of the specimen, unless another value is required.

The test is relatively laborious and takes a relatively long time; for low quality concretes the minimum exposure period is 35 days. For higher quality concretes, however, this period must be extended to 90 days or longer.

4.5.2.2 Non-steady State Chloride Migration Coefficient

The most commonly used and worldwide accepted test method is according to NT BUILD 492 for the evaluation of the non-steady state chloride migration coefficient [45]. This method uses an external electrical field, axially applied across the specimen for accelerating chloride penetration. The test gives values of D_{nssm} (non-steady state migration coefficient), in a relatively simple and rapid way, as described below.

The method is applicable to hardened specimens cast in the laboratory or drilled from field structures, where the inner part (not contaminated) of the concrete core should be used for testing. The chloride migration coefficient determined by this method is a measure of the resistance of the tested material to chloride penetration, and according to the final report of RILEM TC 189-NEC [42] it should be used as a reference test. This non-steady-state migration coefficient cannot be directly compared with chloride diffusion coefficients obtained from the other test methods, such as the non-steady-state immersion test or the steady-state migration test. Correction factors, which enable the correlation between non-steady state migration coefficient with other diffusion coefficients, were suggested in several project reports [43, 47–49].

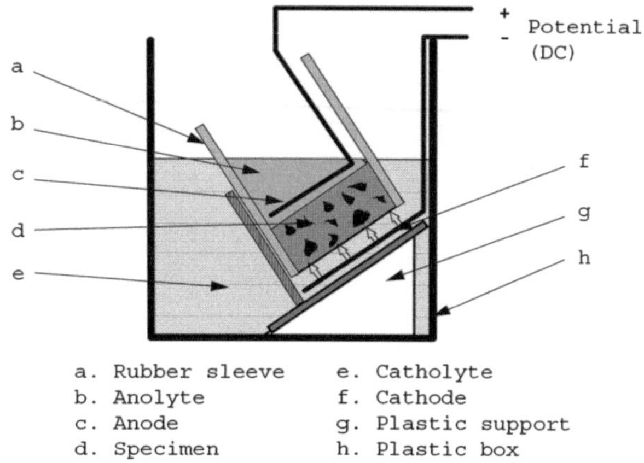

Fig. 4.24 Setup for chloride migration test [45]

Usually specimens of diameter 100 mm and a thickness of 50 mm are tested, with the cut surface as the test surface. If a drilled core is used, the outermost approximately 10–20 mm thick layer should be cut off (depending on the chloride contamination) and the next 50 ± 2 mm thick slice should be cut as the test specimen. The end surface that was nearer to the outermost layer is the one to be exposed to the chloride solution. All specimens should be vacuum saturated with saturated lime-water before testing. Testing procedures include imposing a 10–60 V external potential across the specimen with the test surface exposed to the 10 % *NaCl* solution (catholyte) and the oppose surface in the 0.3 M *NaOH* solution (anolyte) for a certain duration (6–96 h depending on the quality of concrete, in most cases 24 h), Fig. 4.24, then splitting the specimen and measuring the penetration depth of chlorides using a colorimetric method.

From the chloride penetration depth the non-steady state migration coefficient can be calculated using the following equation:

$$D_{nssm} = \frac{RT}{zFE} \frac{x_d - \alpha\sqrt{x_d}}{t} \tag{4.23}$$

where E and α are given by:

$$E = \frac{U - 2}{L} \tag{4.24}$$

$$\alpha = 2\sqrt{\frac{RT}{zFE}} \cdot erf^{-1}\left(1 - \frac{2c_d}{c_0}\right) \tag{4.25}$$

where D_{nssm} is the non-steady state migration coefficient in 10^{-12} m^2/s, U is the absolute value of the applied voltage in V, T is the average value of the initial and final temperature in the anolyte solution in °C, L is the thickness of the specimen in mm, x_d is the average penetration depth in mm, t is the test duration in s, erf^{-1} is the inverse of the error function, c_d is the chloride concentration at which the colour changes (for OPC concrete, $c_d \approx 0.07$ N), and c_0 is the chloride concentration in the catholyte solution ($c_0 \approx 2$ N).

Since:

$$erf^{-1}\left(1 - \frac{2 \times 0.07}{2}\right) = 1.28 \tag{4.26}$$

Equation (4.23) can be simplified:

$$D_{nssm} = \frac{0.0239 \cdot (273 + T) \cdot L}{(U - 2) \cdot t}\left(x_d - 0.0238\sqrt{\frac{(273 + T)L \cdot x_d}{U - 2}}\right) \tag{4.27}$$

At present, no quantitative performance requirements regarding the chloride migration coefficient are included in the European standards, but there are some recommended/required limiting values used in national standards or recommendations [50]. In Switzerland for the exposure classes XD2 and XD3 according to standard EN 206, a maximum chloride migration coefficient of 10×10^{-12} m^2/s is required [29, 51]. In Germany, for XD1 and XD2 a maximum mean chloride migration coefficient of 10×10^{-12} m^2/s and for XD3 of 5×10^{-12} m^2/s is required by the Federal Waterways Engineering and Research Institute (BAW) for the approval of cements that are not included in relevant guidelines [52].

However, most of these limiting values are still given as deterministic values. Concrete is an inherently variable and heterogeneous material, and it is very important that the criteria nominated are set and assessed on a statistical basis that balances the clients risk of accepting defective concrete against the suppliers risk of having compliant concrete rejected [53]. The example (Fig. 4.25) shows a statistical distribution of chloride migration coefficient tested on 51 specimens of the same concrete [54]. Specimens were taken during concreting of elements from different

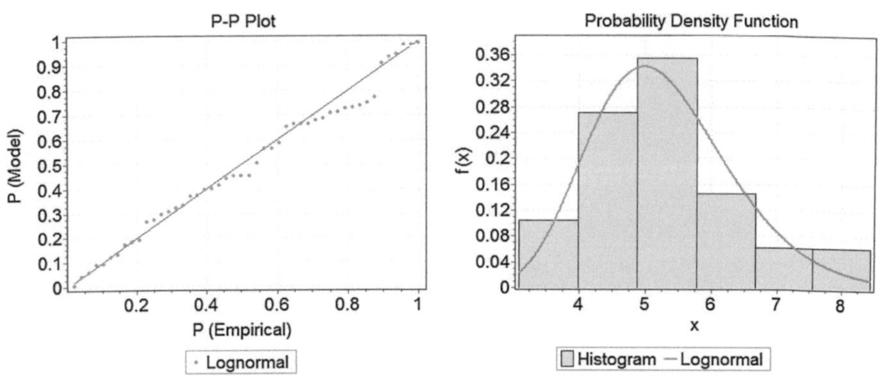

Fig. 4.25 Statistical analysis of chloride diffusion coefficient tested during quality conformity procedure on site [54]

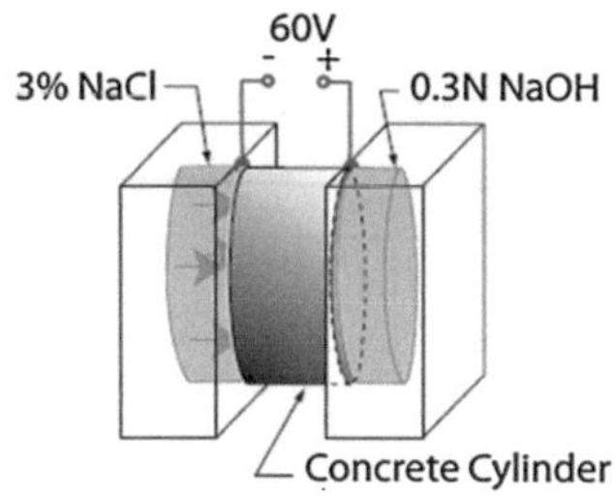

Fig. 4.26 ASTM C 1202 setup for chloride migration test [56]

Table 4.7 Criteria for concrete quality based on chloride penetration resistance [55, 78]

Nordtest method BUILD 492, migration coefficient (m^2/s)	RCPT ratings, charge passed, coulomb	Concrete quality
$<2 \times 10^{-12}$	<100	Very good
$2–8 \times 10^{-12}$	100–1000	Good
$8–16 \times 10^{-12}$	1000–2000	Normal
$>16 \times 10^{-12}$	2000–4000	Poor
	>4000	Very poor

Criteria here are given solely as an overview, with the purpose of highlighting the importance of indicating the test method in the performance-based design approach to avoid misinterpretation due to different test methods

batches of the same concrete quality required by the project. In the design project of the structure a chloride migration coefficient lower than 6×10^{-12} m^2/s was specified, since the structure is exposed to an aggressive marine environment. On the tested 51 specimens, the obtained mean value of chloride diffusion coefficient was 5.31×10^{-12} m^2/s, with a standard deviation of 0.80×10^{-12} m^2/s, and a coefficient of variation of 15 %, Fig. 4.25.

4.5.2.3 Electrical Indication of Concrete's Ability to Resist Chloride Ion Penetration

The rapid chloride permeability (RCP) test is widely accepted as a US standardized testing method [55] for electrical indication of the concrete's ability to resist chloride ion penetration. In the ASTM C1202 standard test, a water-saturated, 50 mm thick, 100 mm diameter concrete specimen is subjected to a 60 V applied DC voltage for 6 h using the apparatus shown in Fig. 4.26. The one reservoir is filled with a 3.0 % *NaCl* solution and the other contains a 0.3 M *NaOH* solution. The total charge passed is determined and this is used to rate the concrete according to the criteria included in Table 4.7.

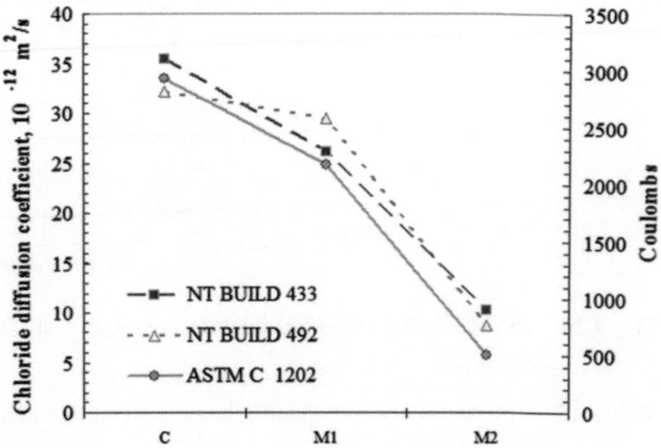

Fig. 4.27 Results of testing chloride diffusion on three mixtures of different quality, NT BUILD 492, NT BUILD 443, ASTM 1202 [57]

There have been a number of criticisms of this technique, although this test has been adopted as a standard test and is widely referred to in the literature [58–61]. The main criticisms are [40, 62–65]:

1. the current passed is related to all ions in the pore solution not just chloride ions
2. the measurements are made before steady-state migration is achieved especially for low quality concretes, which further increases the charge passed.

Lower quality concretes heat up more as the temperature rise is related to the product of the current and the voltage. The lower the quality of concrete, the greater the current at a given voltage and thus the greater the heat energy produced. This heating leads to a further increase in the charge passed, over what would be experienced if the temperature remained constant. Thus, poor quality concrete appears even worse than it would otherwise.

Nevertheless in the US this method is widely used and accepted as a reference durability testing method. Experimental testing also shows a good correlation between this method and NT BUILD 492 and NT BUILD 433, Fig. 4.27. The main shortcoming is the fact that the method gives qualitative results on concrete quality and not the diffusion coefficient, upon which most service life models are based. In the paper by Obla et al. [56] the statistically-based acceptance criteria have been established and proposed with examples of practical applications, in order to improve performance-based specifications.

4.5.2.4 Multiregime Method

The multiregime method, developed and standardized in Spain [43, 66, 67] is a method for the determination of the steady and non-steady state chloride diffusion coefficients by monitoring the conductivity of the electrolyte in the anolyte chamber in a migration experiment. The test is relatively simple due to the indirect measurement of chloride concentration through a simple conductivity measurement.

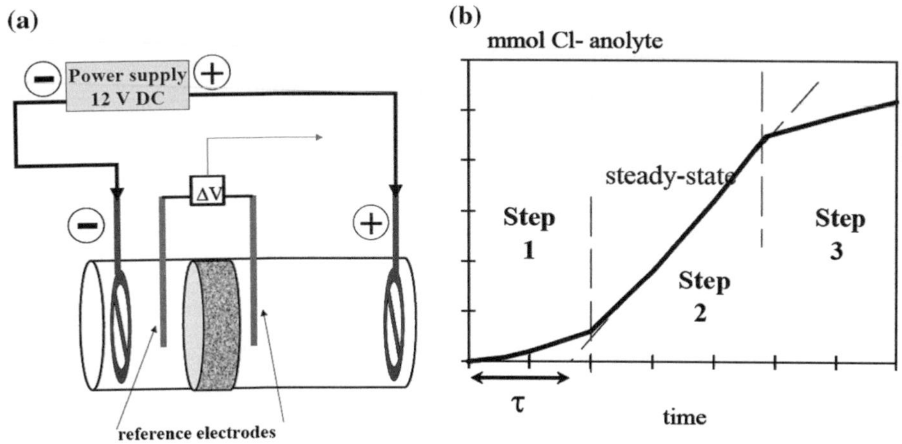

Fig. 4.28 **a** Test set up for the multiregime method, and **b** schematic representation of the evolution of the conductivity and amount of chlorides in the anolyte during the test [66, 67]

The method is applicable to hardened specimens cast in the laboratory or drilled from field structures. The method requires cylindrical specimens of any diameter and a thickness of 15–20 mm, sliced from cast cylinders or drilled cores.

Specimens need to be pre-conditioned with vacuum saturation with demineralised water. At least 3 specimens, with the cut surface as the test surface should be tested. The testing set up is schematically presented in Fig. 4.28a, and the testing procedure is as follows: 12 V of external potential is imposed across the specimen with the test surface exposed to the 1 M *NaCl* solution (upstream cell) and the opposed surface in the demineralised water (downstream cell). The test is based on measuring the amount of chlorides arriving in the downstream cell (anolyte) by means of measuring the conductivity of that solution. The steady-state coefficient is calculated from the flux of chlorides through the specimen calculated from the measurement of the conductivity of the anolyte in the anodic compartment. The calculation of the non-steady-state diffusion coefficient is made from the time necessary for the chloride ions to establish a constant flux, so called time-lag, Fig. 4.28b. The duration of the test depends on the quality of concrete, and varies from a few days up to about 2 weeks.

The test gives values of steady state migration coefficient D_s from the flux and non-steady state migration coefficient D_{ns} from the time-lag. D_s is calculated from the slope of the constant portion of the concentration-time curve (the constant flux), according to the modified Nernst-Planck equation [40, 66, 68]:

$$D_s = \frac{J_{Cl} RTl}{zFC_1 \gamma \Delta \Phi} \tag{4.28}$$

where γ is the activity coefficient of the catholyte solution. The non-steady state migration coefficient D_{ns} is calculated from the intersection on the time-axis of the constant portion of the concentration-time curve (time-lag τ, step 1 in Fig. 4.28b) according to the following equation:

$$D_{ns} = \frac{2l^2}{\tau v^2}\left(v\coth\frac{v}{2} - 2\right) \tag{4.29}$$

where τ is the time-lag in the migration test in s, l is the thickness of the specimen in cm, and v is given by Eq. (4.30):

$$v = \frac{ze(\Delta\varphi)t}{kT} \tag{4.30}$$

where k is Boltzmann's constant, T is the average temperature during the test, and $(\Delta\varphi)t$ is the averaged effective voltage in V through the specimen from the beginning of the test until the time lag.

For the range of values of voltage drop usually applied in the test, Eq. (4.29) can be simplified into:

$$D_{ns} = \frac{2l^2(v-2)}{\tau v^2} \tag{4.31}$$

4.5.2.5 Integral Corrosion Test (UNE 83992-2 EX-2012)

The method is based on an acceleration of the chloride penetration by means of an electrical field (migration) [69, 70]. The test consists of exposing a concrete specimen (prismatic shape, recommended size 10 mm or 15 mm) with a transversally embedded steel bar (8 or 10 mm in diameter, concrete cover depth of around 30 mm) to an electric current generated by two electrodes, one positioned in a chloride solution (0.6 M $NaCl$ and 0.4 M $CuCl_2$), in contact with one face of the

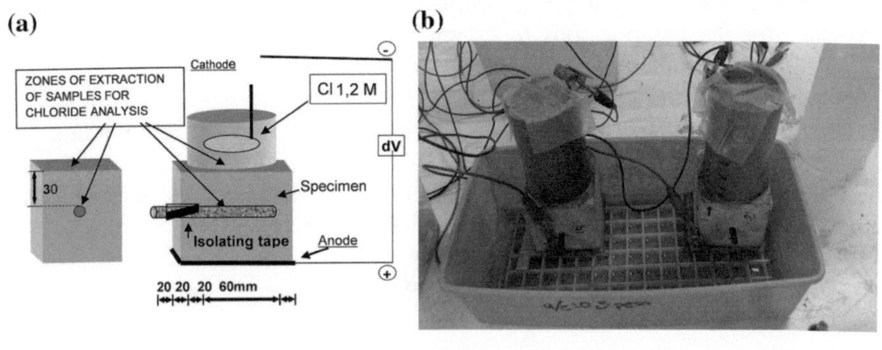

Fig. 4.29 **a** Accelerated chloride migration and corrosion test setup, and **b** the arrangement during the experimentation. The pond is covered by a transparent film to prevent the chloride solution from evaporating

specimen, and the other in contact with the opposite face, Fig. 4.29. A potential drop of 12–30 V is applied by means of a potential source through electrodes placed inside the pond (the cathode—a copper plate) and at the opposite face through a sponge (the anode—a stainless steel mesh or plate). The current is recorded for the subsequent calculations. The electrical field induces the accelerated penetration of chloride ions throughout the concrete. The chloride ions corrode the steel bar placed in their path enabling to study not only the time for steel depassivation, but also the actual chloride threshold developing active corrosion and the corrosion rate produced.

The time lag from the initiation of the experiment when the electrical field is applied and the steel depassivation enables the measurement of the non-steady state diffusion coefficient D_{ns} also named the apparent diffusion coefficient D_a. The current is subsequently disconnected and the specimen cracked open to obtain samples of the concrete on the steel surface on the side closest to the chloride container and on the specimen surface. The critical chloride concentration that initiated corrosion and the chloride concentration on the specimen surface are then obtained from chemical analysis. The test may continue on a second specimen, in which the current is either disconnected or maintained to induce a certain degree of steel corrosion and obtain the mean $I_{corr.mean}$ or total $I_{corr,complete}$ corrosion current density values, respectively, for the embedded bar in the particular concrete.

The non-stationary diffusion coefficient is calculated by means of the following expressions [70]:

$$D_{ns} = \frac{e^2}{2t_{lag}\varphi} \tag{4.32}$$

where D_{ns} is the natural non-stationary diffusion coefficient in cm^2/s, t_{lag} is the "time lag" or time until depassivation is noticed in s, e is the cover thickness in cm, and φ is given by Eq. (4.33):

$$\varphi = \frac{zF}{RT}\Delta\phi = 40\,for\ 22\,^{\circ}C = acceleration\,factor\ of\ the\ electrical\,field \tag{4.33}$$

where R is the gas constant, F is Faraday's constant, T is temperature in K, z is the ion valence (which is 1 for chlorides), and $\Delta\phi$ is the normalised electric field in V given by Eq. (4.34):

$$\Delta\phi = \frac{\Delta V}{L} \tag{4.34}$$

where ΔV is the potential voltage drop applied in V, and L is the distance between electrodes (specimen thickness) in cm.

In order to detect the onset of corrosion, the corrosion potential of the bar is periodically measured by voltmeter; first switching off the potential drop during 30–180 min. The depassivation can be also detected by measuring the polarization

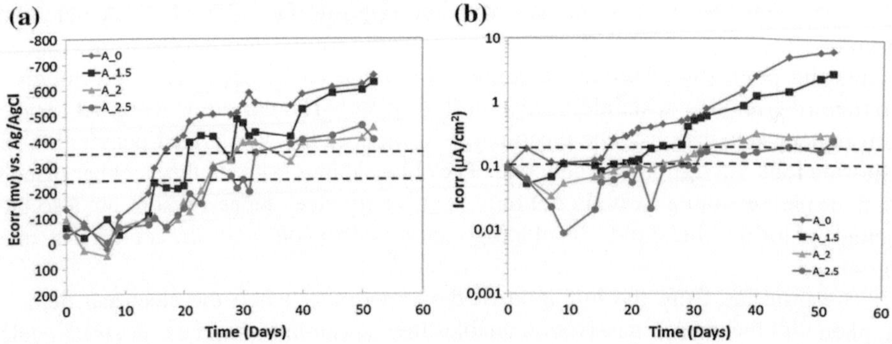

Fig. 4.30 **a** E_{corr}, and **b** I_{corr} values with time for concrete A (*upper part*) with no inhibitor and with three proportions of an inhibitor

resistance through a potentiostat. The R_p value is measured by a sweep rate of 10 mV/min [71]. A constant B of 26 mV is used for the calculation of the corrosion rate I_{corr} of the bar during the experiment. The depassivation is detected when the potential shifts below around −350 mV_{SCE} or the I_{corr} is higher than 0.2 μA/cm^2. The accumulated corrosion P_{corr} is calculated from the corrosion rate I_{corr} through:

$$P_{corr} = 0.0116 \times I_{corr} \times t \tag{4.35}$$

where P_{corr} is the corrosion penetration in mm, I_{corr} is the corrosion rate in μA/cm^2, and t is the time in years. P_{corr} can be also obtained from the integration of the I_{corr}-time plot.

The test enables the determination of all service life parameters in a single accelerated procedure: the diffusion coefficient, the chloride threshold and the corrosion rate (at different corrosion degrees). It serves for evaluating different concrete mixes, galvanized and bare steel or the efficiency of corrosion inhibitors.

An example [72] of the evolution of the corrosion potential and corrosion rate is shown in Fig. 4.30, where a concrete mix without (A_0) and with three proportions of an inhibitor (A_1.5, A_2, A_2.5) are tested. It is clearly deduced from the experiment that the delay in depassivation is due to the increasing amount of inhibitor.

The diffusion coefficients calculated from these figures through the equations given above are shown in Fig. 4.31a. It has to be noted that the lower D_{ns} values found when the inhibitor is present have to be understood as a "fictitious" value in the sense that the concrete is almost the same (except small changes in the rheology due to the presence of the inhibitor) and then the D_{ns} should have been the same. They are different due to the longer times taken for the depassivation. In spite they are "fictitious" values however, they serve for the practical calculation of the corrosion initiation period.

After depassivation is noticed, the specimen is broken in order to find out the chloride threshold, Fig. 4.31b. For it a small sample (about 2 g) near the bar over

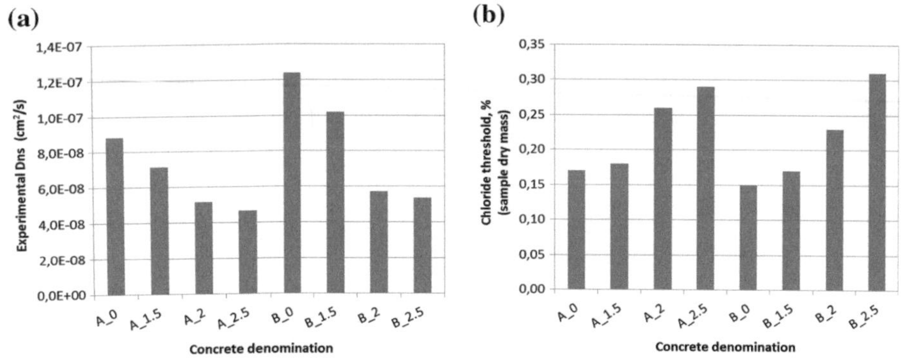

Fig. 4.31 **a** Diffusion coefficients calculated from the time to depassivation, and **b** critical chloride contents in the steel/concrete interface of the specimens after depassivation

the corroded zone is extracted by means of a sharp pointed tool and a hammer. The chloride concentration values are given by concrete mass. It can be seen that the increasing chloride thresholds are a function of the increasing amount of inhibitor.

Using the same experimental setup, the corrosion rate after depassivation can be measured. In that case the specimen is left to corrode naturally by switching off the potential applied during around 15–30 days. If the I_{corr} in much corroded conditions is of interest, then the potential can be left applied until the bar is corroded up to a level of interest and then the polarization resistance can be measured [72].

4.5.2.6 Permit Ion Migration Test

The Permit Ion Migration Test (PERMIT) is a non- invasive test which is capable of determining the chloride migration coefficient of cover concrete [73–75]. The test is designed to function on the concrete surface itself. The migration coefficient obtained from PERMIT can be used for assessing the quality of concrete cover or for modelling the ingress of chloride ions at any given time.

The main disadvantage of the test is that it introduces chloride ions into the test surface. However, the chloride ions only will affect an area of 160 mm diameter and to a depth of up to 50 mm from the surface for a high permeability concrete. Chloride ions within the test area can be effectively removed by reversing the test procedure, which may take up to 4 days. It is also to be noted that the test area will be slightly stained by deposition of the ferrous by-product of the electrochemical reaction. However, if this deposit is washed off immediately after the test, the extent of the staining can be reduced.

The schematic diagram of the test apparatus (PERMIT body) is presented in Fig. 4.32. It consists of two cylinders, of different diameters, concentrically placed on the concrete surface. A rubber seal at the end of the cylinders prevents flow of liquid/ions between the cylinders or to the outside environment.

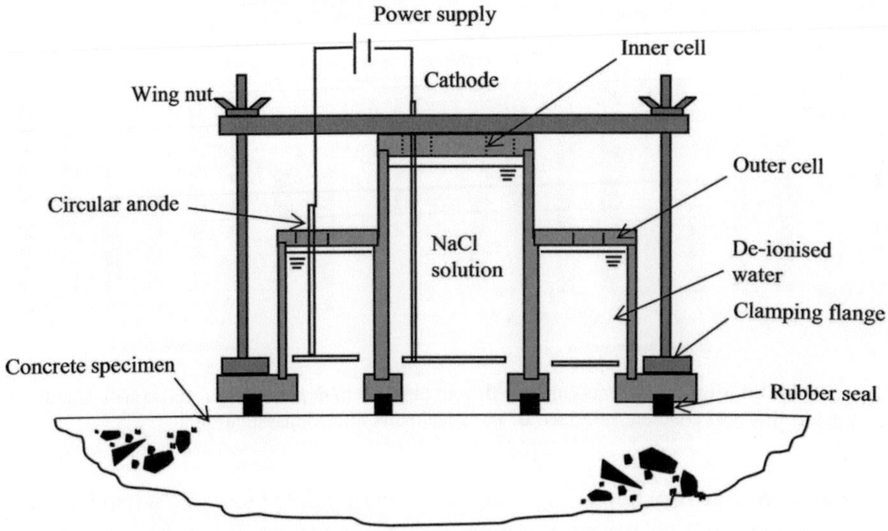

Fig. 4.32 Schematic diagram of permit ion migration test

The inner cell contains a 0.55 solution of *NaCl* and the outer cell contains distilled/de-ionised water. The inner cell accommodates a circular stainless steel mesh anode and the outer cell accommodates an annular mild steel perforated plate cathode. In order to carry out the test, the apparatus is fixed on the surface of the specimen and both inner cell and outer cell are filled with the respective solutions. A potential difference of 60 V DC is applied between the anode and the cathode. This forces chloride ions to travel from the inner cell to the outer cell through the concrete. A steady flow of chloride ions can be achieved in the test after 6–10 h for normal concrete and after 24 h for high performance/low permeability concretes. The change in concentration of the outer cell which initially contained distilled water is monitored using conductivity sensors. This change in conductivity is utilised to identify the steady rate of flow of chloride ions. A migration coefficient can be determined based on the rate of flow of chloride ions, the cell geometry and the test variables such as voltage applied, concentration of inner cell solution, etc. The coefficient obtained is termed as in situ migration coefficient and is expressed in m^2/s. Values for in situ migration coefficients range from 0.01×10^{-12} to 4×10^{-12} m^2/s.

The in situ migration coefficient obtained from PERMIT has been found to correlate well with other established laboratory based tests (see Figs. 4.33 and 4.34).

Research is on-going in identifying an alternative ionic solution in PERMIT so that the test surface will not be contaminated by chloride ions. Also research is on-going on developing a more rapid version of PERMIT for high performance low permeability concretes.

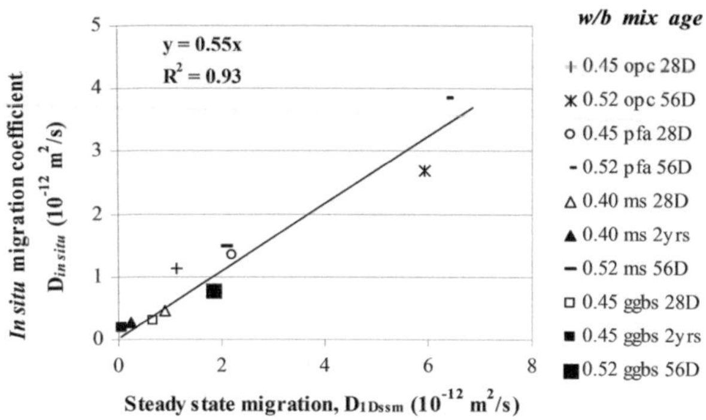

Fig. 4.33 Correlation between conventional laboratory-based steady state migration coefficient and in-situ chloride migration coefficient from PERMIT

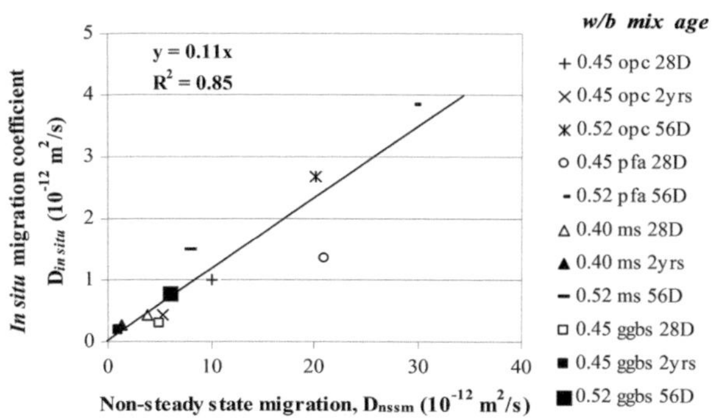

Fig. 4.34 Correlation between non-steady state migration coefficient from Nordic Test NT Build 492 and in-situ chloride migration coefficient from PERMIT

4.5.2.7 Chloride Profiling Method

When the structure has been exposed to chlorides for certain period of time, the chloride profiling method is very suitable for the assessment of concrete resistance to chloride penetration. Concrete dust specimens can be either prepared from slices from drilled cores or from concrete dust, drilled from the concrete surface with a drill hammer, as shown in Fig. 4.35.

The concrete dust specimens are tested in the laboratory, usually by a potentiometric titration method, to determine the total amount of chlorides by mass of concrete. After that the solution to Fick's 2nd law is fitted to measured chloride profiles. In such curve-fitting, if the exposure time t is inserted, the best curve-fitting gives two regression parameters, D_{app} and C_{sa}. The index a or app means achieved or apparent, after a certain exposure time and assuming that the diffusion coefficient and boundary condition were constant during the whole exposure [42, 43, 47].

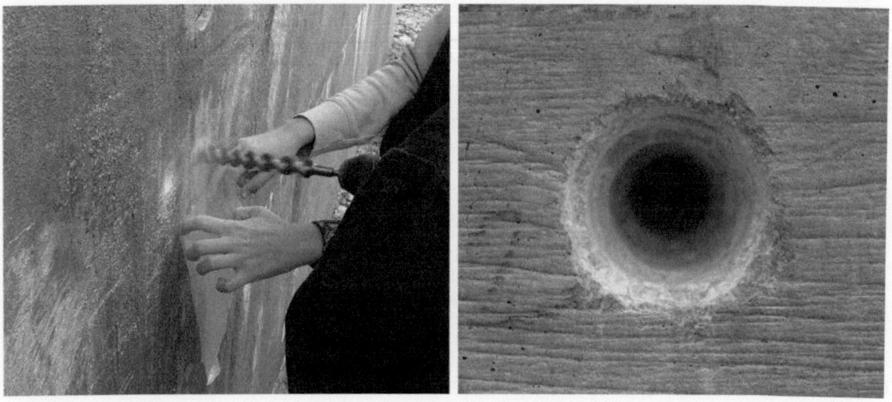

Fig. 4.35 Drilling concrete dust specimens for chloride profiling

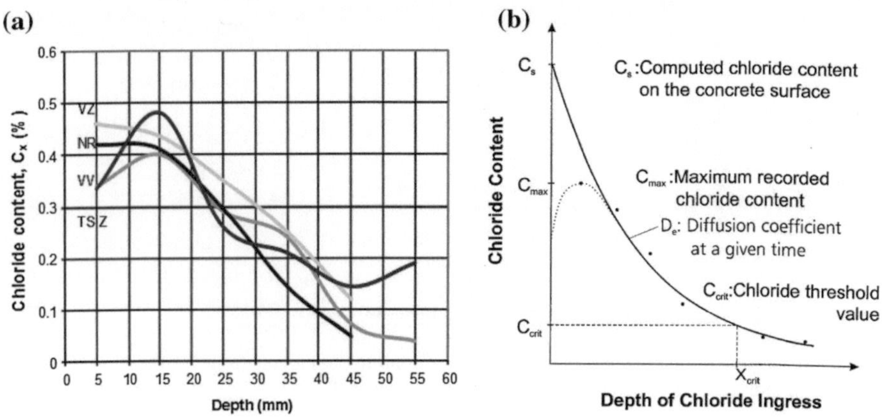

Fig. 4.36 **a** Typical chloride profiles from a concrete structure exposed to sea water [76], and **b** results of regression analyses of data measured in the structure by curve fitting to *erfc*-solution [77]

The calculated C_s from a measured chloride profile is the representative chloride concentration at the concrete surface, Fig. 4.36.

4.5.3 Overview and Criteria for Evaluation of Concrete Quality

From different methods of testing chloride penetration into concrete, the concrete quality can be evaluated using the criteria listed in Table 4.7.

Table 4.8 presents an overview of different testing methods and their advantages and limitations for practice. Advantages and limitations are relative and based on the experience of the authors with the methods considered in this chapter.

4.6 Concrete Resistivity and Conductivity

4.6.1 Principle and Mechanism

Resistivity techniques constitute another type of method of assessing the ability of chlorides to penetrate concrete. Concrete resistivity is the ratio between applied voltage and resulting current in a unit cell that is a specific geometry-independent material property, which describes the electrical resistance. The dimension of resistivity is resistance multiplied by length; usually Ωm. Electrical conductivity is the inverse of electrical resistivity; if the latter is expressed in Ωm, its inverse, the conductivity is expressed in Siemens per unit length (S/m). In concrete electrical current is carried by ions dissolved in the pore liquid. The resistivity of the concrete increases when the concrete is drying out and when the concrete carbonates, while the resistivity significantly decreases in wet concrete with more pore water and with larger and connected pores. In synthesis, concrete resistivity and conductivity are functions of porosity, the chemical composition of the solution in the pores and the number and distribution of pores as a result of the reaction with the environment [68, 79].

4.6.2 Test Methods

4.6.2.1 Direct Resistivity Test According to UNE 83988-1

Direct current resistivity can be measured by applying a voltage between two electrodes with the concrete sandwiched between them, as shown in Fig. 4.37. It is an indirect measurement of the transport property of concrete, because the electrical resistance of concrete is related to the pore structure and ionic strength in the pore solution.

The method is applicable to hardened specimens cast in the laboratory or drilled from field structures. If drilled cores are used, the outermost approximately 10–20 mm thick layer should be cut off and the next (50 ± 2) mm thick slice should be cut from each core as the test specimen. If cast cylinders are used, the central portion of each cylinder should be prepared as the test specimen (50 ± 2 mm thick slice). Specimens need to be pre-conditioned in the vacuum saturation container with saturated lime-water. The testing procedure is based on imposing a constant alternating current across the specimen and measuring the potential response for

Table 4.8 Overview of different methods and their advantages and limitations

Advantages/limitations	Test method						
	Nordtest method BUILD 443	Nordtest method BUILD 492	RCPT ratings ASTM C1202	Multiregime	Integral corrosion test	Permit ion migration Test	Chloride profiling
Non-invasive	–	–	–	–	–	–	–
Applicable on-site	–	–	–	–	–	+	+
Applicable on drilled cores	+	+	+	+	+	–	+
Performance criteria correlated to exposure classes	+	+	–	+	–	–	+
Standardised	+	+	+	+	+	–	–
Used in service life models	+	+	–	+	+	–	+
Easy to use on site	–	–	–	–	–	–/+	+
Natural diffusion	+	–	–	–	–	–	+

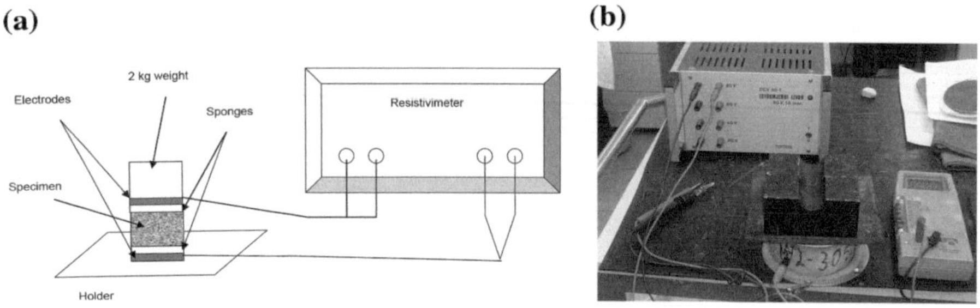

Fig. 4.37 **a** Test set up for concrete resistivity measurements [47], and **b** photograph of the setup for concrete resistivity measurements

calculating the resistance using Ohm's law. The test lasts for a few seconds or minutes for the measurements [47, 67].

4.6.2.2 Concrete Resistivity Wenner Probe

The Wenner array probe is a technique for determining resistivity on concrete in situ. It consists of a set of four points, each a constant distance apart, *a*. The current is applied between the two outer points, while the inner two points measure the potential, Fig. 4.38. This has the advantage of eliminating the influence of polarization as the actual potential is measured across an inner region.

For a semi-infinite region (where the thickness is much greater than the distance between the points) the resistivity can be calculated as [80]:

$$\rho = 2\pi a \frac{P}{I} \tag{4.36}$$

where ρ is the resistivity in Ω, a is the distance between points in m, P is the measured potential in V, and I is the applied current in A.

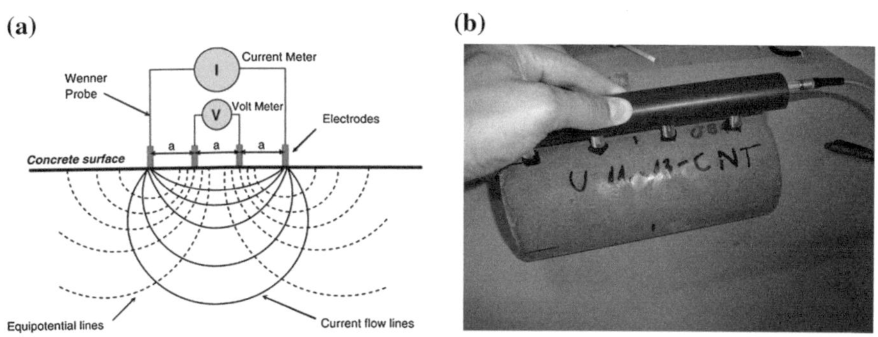

Fig. 4.38 **a** 4-point Wenner probe, and **b** application of test method on cored specimen [54]

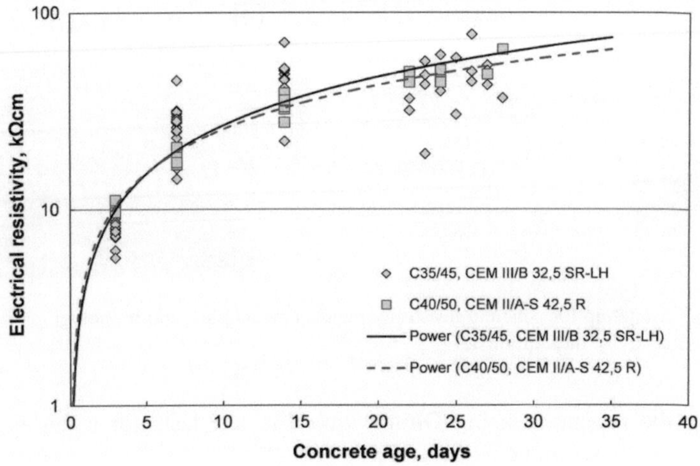

Fig. 4.39 Evolution of electrical resistivity of concrete in time [54]

If the thickness is not much greater than the distance between two points, then correction factors must be applied, and have been developed by Morris et al. [80]. The example here shows monitoring of concrete electrical resistivity during aging, as a part of concrete conformity procedure during construction, using a 4-point Wenner probe, Fig. 4.39 [54].

Resistivity techniques have the advantage of speed and represent methods already familiar to many concrete researchers. These techniques also provide a value that may be useful when determining corrosion rates in concrete. Resistivity tests avoid heating of the concrete as the voltage can be low, usually in the range of 10 V or lower [80], and is only applied for short times.

However it should be noted that rebars conduct current much better than concrete and they will disturb a homogeneous current flow. The measured result may be artificially low or high, if one or more electrodes are placed above or near rebars. To minimise this effect, none of the measuring electrodes should be placed above or near rebars, which is quite difficult to achieve in reality. Therefore the suggested solution for electrode spacing is between 30 and 50 mm. Further recommendations can be found in the literature [79, 81–84].

4.6.2.3 Chloride Conductivity Index Test (South Africa)

The South African chloride conductivity test apparatus (Fig. 4.40) consists of a two cell conduction rig in which concrete core specimens are exposed on either side to a 5 M *NaCl* chloride solution [85]. The core specimens are preconditioned before testing to standardise the pore water solution (oven-dried at 50 °C followed by 24 h

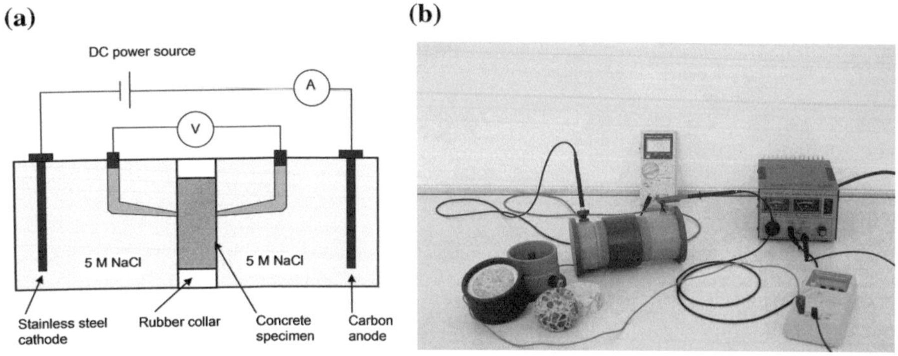

Fig. 4.40 **a** Schematic of chloride conductivity cell, and **b** photograph of the chloride conductivity cell

vacuum saturation in a 5 M *NaCl* chloride solution). The specimen diameter is typically 70 mm and the thickness is 30 mm, representing the cover concrete. The movement of chloride ions occurs due to the application of a 10 V potential difference. The chloride conductivity is determined by measuring the current flowing through the concrete specimen. The apparatus allows for rapid testing under controlled laboratory conditions and gives instantaneous readings. The test equipment and test procedures are described in detail in the literature [23–25].

Chloride conductivity decreases with the addition of fly ash, slag, and silica fume in concrete, extended moist curing and increasing grade of concrete. Portland cement concrete for instance generally has high conductivity values with only high-grade material achieving values below 1.0 mS/cm. Slag or fly ash concretes in contrast have significantly lower chloride conductivity values. While the test is sensitive to construction and material effects that are known to influence durability, results are specifically related to chloride ingress into concrete. Correlations between 28-day chloride conductivity results and diffusion coefficients after several years of marine exposure have shown to be good over a wide range of concretes [86].

An international study revealed that the chloride conductivity test is able to detect differences in w/b ratio, binder type, and curing condition on a statistically highly significant level and that test results obtained with the chloride conductivity method generally correlate well with results obtained from other accepted test methods for chloride resistance [1].

The Chloride Conductivity Index test method is used for performance-based design and quality control of concrete structures exposed to the marine environment in South Africa. Limiting chloride conductivity values were developed based on scientific and empirical correlations between concrete conductivity and chloride ingress for various environmental conditions and binder types (compare Sect. 8.8).

Table 4.9 Criteria for concrete quality based on concrete resistivity

Resistivity, Ωm	Chloride conductivity	Concrete quality
	<0.75	Very good
>1000	0.75–1.50	Good
500–1000	–	Normal
100–500	1.50–2.50	Poor
<100	>2.50	Very poor

Criteria here are given solely as an overview, with the purpose of highlighting the importance of indicating the test method in the performance-based design approach to avoid misinterpretation due to different test methods

Table 4.10 Overview of different methods and their advantages and limitations

Advantages/limitations	Test method		
	Direct resistivity test according to UNE 83988-1	Concrete resistivity Wenner probe	Chloride conductivity test
Non-invasive	–	+	
Applicable on-site	–	+	–
Applicable on drilled cores	+	+	+
Performance criteria correlated to exposure classes	–	–	–
Standardised	+	+	+
Used in service life models	+	+	+
Easy to use on site	–	+	–

4.6.3 *Overview and Criteria for Evaluation of Concrete Quality*

From different methods of testing concrete resistivity, concrete quality can be evaluated using criteria listed in Table 4.9 [23, 87].

An overview of different methods and their advantages and limitations for practice is presented in Table 4.10. Advantages and limitations are relative and based on the experience of the authors with the methods considered in this chapter.

4.7 Final Remarks

Before preparing this chapter, a survey was performed among different members of RILEM TC 230 PSC. They were asked to list the methods for concrete durability indication used and/or standardized in their country. Through this survey, a list of methods, presented in this chapter, was obtained. Methods that were presented here are among those most frequently and widely used in today's research and practice.

Beside the short description and examples of use of different test methods, a list of advantages and limitations was given. These limitations are based on the experience of the authors with a certain method. The reader is advised to consult the referenced literature for further details and recommendations for a method they are interested in. Furthermore, limiting values for certain properties and correlating concrete quality are given here merely as an overview, mostly based on the available literature. Ranges of limiting values are given here with the purpose of highlighting the importance of prescribing both test method and limiting value, in order to ensure that the required quality of concrete will be achieved.

The keys for successful implementation of performance-based design are well established and standardized limiting values and reproducible and repeatable methods of testing specific concrete properties. It is also crucial to establish a link between properties required for a specific environmental class and the required service life of a structure. Also, further efforts have to be aimed at setting in the design phase and assessing in the execution phase durability indicators on a statistical basis, similar to the procedure with concrete compressive strength. That way, the balance between the clients' risk of accepting defective concrete and the suppliers' risk of having compliant concrete rejected will be achieved.

References

1. RILEM report 40. Non-destructive evaluation of the penetrability and thickness of the concrete cover. In: Torrent R, Fernández Luco L, editors. State of the art report (STAR): RILEM TC 189-NEC. RILEM Publication S.A.R.L.; 2007. 246 pp.
2. Lawrence CD. Transport of oxygen through concrete. In: Chemistry and chemically-related properties of cement. London: British Ceramic Society; 1984.
3. RILEM report 12. Performance criteria for concrete durability. In: Kropp J, Hilsdorf HK, editors. London: E&FN SPON; 1995. 327 pp.
4. Parrott L. Design for avoiding damage due to carbonation-induced corrosion. In: Durability of concrete: third international conference. France: Nice; 1994. pp 283–298.
5. Neville AM. Properties of concrete. 4th ed. Essex: Longman Group Limited; 1995.
6. Basheer PAM. Near-surface testing for strength and durability of concrete. In: Fifth CANMET/ACI international conference on durability of concrete, Barcelona, Spain; 2000.
7. Schönlin K, Hilsdorf H. Evaluation of the effectiveness of curing of concrete structures. In: Scanlon JM, editors. Concrete durability: Katharine and Bryant Mather international conference, SP-100. ACI; 1987. pp 207–226.
8. Torrent RJ. A two-chamber vacuum cell for measuring the coefficient of permeability to air of the concrete cover on site. Mater Struct. 1992;25(150):358–65.
9. Figg JW. Methods of measuring air and water permeability of concrete. Mag Concr Res. 1973; 25(85):213–219 (London).
10. Kasai Y, Matsui I, Nagao M. On site rapid air permeability test for concrete, ACI SP-82, pp 501–524. In: Malhotra VM, editors. In-situ/non-destructive testing of concrete, ACI, Detroit; 1984.
11. Imamoto K, Shimozawa K, Nagayama M, Yamasaki J, Nimura S. Air permeability of concrete cover and its relationship with carbonation progress under long-term exposure test in Japan.

In: Alexander MG, Bertron A, editors. Concrete in aggressive aqueous environments—performance, testing, and modeling; 2009.
12. Berissi R, Bonnet G, Grimaldi G. Mesure de la porosite ouverte des betons hydrauliques. Bull Liaison Labor Ponts Chauss. 1986;142:59–67 (in French).
13. Gabrijel I, Mikulić D, Bjegović D, Stipanović Oslaković I. In-situ testing of the permeability of concrete, SACoMaTiS 2008, Varenna, Italy, 2008.
14. Bjegović D, Serdar M, Kožoman E. The influence of controlled permeability formwork on the mechanical and durability properties of concrete cover. eGFOS Electron J Fac Civ Eng Osijek. 2012; 1, 2:62–73.
15. SIA 262/1:2013. Concrete structures—complimentary specifications, Swiss Standards; 2013.
16. Jacobs F, Denarie E, Leemann A, Teruzzi T. Recommendations for the quality control of concrete with air permeability measurements, VSS Report 641, Bern, Switzerland; 2009.
17. Bjegović D, Serdar M, Baričević A, Šimunović T. Air permeability as a parameter of concrete quality compliance. In: Knežević M, Šćepanović B, editors. Civil engineering—research and practice. Podgorica: Štamparija Pavlović; 2012; pp 1263–1269.
18. Paulini P. A laboratory and on-site test method for air permeability of concrete. In: 2nd international symposium on service life design for infrastructure, Delft, The Netherlands; 2010.
19. Paulini P, Nasution F. Air permeability of near-surface concrete. In: Concrete under severe conditions environment & loading, CONSEC'07 Tours, France; 2007.
20. Kollek J. CEMBUREAU recommendation—the determination of the permeability of concrete to oxygen by the Cembureau method. Mater Struct. 1989;22:225–30.
21. RILEM TC 116-PCD. Permeability of concrete as a criterion of its durability recommendations tests for gas permeability of concrete. Mater Struct. 1999; 32:174–179.
22. Abbas A, Carcassés M, Ollivier J-P. Gas permeability of concrete in relation to its degree of saturation. Mater Struct. 1999;32:3–8.
23. Alexander MG, Mackechnie JR, Ballim Y. Use of durability indexes to achieve durable cover concrete in reinforced concrete structures. In: Skalny JP, Mindess S, editors. Materials science of concrete, American Ceramic Society; 2001. vol. VI, pp 483–511.
24. Mackechnie JR, Alexander MG. Durability predictions using early age durability index testing. In: Proceedings of 9th durability and building materials conference, 2002, Australian Corrosion Association, Brisbane; 2002. 11 pp.
25. Beushausen H, Alexander MG, Mackechnie J. Concrete durability aspects in an international context. In: Concrete plant and precast technology BFT, Germany; 2003. vol 7, pp. 22–32.
26. Bjegović D, Serdar M, Stipanović Oslaković I. Assessment of concrete properties in structures. In: Kim S-H, Ann KY, editors. Handbook of concrete durability. Korea: Middleton Publishing Inc.; 2010. pp. 131–179.
27. Basheer PAM. A brief review of methods for measuring the permeation properties of concrete in-situ. Proceeding of ICE structures & bridges. 1993;99:74–83.
28. Gomes AM, Costa JO, Albertini H, Aguiar JE. Permeability of concrete: a study intended for the "in-situ" valuation using portable instruments and traditional techniques. In: International symposium (NDT-CE 2003) non-destructive testing in civil engineering, 2003. http://www.ndt.net/article/ndtce03/papers/v017/v017.htm.
29. Jacobs F, Lemann A. Concrete properties according to SN EN 206-1, VSS report 615, 2007 (in German).
30. Jacobs F. Permeability and porous structure of cementitious materials. In: Building materials report no. 7. Aedificatio Publishers, IRB Verlag, 1994. 164 pp (in German).
31. EN 12390-8. Testing hardened concrete—Part 8: depth of penetration of water under pressure, 2000.
32. Basheer L, Kropp J, Cleland DJ. Assessment of the durability of concrete from its permeation properties: a review. Constr Build Mater. 2001;15:93–103.
33. Long AE, Henderson GD, Montgomery FR. Why assess the properties of near-surface concrete? Constr Build Mater. 2001; 1565–1579.

34. Glanville WH. The permeability of portland cement concrete. In: Building research establishment, Technical paper, No. 3; 1931. 62 pp.
35. British Standard Institution. Methods for testing hardened concrete for other than strength, BS1881, Part 208; 1996.
36. RILEM CPC11.2, Absorption of water by capillarity. 2nd ed.; 1982.
37. Laurent J. Capillary water transfer in stone materials: theoretical and experimental aspects, Advanced study course, Paris, pp 7–18. In: Lefèvre RA, editor. European Commission; 1998. pp 19–39.
38. Hall C. Water sorptivity of mortars and concretes, a review. Mag Concr Res. 1989;41 (147):51–61.
39. EN 13057:2003. Products and systems for the protection and repair of concrete structures—test methods—determination of resistance of capillary absorption; 2003.
40. Andrade C. Calculation of chloride diffusion coefficients in concrete from ionic migration measurements. Cem Concr Res. 1993;23(3):724–42.
41. Bertolini L, Elsener B, Pedeferri P, Polder RP. Corrosion of steel in concrete. Wiley-VCH; 2004.
42. Kropp J, Alexander M. Transport mechanisms and reference tests. In: RILEM TC 189-NEC state-of-the-art report; 2007. pp 13–34.
43. CHLORTEST. Resistance of concrete to chloride ingress—from laboratory tests to in-field performance, EU-Project (5th FP GROWTH) G6RD-CT-2002-00855, Workpackage 5 Report: "Final evaluation of test methods", 2005.
44. NT BUILD 443. Concrete, hardened: accelerated chloride penetration, Nordtest method, 1995.
45. NT BUILD 492. Concrete, mortar and cement-based repair materials: chloride migration coefficient from non-steady-state migration experiments, Nordtest method, 1999.
46. Stipanović I, Bjegović D, Serdar M. Resistance of green concrete to chloride ingress. In: Ferreira RM, Gulikers J, Andrade C, editors. Integral service life modelling of concrete structures. Bagneux: RILEM Publications S.A.R.L.; 2007. pp 141–148.
47. Vivas E, Boyd A, Hamilton HR. Permeability of concrete—comparison of conductivity and diffusion methods, FDOT UF Project no. 00026899, Final report. http://www.dot.state.fl.us/research-center/Completed_Proj/Summary_SMO/FDOT_BD536_rpt.pdf, 2007.
48. EU—Brite EuRam III, DuraCrete final technical report, Probabilistic performance based durability design of concrete structures, Document BE95-1347/R17, 2000.
49. Beushausen H, Alexander MG. The South African durability index tests in an international comparison. J S Afr Inst Civ Eng. 2008;50(1):25–31.
50. CEN/TR15868:2009. Survey of national requirements used in conjunction with EN 206-1:2000, 2009.
51. Jacobs F, Lemann A. Concrete properties according to SN EN 206-1 (in german), VSS report 615, 2007.
52. Bundesanstalt für Wasserbau, Merkblatt Chlorideindringwiderstan, 2004 (in German).
53. Cement & Concrete Association of New Zealand, Specifying concrete for performance. http://www.ccanz.org.nz/. 2014.
54. Perić V, Serdar M, Bjegović D. Prescribing chloride diffusion coefficient as a durability indicator in the performance-based design procedure. Zagreb: Faculty of Civil Engineering University of Zagreb; 2012. p. 139.
55. ASTM C1202-10. Standard test method for electrical indication of concrete's ability to resist chloride ion penetration, 2010.
56. Obla KH, Lobo CL. Acceptance criteria for durability tests: minimizing the risks of accepting defective concrete or rejecting acceptable concrete. Concr Int. 2007;5:43–8.
57. Stipanović Oslaković I, Bjegović D, Rosković R, Serdar M. Resistance of blended cement concrete to chloride ingress. Concr Plant Int. 2010; 3:102–107.
58. Saito M, Ishimori H. Chloride permeability of concrete under static and repeated compressive loading. Cem Concr Res. 1995;25(4):803–8.
59. Goodspeed CH, Vanikar S, Cook RA. High performance concrete defined for highway structures. ACI Concr Int. 1996;18(2):62–7.

60. Thomas MDA, Jones MR. A critical review of service life modelling of concretes exposed to chlorides. In: Dhir RK, Hewlett PC, editors. Concrete in the service of mankind: radical concrete technology. London: E.&F.N. Spon; 1996. pp 723–736.
61. Samaha HR, Hover KC. Influence of microcracking on the mass transport properties of concrete. ACI Mater J. 1992;89(4):416–24.
62. Zhang MH, Gjorv OE. Permeability of high strength lightweight concrete. ACI Mater J. 1991;88(5):463–9.
63. Malek RIA, Roy DM. The Permeability of chloride ions in fly ash-cement pastes, mortars, and concrete, MRS symposium, vol 113, Materials Research Society, Pittsburgh; 1996. pp 291–300.
64. Roy DM. Hydration, microstructure and chloride diffusion of chloride ions in hardened cement pastes, ACI SP-114, vol 2, American Concrete Institute, Detroit; 1989. pp 1265–1281.
65. Geiker M, Thaulow N, Andersen PJ. Assessment of rapid chloride ion permeability test of concrete with and without mineral admixtures. In: Baker JM, Nixon PJ, Majumdar AJ, Davis H, editors. Durability of building materials. London: E&FN Spon; 1990. pp 493–502.
66. Castellote M, Andrade C, Alonso C. Measurement of the steady and non-steady state chloride diffusion coefficients in a migration test by means of monitoring the conductivity in the anolyte chamber. Comparison with natural diffusion tests. Cem Concr Res. 2001;31:1411–20.
67. UNE 83988-1:2008, Concrete durability—test methods—determination of the electrical resistivity—Part 1: direct test (reference method), 2008.
68. Andrade C, Polder R, Basheer M. Non-destructive methods to measure ion migration. In: RILEM TC 189-NEC state-of-the-art report; 2007. pp 91–112.
69. PNE 83992-2 EX., Concrete durability. test methods. Chloride penetration tests on concrete. Part 2: Integral accelerated method, 2012.
70. Castellote M, Andrade C, Alonso C. Accelerated simultaneous determination of chloride depassivation threshold and of non-stationary diffusion coefficient values. Corros Sci. 2002;44:2409–24.
71. Andrade C, González JA. Quantitative measurements of corrosion rate of reinforcing steels embedded in concrete using polarization resistance measurements. Werkst Korros. 1978;29:515–9.
72. Andrade C, Rebolledo N. Accelerated evaluation of corrosion inhibition by means of the integral corrosion test. In: Proceedings of the 3rd international conference on concrete repair, rehabilitation and retrofitting (ICCRRR), Cape Town, South Africa, 2012.
73. Basheer PAM, Andrews RJ, Robinson DJ, Long AE. 'PERMIT' ion migration test for measuring the chloride ion transport of concrete on site. NDT and E Int. 2005;38(3):219–29.
74. Nanukuttan SV, Basheer PAM, Robinson DJ. Further developments of the permit ion migration test for determining the chloride diffusivity of concrete. In: Forde MC editor. Structural faults and repair-06, Edinburgh, Vol. CD Rom, Engineering Technic Press; 2006. 14 pp.
75. Nanukuttan S, Basheer PAM, Robinson DJ. Determining the chloride diffusivity of concrete in-situ using Permit ion migration test. In: Proceedings of the concrete platform conference, Belfast, 2007; ISBN:978-0-85389-913-6, pp 217–228.
76. Stipanović I, Bjegović D, Mikulić D, Serdar M. Time dependent changes of durability properties of concrete from Maslenica Bridge at the Adriatic coast. In: Ferreira RM, Gulikers J, Andrade C, Bagneux C, editors. Integral service life modelling of concrete structures. RILEM Publications S.A.R.L.; 2007. pp 87–95.
77. Fluge F. Marine chlorides—a probabilistic approach to derive provisions for EN 206-1, DuraNet, third workshop, service life design of concrete Structures—from theory to standardisation, Tromsø, 2001.
78. Tang L. Chloride transport in concrete—measurement and prediction, PhD Thesis, Chalmers University of Technology, Sweden, 1996.
79. RILEM recommendation. Test methods for on-site measurement of resistivity of concrete. Mater Struct. 2000; 33:603–611.

80. Morris W, Moreno EI, Sagues AA. Practical evaluation of resistivity of concrete in test cylinders using Wenner array probe. Cem Concr Res. 1996; 26(12):1779–1787.
81. Streicher PE, Alexander MG. A chloride conduction test for concrete. Cem Concr Res. 1995;25(6):1284–94.
82. Ewins AJ. Resistivity measurements in concrete. Brit J NDT. 1990;32(3):120–6.
83. Millard SG. Reinforced concrete resistivity measurement techniques. In: Proc. Institution Civil Engineers, 1991. part 2, pp 71–88.
84. ASTM G57-95a, Standard test method for field measurement of soil resistivity using the Wenner four-electrode method, 2001.
85. Alexander MG, Ballim Y, Mackechnie JM. Concrete durability index testing manual, Research Monograph no. 4, Departments of Civil Engineering, University of Cape Town and University of the Witwatersrand, 1999.
86. Mackechnie JR. Predictions of reinforced concrete durability in the marine environment, Research Monograph No. 1, Department of Civil Engineering, University of Cape Town, 2001. 28 pp.
87. Andrade C, Alonso C, Gulikers J, Polder R, Cigna R, Vennesland O, Salta M, Raharinaivo A, Elsener B. Test methods for on-site corrosion rate measurement of steel reinforcement in concrete by means of the polarization resistance method, RILEM TC 154-EMC Electrochemical techniques for measuring metallic corrosion. Mater Struct. 2004;37:623–43.

Chapter 5
Principles of the Performance-Based Approach for Concrete Durability

H. Beushausen, M.G. Alexander, M. Basheer, V. Baroghel-Bouny, R. d'Andréa, A. Gonçalves, J. Gulikers, F. Jacobs, M. Khrapko, A. V. Monteiro, S.V. Nanukuttan, M. Otieno, R. Polder and R. Torrent

5.1 Introduction

In general, design approaches for durability can be divided into prescriptive design concepts and performance-based design concepts. As discussed in Chap 3, prescriptive concepts result in material and cover depth specifications from using factors such as exposure classes and compressive strength. Following this approach, durability specifications in most existing codes and standards are based primarily on establishing constraints to the material and mix proportions of the concrete, such as maximum water binder (w/b) ratios, and total minimum cementitious materials content, as a function of the severity of the anticipated exposure (e.g. [1]). Design for durability includes the correct choice of exposure class and compliance with material requirements and concrete cover requirements, as well as with placing, compacting and curing procedures.

H. Beushausen (✉) · M.G. Alexander
University of Cape Town, Cape Town, South Africa

M. Basheer
University of Leeds, Leeds, UK

V. Baroghel-Bouny
Laboratoire Central des Ponts et Chaussées (LCPC), Chelles, France

R. d'Andréa
Eduardo Torroja Institute (IETcc-CSIC), Madrid, Spain

A. Gonçalves
Laboratório Nacional de Engenharia Civil—LNEC, Lisboa, Portugal

J. Gulikers
Rijkswaterstaat Bouwdienst, Utrecht, The Netherlands

F. Jacobs
TFB, Wildegg, Switzerland

H. Beushausen and L. Fernandez Luco (eds.), *Performance-Based Specifications and Control of Concrete Durability*,
RILEM State-of-the-Art Reports 18, DOI 10.1007/978-94-017-7309-6_5

However, many researchers and engineers argue that durability is a material performance concept for a structure in a given environment and that as such it cannot easily be assessed through simple mix parameters [2–5]. The prescriptive approach ignores, to a certain extent, the different performance of the various types of cement and mineral additions or to the concrete itself, as well as the type of aggregate, and does not allow to take into account the influences of on-site practice during the construction process. It also cannot explicitly account for a specific service life requirement. Furthermore, the prescriptive approach is a barrier to the use of new or recycled materials.

Performance-based design concepts, on the other hand, are based on quantitative predictions for durability (or service life) from prevailing exposure conditions and measured material parameters. The resistance of the structure against deterioration, measured through durability indicators of the actual concrete used, is compared against the environmental load. On this basis, deterioration of a structure during its lifetime is quantified using appropriate deterioration models. In this concept the concrete composition is only important to the extent that it controls the concrete properties.

To move from prescriptive to performance-based approaches for concrete durability, a framework guide is needed. One such framework has been suggested by [6], which considers seven steps for development:

- Define exposure classes related to the mechanism(s) of deterioration
- Derive a quantitative design methodology, including definition of end of service life
- Develop test methods that relate to the input parameters of the design method
- Produce provisional compliance criteria and calibrate against traditional solutions
- Establish limitations of test applicability
- Ensure production control and acceptance testing
- Conduct full-scale trials and long-term monitoring to confirm compliance requirements.

M. Khrapko
CBE Consultancy Ltd, Auckland, New Zealand

A.V. Monteiro
National Laboratory for Civil Engineering, Lisboa, Portugal

S.V. Nanukuttan
Queen's University Belfast, Belfast, UK

M. Otieno
University of the Witwatersrand, Johannesburg, South Africa

R. Polder
TNO Built Environment and Geosciences, TU Delft, Delft, The Netherlands

R. Torrent
Materials Advanced Services, Buenos Aires, Argentina

For the practical application of a performance-based approach in durability specifications and for service life assessment, the following elements need to be developed [2]:

- Limit state criteria
- A defined service life
- Deterioration models
- Compliance tests
- A strategy for maintenance and repair
- Quality control systems.

Limit state criteria for concrete durability should be quantifiable and preferentially have a clear physical meaning. The deterioration models generally comprise mathematical expressions and should include parameters that are directly or indirectly linked to the performance criteria. However, in some instances, deterioration modelling may not be necessary for a performance-based approach to be implemented. For example, in the case of freeze thaw resistance evaluation, the performance test leads to the acceptance of the concrete mix if for example the loss of mass is lower than a certain value. Here, no appropriate model exists to integrate the result, but the long-term experience shows that such a mix will perform satisfactorily in most real exposures.

Different levels of sophistication may be applied to performance-based design for durability, including the use of durability indicators, the application of analytical deterioration models, and full probabilistic approaches. Depending on the required level of sophistication, the necessary tools for performance-based service life design may incorporate appropriate service life models, including end of service life criteria, and test methods for the verification of material characteristics. Performance-based as-built compliance assessment must incorporate testing of relevant properties of the concrete cover layer, which can be used to assess the expected ingress of harmful substances such as chlorides and carbon dioxide. Various performance-based test methods have been developed in different parts of the world, as discussed in Chap. 4.

The performance-based approach for concrete durability can be considered to be an important advance in the design of reinforced concrete (RC) structures. Current limitations to this approach link to the circumstances that the various deterioration processes affecting RC structures are presently not fully understood in all necessary details, test methods used in the laboratory do not always reflect real-life conditions, and the variation in concrete quality across the structure or single element is not sufficiently known. However, the problem of the difference between site and laboratory conditions can to some extent be overcome by the use of physical models that take testing conditions into account and are based on ageing phenomena, microstructural changes, etc. The larger the difference between the site and laboratory conditions (e.g. 5 °C instead of 20 °C), the larger is the uncertainty of the performance of the site concrete.

5.2 Pre-qualification Versus Compliance Control

5.2.1 Principles

If solely used for the purpose of pre-qualification, performance-based design refers to the performance evaluation of a particular concrete mix that is intended to be used later in construction. In such a case, EN-206:2013 [1] refers to "initial testing". With the measured durability indicators, the expected service life of the structure can be evaluated using appropriate service life models. Based on the measured durability indicators of the concrete prior to construction it is assumed that the as-built structure too will be inherently durable, but no further tests are performed on the as-built structure to verify the actual in situ quality.

Compared to traditional prescriptive design approaches, the performance-based design approach has the advantage that the influence of constituent materials on the durability of the concrete can be directly assessed, accounting for locally available materials and design innovations. However, the influence of construction procedures including the environmental conditions on concrete quality cannot be assessed in the pre-qualification stage and needs to be checked on the as-built structure.

As a complement to the above, performance-based compliance control includes the evaluation of the as-built structure using appropriate test methods. The actual in situ quality of the structure can then be compared to design specifications and appropriate measures can be implemented if the structure does not conform to limiting design values for concrete durability. This method allows evaluating not only mix design parameters but also construction-related influences and will therefore give a much better indication of the expected durability of the as-built structure. The principles of the above two approaches are discussed further in the following sections.

5.3 Performance-Based Design and Specification (Pre-qualification)

5.3.1 Principles

Performance-based design for durability of concrete structures includes the measurement of relevant concrete properties in the design stage in order to assess the resistance of the material against deterioration. Various performance-based service life design models have been developed in different parts of the world, as discussed in Chap. 2. For example, the European performance-based design approach "DuraCrete" [7] was developed to model both chloride ingress into concrete, and carbonation. The models were slightly revised in the research project DARTS and are described in the fib Model Code for Service Life Design [8]. Other models dealing with chloride ingress include the South African chloride prediction model [9] and the Scandinavian model "Clinconc" [10].

Using these models for the prediction of chloride ingress or carbonation, the onset of the corrosion propagation period is predicted. Durability indicators of concretes in relation to constituent materials and mix proportions, which are needed as input parameters in the service life models, are determined through experiments, usually using laboratory-cured concrete at an age of 28 days. Longer curing periods may be needed if pozzolanic or latent hydraulic materials (such as for example fly ash, natural pozzolans, slag) are used. However, it should be checked to which extent such longer curing periods replicate site exposure conditions.

The performance-based design approach allows to directly assess the influence of various mix constituents on concrete durability for the investigated environmental conditions. The use of higher quality cements, supplementary cementitious materials or durability enhancing admixtures, for example, could, if properly selected and applied, result in a more durable concrete. Testing the specific concrete in the design stage therefore allows an optimization between mix design properties and cover depth specifications. The results obtained with the design mix can be used as input parameters for service life models or for the prediction of the reinforcement corrosion initiation period. The underlying principles of performance-based design of concrete structures are illustrated in Fig. 5.1.

Note that performance-based design done in this way may need to include prescriptive requirements for material parameters (mix proportions and mix constituents) since the interpretation of the measured durability indicators may depend on the type of material used. For example, concretes made with different binder types may have the same values for durability indicators (e.g. the same transport properties such as permeability, electrical resistivity, etc.) but different durability since the latter depends on additional factors such as chloride binding mechanisms, amount of carbonatable material in the concrete, etc., and hence on type and amount of constituent materials used.

In performance-based design for concrete durability, the variation of the concrete properties over the relevant production period has to be known and considered. The variations will be caused by e.g.:

- variation in the properties of the constituent materials
- variation in the mix composition
- variation in production, placing, compacting and curing
- variation in the environmental conditions.

Up to now, this data is only rarely available and the implementation of performance-based approaches can help to improve this situation.

Performance-based design approaches assist in determining optimum mix constituents and mix proportions for the particular exposure environment, which in turns assists in providing RC structures with adequate durability. However, as already mentioned, performance-based approaches that rely purely on the pre-qualification stage cannot account for the important influence of construction procedures, such as on-site quality control, batching and mixing procedures,

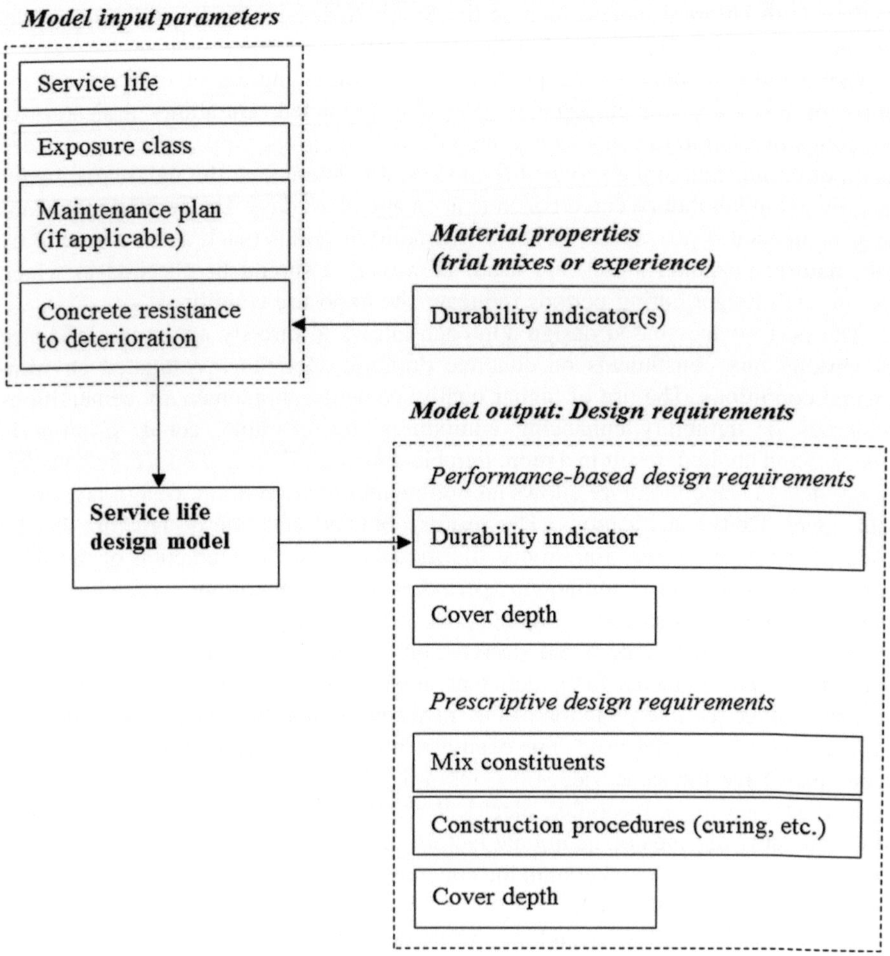

Fig. 5.1 Principles of performance-based design of concrete structures

concrete transport and placement, compaction, curing, environmental influences, etc. The combined influence of mix parameters and construction processes can be evaluated using a scheme for performance-based quality assessment and compliance control on the as-built structure, as discussed in Sect. 5.4.

As indicated in Fig. 5.1, modern service life design should ideally also include a maintenance plan, the extent of which needs to be considered when specifications are drawn up. This implies that the owner or manager of the structure needs to be given a "manual" by which to best manage the structure for optimum service life. Maintenance is defined in EN 1504-9 [11] as the "recurrent or continuous measures that provide repair and/or protection". In the current context, the maintenance plan should include routine inspection of the structure, which may (or may not) result in specification of appropriate protective measures or repair, if needed.

5.3.2 Test Methods, Procedures and Variables

Performance-based design of concrete structures is potentially most reliable if performance simulation tests are used in the assessment of the concrete's resistance to deterioration. Such test methods measure concrete deterioration characteristics under conditions that simulate accelerated exposure to the relevant deteriorating agent. The most common performance simulation tests applied for the assessment of the concrete quality necessary for reinforcement corrosion resistance are the Bulk Diffusion [12–14] and accelerated carbonation tests [15, 16]. The advantage of these test methods is that they attempt to simulate the real deterioration process, exposing concrete to a high concentration of the deleterious species. Another feature of these test methods is that certain pre-conditioning procedures are chosen to facilitate the deterioration process (e.g. saturation of specimens prior to the determination of the concrete resistance against chloride ingress). The Bulk Diffusion test measures chloride ingress into concrete specimens exposed to a concentrated chloride solution. Results are typically analysed to determine a bulk chloride diffusion coefficient, which can be used as an input parameter in deterioration models using Fick's laws of diffusion.

The experimental set-up of the different tests practised can vary largely. Therefore, it is important to understand that the durability property measured is dependent on the experimental set-up. For example, the bulk chloride diffusion coefficient will differ from a diffusion coefficient obtained from EN 13396 [17], which employs a low concentration of chloride solution, compared to NT Build 443 [12]. The deterioration models need to take this variation into account.

The results of accelerated carbonation tests can be used to predict carbonation in real structures using established relationships between test results and expected field performance, as is the case for the model proposed in the fib Model Code [8].

The main problems with these performance simulation methods, however, are that they are time-consuming, as test specimens need to be monitored for several weeks or months. This may not be practical for certain projects if they are bound by short bid times and quick project implementations. Additionally, the curing of the specimens deviate significantly from on-site conditions. As an alternative, chloride ingress can be accelerated by applying an electrical voltage to achieve a test result within hours or days. Other test methods, usually based on the evaluation of transport mechanisms such as permeability and resistivity, can also be applied for performance-based design procedures. The application of such test methods is discussed further in Sect. 5.4.

It needs to be noted that both performance simulation tests and tests that yield transport properties are commonly carried out using laboratory samples at relatively young age. As such, all these tests can only give a more or less rough indication (or 'simulation') of the potential long-term performance of the actual in situ concrete. It is therefore imperative to undertake, preferably long-term, natural exposure tests to calibrate and verify short-term test results obtained in the laboratory. Because such an undertaking is time-consuming it can be expected that test methods, and the interpretation/calibration of test results, will be continuously improved over time.

Table A1 in Appendix A gives an overview of the application of different test methods for performance-based design in various countries. The principles of relevant test methods are discussed in Chap. 4.

Test procedures for the assessment of concrete mixes need to be properly designed and executed, following clearly defined guidelines for the following aspects:

- Preparation of specimens (curing method and duration, preconditioning procedures, etc.)
- Specimen age at testing
- Number of test specimens required
- Sequence of testing
- Analysis of test results (statistical evaluation, acceptance limits, etc.).

The above needs to be based on the specific requirements of the project and previous experience with the test procedures. The principles of test method application in view of the above aspects can therefore best be explained using practical examples, which are presented in Chap. 8.

5.4 Performance-Based Quality Control and Compliance Assessment

5.4.1 Principles

Performance-based quality control and compliance assessment procedures provide the means to assess the quality of the as-built structure and establish if design specifications have been met. The philosophy of performance-based compliance control involves the understanding that durability will be improved only when measurements of appropriate cover concrete properties can be made. Such measurements must reflect the in situ properties of concrete, influenced by the dual aspects of material potential and construction quality. Performance-based quality control needs to be performed either in situ or on cores removed from the in situ structure. The quality of the concrete is then assessed and compared to the design requirements.

In order to compare as-built concrete quality to design specifications, it is best to use the same test methods for design and quality control. This sets certain limits on the choice of test methods as practical considerations commonly dictate time constraints. Possible test methods for the resistance against chloride penetration and carbonation are the performance simulation tests discussed in Sect. 5.3.2. However, these tests require several weeks or months before useful data can be obtained. For most structures this is not suitable for a quality control procedure where remedial measures may have to be designed timeously if performance specifications have not been met. Another situation that often occurs in practice is that mix designs or mix constituents are adjusted during construction, resulting in the need for a short-term assessment of the quality of the new mix. Rapid test methods are therefore required for performance-based design and quality control procedures.

Depending on country-specific testing procedures it may also be practical to design concrete mixes using the more time-consuming performance-simulation tests but perform compliance assessment based on transport properties. For such an approach, reliable correlations between the different test methods need to be established.

Appropriate test methods for the control of in situ durability properties of concrete have been developed in different parts of the world. An evaluation of the principles, merits and limitations of such methods was presented by RILEM TC 189-NEC (Non-Destructive Evaluation of the Covercrete) [18, 19]. One important conclusion of the work of TC 189-NEC was that suitable devices exist, with which the quality of the concrete cover can be assessed in situ or on cores that are removed from the structure. The test procedures investigated included methods to assess gas permeability, capillary suction and electrical conductivity (or alternatively electrical resistivity) of concrete.

The results obtained with such test methods are used to assess the resistance against the ingress of deleterious agents, commonly carbon dioxide and chlorides, but do not necessarily directly measure the actual ingress of these substances. Test methods for permeability, capillary suction and conductivity measure transport mechanisms that principally reflect the pore structure, pore connectivity and pore fluid chemistry of the concrete. These transport mechanisms can be linked to deterioration processes through calibration and modelling and thus provide a useful and practical measure of the durability of the concrete.

If new types of concrete or new mix constituents are used, it is important to check if previously established correlations between transport properties and actual performance of the concrete still apply (and make adjustments to the interpretation of the test results, if necessary).

A framework for performance-based service life design and quality control is illustrated in Fig. 5.2. Limiting values for durability indicators should ideally be based on a probabilistic approach, i.e. they should be calibrated against a comprehensive statistical evaluation of relevant influencing factors (environment, expected scatter in test results, measurement uncertainties, etc.), applying relevant service life models. Acceptance criteria for on-site concrete quality and sampling requirements need to be clearly defined, including guidelines for remedial measures if compliance requirements have not been met.

5.4.2 Test Methods and Test Parameters

Since the service life of RC structures depends largely on the thickness and the quality of the cover zone, it is necessary to have reliable tests to measure both of these aspects. Appropriate test methods for concrete durability are discussed in

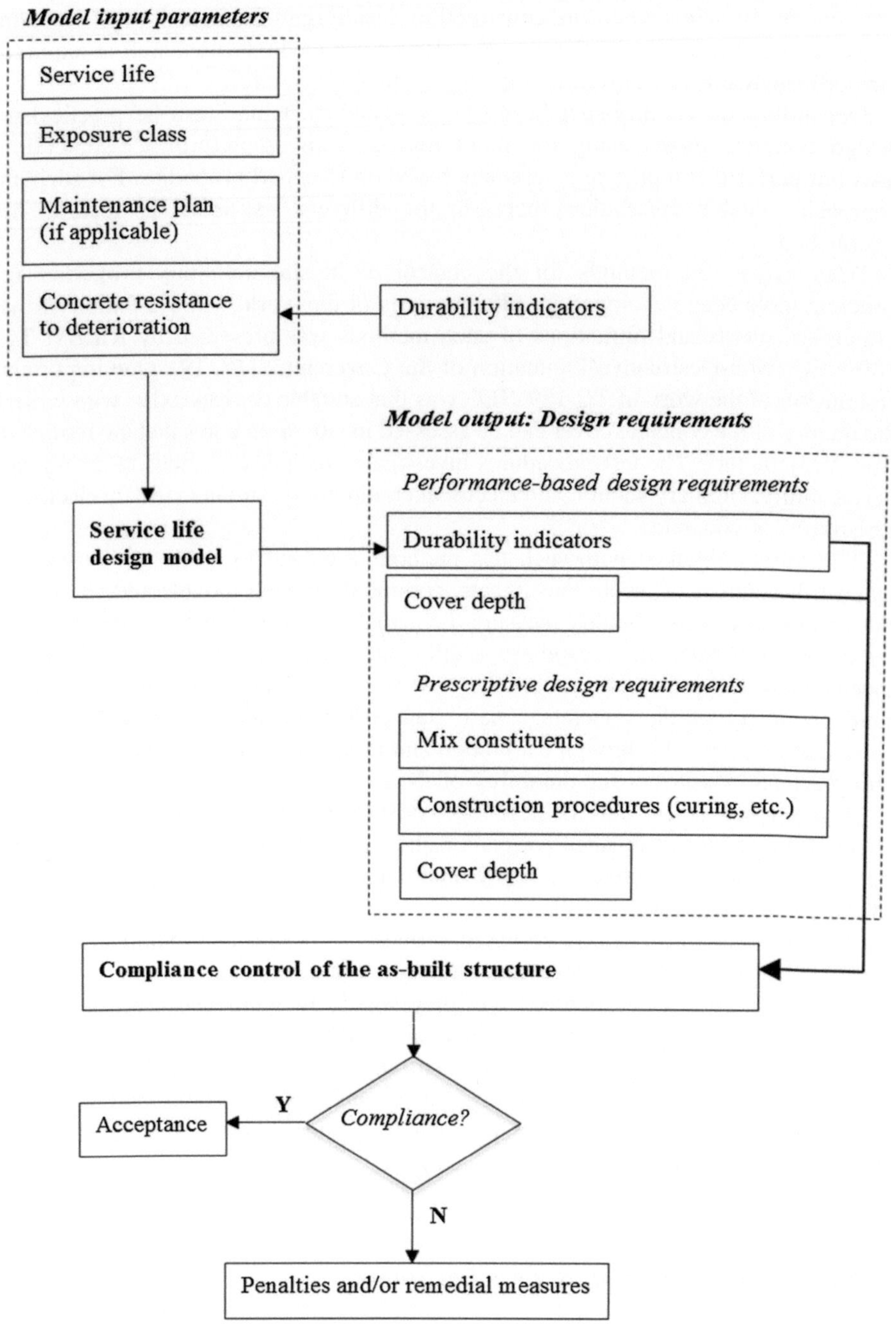

Fig. 5.2 Principles of performance-based design and compliance control of concrete durability

Chap. 4. These methods include accelerated tests, which have been developed to provide useful design information in time periods appropriate for project specifications. Since it is generally too time-consuming to test the ingress of deleterious substances such as chlorides or carbon dioxide, accelerated test methods commonly evaluate the pore structure and, to a certain extent, the pore fluid chemistry of the concrete, with the most common test methods measuring permeability or resistivity (or the inverse, conductivity) characteristics. Using deterioration prediction models or empirical relationships, the test values for permeability and resistivity can be linked to relevant deterioration mechanisms.

The test methods used for the evaluation of the as-built quality of the structure must have proven to provide reliable measures of durability. Clear guidelines for the testing and sampling procedures, as well as the recording and interpretation of test results need to be developed, as already discussed in Sect. 5.3.2.

Test methods for the evaluation of as-built quality control include non-invasive or semi-invasive methods that can be applied directly on the structure, as well as laboratory-based methods that make use of concrete cores removed from the in situ structure. Non-invasive methods have the obvious advantage that test results can be obtained without impairing the aesthetics of the concrete surface. The disadvantage of methods applied on the in situ structure is that the test results commonly need to be calibrated against moisture content and temperature of the concrete, since the measured transport properties depend largely on these factors. Testing of cores removed from the structure has the advantage that the concrete quality can be assessed on pre-conditioned samples under controlled conditions, which eliminates the influence of daily or seasonal fluctuations in the in situ concrete moisture content and temperature. The choice of test method for compliance control normally depends on country-specific experiences.

Depending on project requirements and preferred test methods, it may be practical to combine non-invasive and semi-invasive tests for the determination of the in situ concrete cover quality. Some of the non-invasive tests (for example resistivity measurements or permeability measurements using, respectively, the Wenner probe and Torrent Tester, compare Chap. 4) are relatively easy and fast to perform and thus provide an overall indication of the quality of a certain area of concrete. In countries, where non-invasive methods are usually preferred, a limited number of cores can be extracted as a reference and/or when in doubt to verify the findings.

Table A1 in Appendix A gives an overview of test methods for performance-based specification and compliance control used in different countries. Examples for the application of various test methods for durability specification and compliance control are presented in Chap. 8.

5.4.3 Sampling Criteria and Sampling Procedures

The determination of a sample size for concrete durability testing, either in the laboratory or in the field (in situ), is an important step towards making valid conclusions with respect to the durability performance of a RC structure in a given exposure environment. General principles regarding the determination of a representative sample size indicate that the sampling process should be governed mainly by:

- The degree of (expected) variability in the population (temporal, spatial and parameter variability)
- The desired trueness (i.e. how close the measured values are to the actual/true values)
- Precision (i.e. how close the measured values are to each other)
- The nature of the analysis to be performed
- The number of variables to be examined simultaneously.

In the case of RC durability, all the above factors are important and should be considered a priori. Most importantly, sampling should be done in a manner that does not create bias in favour of any observation. Some of the common sampling criteria/procedures include [20]:

- Simple random (haphazard) sampling: where any observation has equal probability of being collected. The disadvantage of this method is that it is less effective if there is heterogeneity in the population or in the estimation of parameters at a range of spatial and temporal scales, which is common in RC structures
- Stratified sampling: where the population is divided into levels or strata that clearly define groups or units within the population. The sampling is done randomly and independently. This method takes into consideration the heterogeneity in the population
- Systematic sampling (also called non-targeted or grid sampling): where the sample is chosen from the population at a regular/systematic interval, either spatially or temporally
- Adaptive sampling: where the sampling method (or the sample size) is modified depending on preliminary estimates e.g. of variance.

With respect to sampling and sample size determination for durability assessments in RC structures, important factors that should be kept in mind are:

- The process of sampling and sample size determination is critical
- Inasmuch as sampling principles provide clear guidance on how to carry out the process, long-term experience may also provide the basis of a successful process
- Concrete is a highly heterogeneous material and, even with good sampling techniques, the test results should be considered only as approximations to the true state of the structure.

5.4.4 Actions in Case of Non-compliance

The performance-based approach for quality control and compliance assessment enables owners and engineers to specify certain durability parameters in relation to the anticipated service life, environmental conditions, binder types and cover depth requirements. Durability specifications commonly comprise limiting values for the thickness and penetrability of the concrete cover. When limiting values, obtained on the as-built structure, meet the specified requirements, the structure is considered to be inherently durable.

However, a framework for remedial interventions for concrete structures that do not meet the specified requirements needs to be established. If limiting values for durability indicators have not been achieved, the owner of the structure principally has the following options:

1. Verify the concrete quality by further testing or other test methods with respect to the real exposure of the element; all test methods have a certain scatter, not only depending on the concrete inhomogeneity but for example also depending on specimen preparation
2. Check for any safety margin in the cover depth
3. Protect the structure against the ingress of harmful substances, such as carbon dioxide and chlorides. This could include temporary protection, e.g. a surface treatment that will limit chloride ingress for a period of time
4. Accept that some structural elements have to be later replaced or submitted to anticipated maintenance
5. Accept that the anticipated service life duration may be compromised
6. Demolish and rebuild the structure or the defective element.

For most projects, the last two of the above options will be undesirable, for obvious reasons. The first two options involve a re-evaluation of the original design parameters and may also not be acceptable. In most cases, the third or fourth option will be adopted, i.e. protecting the structure against deterioration to ensure that the design service life can be reached. Such methodology may for example include the application of protective surface treatments [21]. Depending on the discrepancy between desired quality and actual quantity achieved, a once-off application may be sufficient, whereas in other cases a detailed maintenance plan may need to be established, taking monitoring or repeated application of protective measures into account.

The decision of appropriate repair and maintenance strategies needs to be based on an evaluation of the expected service life. For this, the measured durability performance value is used as an input parameter in the service life model, with which the original design parameter was established. This will allow an estimation of the actual service life duration that can be expected. This, in turn, will give the information of how many years of additional service life the protective measure needs to provide.

On this basis, it can for example be argued that a coating, which prevents the ingress of harmful substances over that required duration, presents a suitable protective measure, bringing the structure back to its original service life. For this, a clear philosophy needs to be developed, based on which the design of appropriate protective measures can be carried out. The owner of the structure needs to be given clear guidance on what steps to follow and on what options are available. It needs to be established, which coatings can be used to either prevent or slow down the ingress of contaminants and aggressive agents sufficiently. Information on protective surface treatments is discussed in the literature (some general principles and guidelines are presented in EN 1504 series [22]). Another protection method for RC structures is the application of penetrating corrosion inhibitors.

5.5 Environmental Actions (Quantification)

5.5.1 *General*

The durability performance of concrete structures is closely related to the aggressiveness of the exposure environment. It is therefore important that this aspect is considered in the durability assessment of concrete structures. The aggressiveness of the environment can be classified as either causing chemical or physical attack, or both. Important environmental aspects include, among others, temperature, relative humidity, nature and concentration of the aggressive agents and freeze-thaw cycles, which are usually described in terms of exposure classes. Very often, more than one exposure class is necessary to describe the prevailing environmental actions.

These environmental aspects may vary between different regions in a country or different locations in a region or city. They may also vary between different parts of the same structure, for example between parts exposed to rain and those that are sheltered. Therefore, the relevant environmental aspects should be considered and quantified for a given RC structure.

5.5.2 *Quantification of the Chloride Environment*

5.5.2.1 Marine Environment

Marine chloride-laden environments can be very aggressive to concrete structures, leading to chloride-induced corrosion and deterioration. Chloride-induced corrosion in the marine environment is most aggressive in the tidal and splash/spray zones due to the wetting and drying cycles. Conditions are also different at different heights above average sea level. Maximum chloride contents are found at a height where salt water is frequently supplied to the surface but where the surface intermittently dries out. In the tidal zone the conditions are somewhat similar to that of the splash/spray zone, however, the periods of wetness and intermittent drying are different.

An example of an attempt to classify the marine environment is that in the European Standard EN 206: 2013 [1] (Table 5.1). The classification is based on the aggressiveness of the environment with respect to the expected ingress of chlorides

Table 5.1 EN 206: 2013 XS Environmental sub-classes [1]

Designation	Description
XS1	Exposed to airborne salt but not in direct contact with seawater
XS2	Permanently submerged
XS3	Tidal, splash and spray zones

Table 5.2 South African marine environmental classes (after EN 206: 2013 [1])

Designation	Description
XS1	Exposed to airborne salt but not in direct contact with seawater
XS2a[a]	Permanently submerged
XS2b[a]	XS2a + exposed to abrasion (i.e. heavy seas and pounding waves)
XS3a[a]	Tidal, splash and spray zones
XS3b[a]	XS3a + exposed to abrasion

[a]These sub-clauses have been added for South African coastal conditions

into the concrete and the subsequent degree of deterioration in the structure. However, these environmental classes may not be applicable to all marine exposure environments and should be viewed as guidelines only. Factors that alter the aggressiveness are not explicitly given like water temperature, salinity, and relative ambient humidity, as these may influence chloride transport into the concrete.

The guidelines given by the EN 206 [1] can be modified in its National Annexes, taking the particular site conditions into account. In South Africa, for example, based on local experience and long-term performance-based tests, the EN classes for the marine environment (XS1-XS3) have been modified to suit local exposure conditions (Table 5.2).

5.5.2.2 Exposure to De-Icing Salts

Concrete structures may be exposed to chlorides other than in a marine environment. In many countries it is common to regularly apply chloride-based de-icing salts during freezing periods. Consequently, highway structures and parking decks may experience exposure to chlorides. In addition, concrete structures exposed to industrial water may suffer from chloride ingress and the resulting reinforcement corrosion.

For marine structures the exposure conditions may remain almost constant over time and space. However, for structures exposed to de-icing salts the chloride load may vary significantly over time and its spatial variation may be very high.

In EN 206 [1] exposure class XD is introduced for reinforcement corrosion induced by chlorides other than from seawater (Table 5.3). In this class three levels are distinguished dependent on the humidity level. The moisture conditions given in the class description are those in the concrete cover to reinforcement, but in many

Table 5.3 EN 206: 2013 XD Environmental sub-classes [1]

Designation	Description
XD1	Moderate humidity, e.g. concrete surfaces exposed to spray water containing de-icing salt
XD2	Wet, rarely dry, e.g. concrete in swimming pools
XD3	Cyclic wet and dry, e.g. parts of bridges exposed to splash water containing de-icing salt

cases conditions in the concrete cover can be taken as being the same as those in the surrounding environment.

Exposure class XD1 is exemplified by concrete structures exposed to airborne chlorides whereas swimming pools and structures exposed to industrial waters containing chlorides fall within exposure class XD2. Exposure class XD3 is applicable to parts of bridges exposed to splash water containing chlorides, pavements, and car park slabs.

5.5.2.3 Quantification of CO_2 Environment

Carbonation-induced reinforcement corrosion, although frequently experienced, is often not a major concern for safety or serviceability as it generally proceeds very slowly in natural environments with carbon dioxide concentration at approximately 0.04 % (non-industrial areas). Carbonation of concrete is affected by environmental factors such as relative humidity, temperature and the ambient concentration of CO_2. Optimal conditions for increased carbonation rates include temperatures near 20 °C and a relative humidity between 50 and 80 % [23, 24].

Similar to the classification of the marine environment, it is necessary that exposure conditions for concrete structures subjected to carbonation are taken as guidelines and that adjustments are made to local conditions where necessary. EN 206: 2013 [1] includes an example of a classification of the carbonation environment (Table 5.4). This type of classification does not take into account the variation in carbon dioxide concentration in the surrounding environment. For example, in industrial areas, the carbon dioxide concentration is usually high and hence carbonation rates can increase if there is sufficient moisture to support the process. Parking garages and tunnels may also be subject to high CO_2 contents.

Table 5.4 Environmental classes for carbonation-induced corrosion [1]

Designation	Description
XC1	Permanently dry or permanently wet
XC2	Wet, rarely dry
XC3	Moderate humidity—(exterior concrete sheltered from rain), interior concrete with moderate or high moisture content of the air
XC4	Cyclic wet and dry

5.6 Development of Limiting Values for Specification and Compliance Control

5.6.1 *General Considerations*

Performance-based specification and compliance control for concrete durability must incorporate testing of relevant concrete properties, which can be used to assess resistance against the ingress of harmful substances, such as chlorides and carbon dioxide. Various performance-based test methods have been developed in different parts of the world, as discussed in Chap. 4. The results obtained with these test methods can be used as input parameters for relevant service life models for the prediction of corrosion initiation and propagation.

The test results obtained with these methods must be able to characterise the quality of the cover or surface layer, using parameters that are related to the deterioration processes acting on the concrete. These processes are linked with transport mechanisms, such as gaseous and ionic diffusion, capillary suction, etc.

Design and compliance control for durability requires quantifiable physical or engineering parameters to characterise the concrete at early ages. Such parameters must be sensitive to material, construction, and environmental factors such as binder (cement, mineral additions) type, water/binder ratio and curing, etc. Correlations are required between test results and durability characteristics, and between these two and actual structural performance, such that the durability compliance tests can be used as follows:

- As a means of controlling the quality of the covercrete
- As a means of assessing the quality of construction for compliance with a set of criteria
- As a means of predicting the performance of concrete in the design environment, on an empirical or theoretical basis.

The criteria for suitable limiting values require that the tests:

- Be site- or laboratory-applicable (site-applicability could involve retrieval of core specimens from site for laboratory testing)
- Be linked to fluid and ionic transport mechanisms and have a reasonable and sound theoretical basis
- Have sufficiently low statistical variability
- Be independent of 'executor'.

The obtained durability indicators represent a measure of the potential service life of the structure. The link between test values and service life prediction can be based on empirical correlations and/or the modelling of fundamental relationships between test parameters and durability. The latter involves the application of relevant deterioration and service life prediction models.

To establish limiting values for durability indicators for concrete mixes and/or in situ structures, and evaluate compliance with durability requirements, the following two aspects need to be considered:

- Statistical variability of test results (hence selection of appropriate characteristic values and sufficient number of measurements); and
- Differences between the actual as-built quality (in situ concrete) and the potential quality of the concrete (assessed on laboratory-cured concrete).

The statistical variability of tests can be accounted for by applying different levels of analysis. The two principal approaches for statistical considerations refer to deterministic and probabilistic methods, as discussed in the following sections. The most sophisticated method of designing for concrete durability would appear to be the full probabilistic approach. The design engineer can select the required reliability against failure and assess the durability of the structure using established models for environmental influences and material resistance. However, the full probabilistic approach is time-consuming and requires a very substantial amount of specific knowledge, data of the in situ concrete (which is hardly available), and expertise. The fib Model Code for service life design [8] therefore states that this approach is intended to be used for exceptional structures only. For common structures, a semi-probabilistic approach, using partial safety factors, based on probabilistic models, can be used in durability design.

5.6.2 The Use of Transport Properties in Compliance Control

A range of existing approaches for compliance control of performance-based durability specifications make use of test methods that assess transport properties of the concrete (e.g. permeability, electrical resistivity, absorption, etc.). Empirical or fundamental relationships can be used to link transport properties to the performance of actual structures in service. Using this method, test values obtained on concrete in a structure at a relatively young age, usually at about 28 days, are correlated to the long-term performance of the structure. Such a correlation can be based on previous measurements of relevant parameters such as carbonation depth or chloride ingress on real structures. This approach may however be too time-consuming, as it requires long-term testing. Alternatively, empirical or fundamental relationships can be established between transport properties and performance simulation tests. The application of test values obtained with such performance simulation tests in connection with relevant service life models for durability prediction has been well documented in the literature [25–34]. Using the correlation between transport properties and fundamental durability characteristics such as diffusion coefficients, transport properties can be used as input parameters in service-life models. The principles of using transport properties for performance-based design and

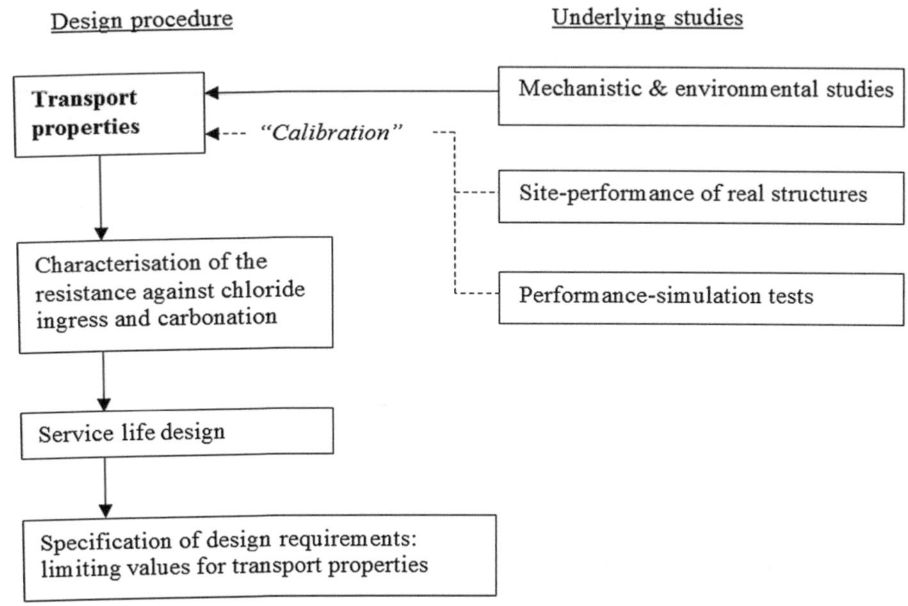

Fig. 5.3 Principles of underlying studies and performance-based design using transport properties as durability indicators

compliance control of concrete durability are illustrated in Fig. 5.3. Most test methods discussed in Chap. 4 are based on this principle.

5.6.3 Principles of the Probabilistic Approach to Statistical Variations

The probabilistic approach allows for the assessment of uncertainties associated with test results as well as those caused by inherent random variations in concrete properties and environmental influences, insufficient data and lack of knowledge on durability parameters. Probabilistic methods can either be full-probabilistic or semi-probabilistic (partial safety factor method) and involve the use of reliability-based design and the limit state methodology.

The concept and principles of reliability based design for durability was introduced by Siemes et al. [35], later developed to operational level in DuraCrete [7]. Nowadays this approach of limit state design for durability has been accepted in some international standards. The limit-states method (LSM) for design was defined in ISO 2394 [36] and consequently adopted by various design standards and codes such as the ISO 13823 [37] and fib Model Code for service life design [8]. Although ISO 2394 includes durability in its principles, the LSM has not been developed for failure due to material deterioration to the extent that it has for failure due to gravity, wind, snow and earthquake loads [37].

The LSM incorporates the use of a service life prediction model, for example that for chloride induced corrosion given by the error function solution to Fick's 2nd law of diffusion, as shown in Eq. (5.1), which describes the ingress of chloride ions in concrete up to the initiation limit state (i.e. initiation of reinforcement corrosion once a certain chloride threshold value has been reached).

$$C(x,t) = C_s - (C_s - C_i)erf\left(\frac{x}{\sqrt{4kD(t)t}}\right) \tag{5.1}$$

where *C(x,t)* is the chloride concentration at distance *x* from the exposed surface, at a certain time *t* in s, in % Cl^- by mass of cement, C_s is the surface chloride concentration in % Cl^- by mass of cement, C_i is the initial chloride concentration in concrete in % by mass of cement, *D(t)* is the chloride diffusion coefficient at time *t* in m^2/s, and *erf* is the mathematical error function.

The change in chloride diffusion coefficient with time could be accounted for by considering the diffusion coefficient at a certain reference time (D_{ref}) as follows:

$$D(t) = D_{ref}\left(\frac{t_{ref}}{t}\right)^m \tag{5.2}$$

where D_{ref} is the diffusion coefficient at t_{ref}, t_{ref} is the reference age at testing (e.g. 28 days), and *m* is the aging coefficient.

The variables in Eqs. (5.1) and (5.2) are stochastic in nature. This implies that for the fullest representation of the problem, it is necessary to use a reliability-based design methodology for analysing the mathematical model at the initiation limit state.

To carry out a reliability analysis of Eq. (5.1), the parameters in the model are characterised further as either action effect *S(t)* or resistance effect *R(t)* (alternatively an initiation limit S_{lim}) [37]. Corrosion initiation is assumed to occur at any time *t*, when the condition given by Eq. (5.3) occurs [37]:

$$S(t) \geq R(t) \tag{5.3}$$

R(t) is taken to be the critical chloride content and *S(t)* (which in this case is the chloride concentration at the cover depth) is represented by Eq. (5.1). The parameters in the functions *S(t)* and *R(t)* are statistically quantified using data obtained from in situ and laboratory tests to give their respective distribution types, mean values and variability. The statistical information of each relevant parameter is then exploited to provide improved uncertainty estimates in the output, which is usually stated in terms of the probability that the condition represented by the so-called Limit State Function (LSF) occurs. The probability of this occurring during the design service life of the structure is termed the probability of failure (P_f) [37].

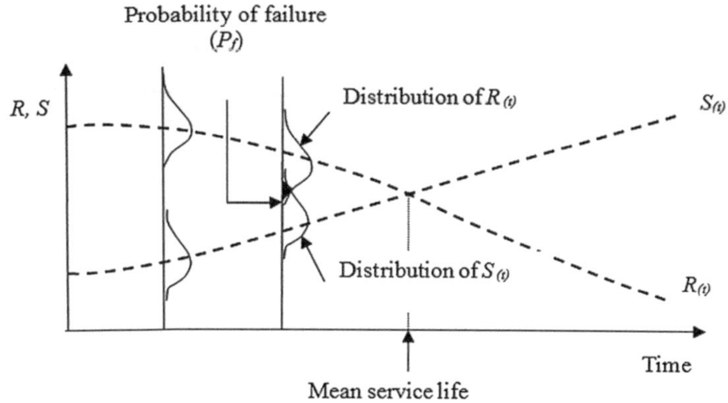

Fig. 5.4 Illustration of the deterministic and probabilistic approach [7, 38, 39]

The time dependent natures of both *S(t)* and *R(t)* are shown in Fig. 5.4. The point in time when their characteristic values intersect is the service life in a deterministic approach (see below). The stochastic nature of *S* and *R*, expressed in statistical distributions, allows the ability to calculate the probability of failure at any point in time. The failure probability is given by the amount of overlap of the distributions for *S* and *R*, as indicated in Fig. 5.4. Examples of the application of probabilistic approaches for concrete durability design and service life prediction can be found in the literature [7, 8, 39–43].

Examples of the application of probabilistic approaches for concrete durability design and service life prediction can be found in the literature [7, 8, 39–43].

5.6.4 Semi-probabilistic Approach to Statistical Variations

Full probabilistic calculations require (at least) special software and expertise. Consequently, simplification is commonly desired. As for structural design, it is possible to "translate" a full-probabilistic service life design method into a semi-probabilistic format by making deterministic calculations based on mean or characteristic values and applying safety factors. Doing so, Eq. (5.3) would change into:

$$S^*(t) \cdot \gamma_S \geq \frac{R^*(t)}{\gamma_R} \tag{5.4}$$

where $S^*(t)$ is the mean value of *S* at time *t*, γ_S is the safety factor for the load, $R^*(t)$ is the mean value of *R* at time *t*, and γ_R is the safety factor for resistance.

In more practical terms, with respect to chloride-induced corrosion the critical chloride threshold value would be reduced and the chloride content at the steel surface would be increased, such that the probability of them being equal reduces from approximately 50 % (deterministic case when mean values are used as input) to the acceptable probability level.

Determining the safety factors in Eq. (5.4) requires information on variability; they should be "calibrated" with respect to a large number of cases (experiments, structures, etc.).

Dutch Committee CUR VC81 has proposed a semi-probabilistic simplification of its service life design method, based on a safety margin on the concrete cover [43, 44]. The safety margin is based on calibration with a set of full-probabilistic calculations. It was criticised for causing inconsistencies [45–47] and a safety factor approach was proposed as an alternative. Apparently this method needs further work. Equation (5.4) may also be written in terms of service life [39], multiplying the intended (target) service life, t_g, by a safety factor, γ_t.

$$t_d = \gamma_t \cdot t_g \tag{5.5}$$

where t_d is the design service life. The semi-probabilistic approach on one hand appears to offer a useful simplification, on the other hand requires careful calibration and developing consensus before it can be widely accepted.

5.6.5 Principles of the Deterministic Approach to Statistical Variations

The deterministic approach to service life design takes a similar format to the probabilistic approach in that it utilizes a mathematical model to carry out design calculations. The solution to the deterministic approach can be obtained using the limit state format, similar to the probabilistic approach. As aforementioned, the limit state format characterises the model parameters as either action effect, $S(t)$ or as a resistance R (or initiation limit, S_{lim}) [37].

The deterministic format has been applied to service life prediction models such as the North American Life 365 model [48]. The model produces only a single deterministic time to corrosion initiation, neglecting that concrete properties are quite variable both throughout the structure and in terms of quality of construction and materials used and in time [49]. Others aspects that are neglected include uncertainties in the environmental actions and the error in the model. The latter is often not included in a probabilistic format. In contrast, a probabilistic model is able to predict a range of expected times to corrosion initiation rather than a single value so as to allow owners to make an easier and more accurate selection of durability parameters and economical decisions of structural concrete. However, owners, engineers and researchers often lack sufficient information and/or experience to select relevant probability parameters in terms of nature and extent of expected damage and economic implications. In practice, the selection of relevant model parameters may therefore be difficult.

A deterministic approach utilizes only the mean or (at best) characteristic values of the model parameters, as single parameters, unlike the probabilistic model which takes into account the range of possible values for each model parameter. At present, there is no practical application of the characteristic value approach to deterministic service life prediction.

References

1. BS EN 206, Concrete: specification, performance, production and conformity, British Standards Institution, 2013.
2. Walraven J. Design for service life: how should it be implemented in future codes. In: International conference on concrete repair, rehabilitation and retrofitting, Proceedings ICCRRR 2008, Cape Town, 24–26 Nov 2008. pp. 3–10.
3. Alexander MG, Mackechnie JR, Ballim Y. Use of durability indexes to achieve durable cover concrete in reinforced concrete structures. In: Skalny JP, Mindess S, editors. Materials science of concrete, vol. 6, American Ceramic Society, 2001. pp. 483–511.
4. Simons B. Concrete performance specifications: New Mexico experience. Concr Int. 2004;26 (4):68–71.
5. Day KW. Perspective on prescriptions. Concr Int. 2005;27(7):27–30.
6. Harrison TA. Framework for durability performance specifications for concrete used in normal construction. In: Proceedings: XIth European ready-mixed concrete congress, Turkey, 1995.
7. DuraCrete R17, Final technical report. DuraCrete—probabilistic performance based durability design of concrete structures, The European Union—Brite EuRam III, May 2000, Document BE95-1347/R17, 2000.
8. fib bulletin 34, Model code for service life design, Switzerland, 2006, 110 pp.
9. Alexander MG, Mackechnie JR, Ballim Y. Guide to the use of durability indexes for achieving durability in concrete structures. Research monograph no 2, Department of Civil Engineering, University of Cape Town, 1999, 35 pp.
10. Tang L. Chloride transport in concrete-measurement and prediction. Building materials, Gothenburg: Chalmers, 1996.
11. EN 1504-9, Products and systems for the protection and repair of concrete structures—definitions, requirements, quality control and evaluation of conformity—Part 9: general principles for the use of products and systems, 2008.
12. Nordtest, Concrete, hardened: accelerated chloride penetration, NT Build 443, Espoo, Finland, 1995.
13. ASTM C1556, Standard test method for determining the apparent chloride diffusion coefficient of cementitious mixtures by bulk diffusion, Philadelphia: American Society for Testing and Materials, 2004.
14. XP CEN/TS 12390-11, Testing hardened concrete—Part 11: determination of the chloride resistance of concrete, unidirectional diffusion, January 2010.
15. Kollek JJ. The determination of the permeability of concrete to oxygen by the Cembureau method—a recommendation. Mater Struct. 1989; 22(129):225–230.
16. CEN/TS 12390-XXX, Testing hardened concrete—Part XX: determination of the potential carbonation resistance of concrete: accelerated carbonation method, Draft version, Nov 2009.
17. EN 13396, Products and systems for the protection and repair of concrete structures. Test methods. Measurement of chloride ion ingress, 2004.
18. Torrent R, Fernández Luco L, editors. Non-destructive evaluation of the covercrete, RILEM TC 189-NEC state-of-the-art report, Zurich, Switzerland, 2007. 223 pp.

19. Romer M, Fernández Luco L. RILEM TC 189-NEC: non-destructive evaluation of the concrete cover: comparative test—part I: comparative test of 'penetrability' methods—part II: comparative test of 'Covermeters'. Mater Struct. 2005;38(284):895–911.
20. Quinn GP, Keough MJ. Experimental design and data analysis for biologists. Cambridge University Press, The Edinburg Building, Cambridge CB2 2RU, UK, 2002.
21. Beushausen H, Burmeister N. The use of surface coatings to increase the service life of reinforced concrete structures for durability class XC. Mater Struct. 2015;48:1243–52.
22. EN 1504, Products and systems for the protection and repair of concrete structures—definitions, requirements, quality control and evaluation of conformity, part 1 to 9.
23. Parrott L. A review of carbonation in reinforced concrete. Cement and concrete association. United Kingdom, Waxham Springs, Slough, 1987.
24. Richardson MG. Fundamentals of durable concrete. Modern concrete technology series. London: Spon Press, 2002.
25. Bentz DP, Clifton JR, Snyder KA. Predicting service life of chloride-exposed steel-reinforced concrete. Concr Int. 1996;Dec. 1996:42–47.
26. Thomas MDA, Jones MR. A critical review of service life modelling of concretes exposed to chlorides. In: Dhir RK, Hewlett PC, editors. Concrete in the service of mankind: radical concrete technology, E & FN Spon, London, 1999. pp. 723–736.
27. Berke NS, Hicks MC. Estimating the life cycle of reinforced concrete decks and marine piles using laboratory diffusion and corrosion data. In: Chaker V, editor. Corrosion forms and control for infrastructure, ASTM STP 1137, American society for testing and materials, Philadelphia, 1992. pp. 207–231.
28. Thomas MDA, Bentz EC. Manual: computer program for predicting the service life and life cycle costs of reinforced concrete exposed to chlorides, 2000.
29. Maage M, Helland S, Poulsen E, Vennesland Ø, Carlsen JE. Service life prediction of existing concrete structures exposed to marine environment. ACI Mater J. 1996;93(6):602–8.
30. Papadakis VG. Effect of supplementary cementing materials on concrete resistance against carbonation and chloride ingress. Cem Concr Res. 2000;30(2):291–9.
31. Castellote M, Andrade C. Modelling the carbonation of cementitious matrixes by means of the unreacted-core model, UR-CORE. Cem Concr Res. 2008;38(12):1374–84.
32. Jiang L, Lin B, Cai Y. A model for predicting carbonation of high-volume fly ash concrete. Cem Concr Res. 2000;30(5):699–702.
33. Thiery M, Dangla P, Villain G, Platret G. A prediction model for concrete carbonation based on coupled CO_2-H_2O-ions transfers and chemical reactions. In: 10th DBMC international conference on durability of building materials and components, Lyon, France, 17–20 April 2005.
34. Barya B, Sellier A. 'Coupled moisture—carbon dioxide–calcium transfer model for carbonation of concrete. Cem Concr Res. 2004;34(10):1859–72.
35. Siemes T, Vrouwenvelder T, van den Beukel A. Durability of buildings: a reliability analysis. HERON. 1985;30(3):2–48.
36. ISO 2394, General principles on reliability for structures for durability, 1998. pp. 73.
37. ISO 13823, General principles on the design of structures for durability, ISO TC98/SC2, Final draft, 2008.
38. Rostam S. Service life design of concrete structures-a challenge to designers as well as to owners. Asian J Civ Eng (Build Hous). 2005;6(5):423–45.
39. Sarja A, Vesikari A. Durabilty design of concrete structures. Report of rilem technical committee 130-CSL, E&FN SPON, 1996.
40. DuraCrete, Modelling of degradation, The European Union—Brite EuRam III, Project BE95-1347, Probabilistic performance-based durability design of concrete structures, 1998.
41. Gehlen C. Probabilistische Lebensdauerbemessung von Stahlbetonbauwerken, Deutscher Ausschuss für Stahlbeton 510, Berlin, 2000 [in German].
42. Kirkpatrick TJ, Weyers RE, Anderson-Cook CM, Sprinkel MM. Probabilistic model for the chloride-induced corrosion service life of bridge decks. Cem Concr Res. 2002;32:1943–60.

43. van der Wegen G, Polder RB, van Breugel K. Guideline for service life design of structural concrete—a performance based approach with regard to chloride induced corrosion. HERON. 2012;57(3):153–68.
44. Polder RB. Test methods for on-site measurement of resistivity of concrete—a RILEM TC-154 technical recommendation. Constr Build Mater. 2001;15:125–31.
45. Polder RB, van der Wegen G, van Breugel, K. Guideline for service life design of structural concrete—a performance based approach with regard to chloride induced corrosion. In: Dehn F, Beushausen H, editors. Proceedings of fib workshop performance-based specifications for concrete, Leipzig, 14–15 June 2011. pp. 25–34.
46. Gulikers J. Development of guidelines for durability design of reinforced concrete structures. In: van Breugel K, Ye G, Yuan Y, editors. Proceedings of the 2nd international symposium service life design for infrastructure, 4–6 Oct 2010, Delft, The Netherlands, 2010. p. 359-368.
47. Gulikers J. Practical implications of performance specifications for durability design of reinforced concrete structures. In: Dehn F, Beushausen H, editors. Proceedings of the fib 8.10 workshop on performance-based specifications for concrete, June 2011, Leipzig, Germany, 2011. p. 341–350.
48. Thomas MDA, Bentz EC. Manual: computer program for predicting the service life and life cycle costs of reinforced concrete exposed to chlorides, 2000.
49. Bentz E. Probabilistic modelling of service life for structures subjected to chlorides. ACI Mater J. 2003;Sept–Oct 2003:390–7.

Chapter 6
Statistical Procedures for Performance-Based Specification and Testing

W.A. Stahel, F. Moro and L. Fernandez Luco

6.1 Introduction

a **Statistical approach needed**. Durability specification and control of concrete should be based on performance measures. Chapters 3 and 4 suggest that transport properties provide suitable criteria.
Concrete shows intrinsic variability, and measurements are always subject to some measurement error. Specifications should take these variations into account by relying on statistical methods, both for estimating "true values" of durability indices and for deciding about conformity with prespecified limits of acceptability.

b **Tests**. Compliance assessment is based on rules for deciding if a construction meets the needed quality standard. Such rules can be viewed as statistical tests. They come in different flavors:
Conformity tests assess if a specified limit of a durability criterion is met on the basis of a sample of measurements obtained from a batch of concrete or from a construction. Such a test may be applied to show that the proposed mix design conforms with the specification by **prequalification testing** under laboratory conditions. Subsequently, an **identity test** may assess if the delivered concrete is "the same" as the prequalified mixture. The meaning of "the same" should be understood from a statistical view reflecting variability and measurement error. Finally, another conformity test may use measurements taken on the final construction to judge if the requirements are fulfilled.

W.A. Stahel (✉)
Department of Mathematics, ETH Zurich, Switzerland

F. Moro
Holcim Technology Ltd, Holderbank, Switzerland

L. Fernandez Luco
Engineering Faculty, University of Buenos Aires, Buenos Aires, Argentina

H. Beushausen and L. Fernandez Luco (eds.), *Performance-Based Specifications and Control of Concrete Durability*,
RILEM State-of-the-Art Reports 18, DOI 10.1007/978-94-017-7309-6_6

c **Scope**. This chapter provides an introduction into the statistical foundations of these types of tests. It also describes the analysis of interlaboratory studies, which help to understand the precision of measurements. Finally, statistical regression methods are discussed, since studying relations between different quantities is a prominent theme in several other chapters of this book, mostly for relating quantities that can be measured to the ultimately targeted aspects of durability. In this text, we cannot give a self-contained introduction into probability. We suppose that the reader has been exposed to such an introduction. Sources for reference are textbooks like [1, 2].

Three example data sets that will be used repeatedly to illustrate the methods follow.

d **Example Tunnel**. Concrete properties (compressive strength, permeability and porosity) in selected structural components of a new cut and cover tunnel have been measured and the spatial variability determined [3]. 400 cores were taken from two deck and two wall elements, and different durability indicators were measured.

e **Example Permeability**. The results of a round robin of the air permeability test (SIA Standard 262/1, Annex E) are reported in [4]. On two selected elements of a bridge a regular grid of 75 identically sized square areas was delineated. The 75 areas were randomly assigned to 5 participating teams, so each team had to measure permeability in 15 areas each for both elements.

f **Example Cover Depth**. A third example is taken from Chap. 9 of this volume (Fig. 9.6, Spans 1 and 2). It features a 5×40 grid of cover depth readings obtained in the top reinforcement layer of the deck slab of a freeway viaduct. Table 6.1 re-displays the data.

g **Software**. In order to facilitate the application of the methods described in this chapter, we have collected data sets and functions in a "package" of the R software, see r-project.org. The R system has developed into the dominating language to make statistical procedures available for a general audience. Our contribution is called qmrobust and is currently available from r-forge.r-project.org. We plan to develop it further and move it to the package collection cran.r-project.org.

h **Cross-references**. This chapter uses a special way to provide cross-references. Within subsections, the paragraphs are labelled by letters in the margin, and cross-references contain the appropriate letter. Thus, the reference Sect. 6.2.1 b

Table 6.1 Cover depth readings for a 5×16 grid from the top reinforcement layer of the deck slab of a freeway viaduct

64	56	66	76	52	73	69	71	79	76	79	72	76	79	63	53
61	56	63	69	69	70	74	76	55	72	76	53	57	76	80	28
57	57	37	40	64	56	60	56	56	48	47	57	54	47	56	56
49	51	42	51	57	58	57	54	59	63	64	71	48	72	48	39
58	61	53	49	51	48	55	50	46	49	65	59	63	62	58	47

The data from two spans of the bridge are shown

points to the paragraph entitled "Density" on the next page. It is also used to refer to the equation appearing there. This gives more precise locations and facilitates searching for equations.

6.2 Distributions

6.2.1 Random Variables

a **Histogram**. Figure 6.1 shows histograms of three variables from the three examples introduced in the preceding section.

b **Density**. These are three examples of measurements of continuous variables. Let us call the possible values x. The model of a continuous random variable is helpful to derive conclusions about the "truth" behind them. It consists of a distribution, an idealized histogram, which describes our thinking about such measurements. The distribution of a continuous random variable is characterized by a curve $f(x)$ describing the **probability density**, like the one shown in Fig. 6.2. Its integration over any interval $[a, b]$ measures the probability of getting a value between a and b. Since the probability of getting any of the possible values is 1, the integral of the density over all possible values must be 1. We denote the random variable by a capital letter, often X when the possible values are x. We then denote the probabilities

$$P(a < X \leq b) = \int_a^b f(x)\,\mathrm{d}x.$$

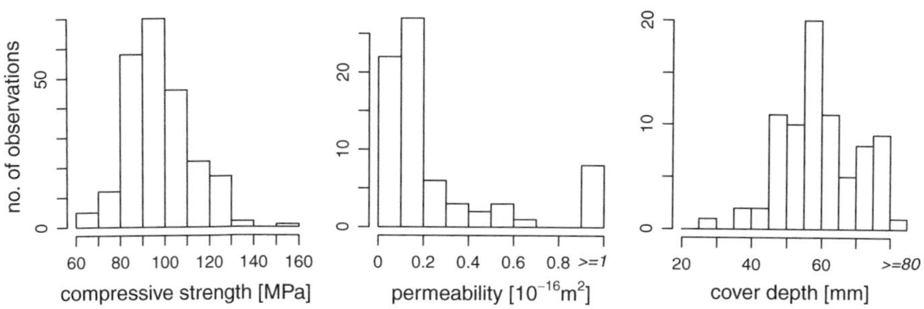

Fig. 6.1 Histograms of the compressive strength of concrete of 233 samples of the deck elements in Example 6.1d [MPa], of 72 permeability values from Example 6.1e [10^{-16} m^2], and of 80 cover depth readings [mm] from Example 6.1

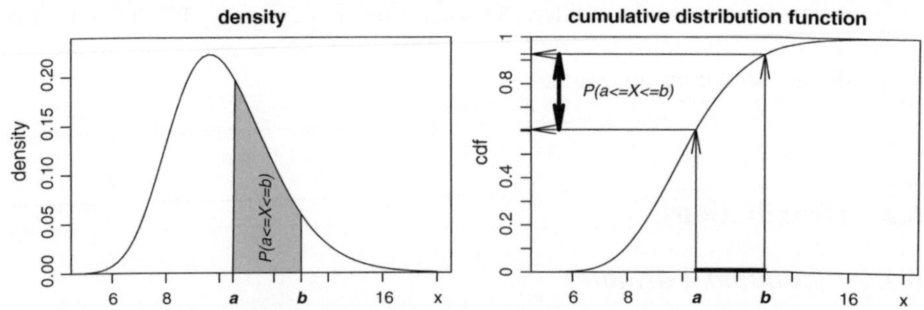

Fig. 6.2 Density and cumulative distribution function. The probability of X falling between a and b is given by the shaded area under the density curve on the *left* or by the length of the interval between the *arrowheads* on the vertical axis in the *right-hand panel*

Note that X is a kind of place holder for the possible values. It stands for the whole distribution, which in turn is characterized by f. Figure 6.2 illustrates the determination of the probability of an interval.

Technical Remark: **Notation**. In order to keep the distinction between random variables and ordinary numbers clear, random variables are denoted by capital letters, often X, whereas lower-case letters denote numbers.

c **Cumulative distribution function (cdf)**. A distribution can be characterized, instead of the density, by its integrating function, which is

$$F(x) = P(X \leq x) = \int_{-\infty}^{x} f(t)\,\mathrm{d}t$$

and is called the cumulative distribution function (cdf). It is an increasing (non-decreasing to be precise) function which goes from 0 to 1. Even though it can be drawn, see Fig. 6.2, it is less suitable for graphical purposes since the shape of the distribution is more difficult to grasp from it even for the expert. It has its merits for theoretical considerations as we will see later.

d **Theoretical and empirical distribution**. A number of values x_i have a "distribution" in an everyday sense. If the values have been obtained under constant circumstances, we call this an **empirical distribution**, and the set of values themselves, a **sample**. They are visualized by a histogram, which resembles a density curve. The histogram, however, is not uniquely defined, since it depends on the chosen width and position of the bars. On the other hand, there is a version of the cumulative distribution function, which is uniquely determined by the sample values. It is called the empirical cdf and defined as

$$\widehat{F(x)} = \frac{1}{n} \#(i | x_i \leq x)$$

see Fig. 6.3 (right hand side). It is a step function with steps of height 1/n at each sample value x_i (if all x_i are different). Its advantage over a histogram is that it allows for identifying the precise sample values, in contrast to the histogram.

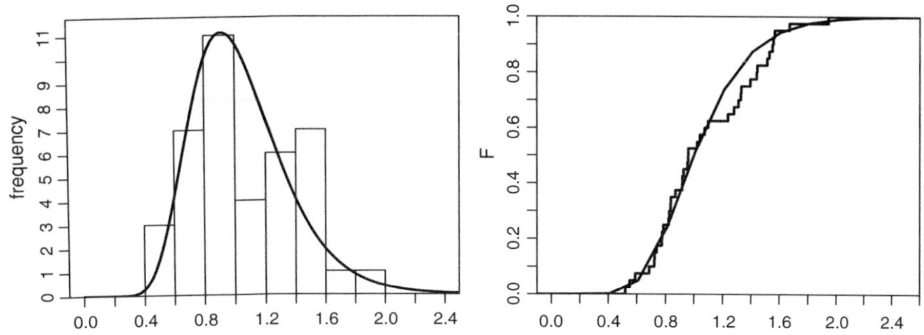

Fig. 6.3 Theoretical and empirical distribution. 40 random numbers according to the given theoretical distribution are used to generate the histogram (*left*) and the empirical cumulative distribution function (*right*)

Thus, an empirical distribution is determined by a sample of values x_i that have been obtained, whereas a theoretical distribution is an idealized model for it, which should help us to judge what values we will get if we follow the same process of obtaining values again. The model cannot be right or wrong, it can be plausible or not, and more or less useful. If it is a good model for the circumstances under which a sample of values has been obtained, we expect the theoretical and empirical distribution functions to be similar, and the histogram to resemble the density curve to some extent. Probability theory can tell us how similar we should expect them to be. We will come back to these thoughts in Sect. 6.3.2 r.

e **Counts, discrete random variables**. When we count the number of cracks in concrete or any other number of "events", the resulting random variable has no density. The distribution of a random variable that models such a count is characterized by the probabilities to get the result 0, 1, 2, ..., x, ..., denoted by $P(X = x)$. The cumulative distribution function is

$$F(x) = P(X \le x) = \sum\nolimits_{k=0}^{x} P(X = k) .$$

More generally, a discrete random variable can take only discrete values x_k and is characterized by the probabilities $P(X = x_k)$ or by the cdf

$$F(x) = P(X \le x) = \sum\nolimits_{x_k \le x} P(X = x_k) .$$

These functions are step functions similar to empirical cdf's, but with steps of height $P(X = x_k)$ at the values x_k.

We could then say that the **empirical distribution**, defined by a sample from a continuous variable, is the distribution of a discrete random variable with possible values x_i and probabilities $P(X = x_i) = 1/n$.

6.2.2 Expectation and Variance

a **Location, expected value**. The distribution of a random variable roughly determines a value around which we expect the realizations to fall (a "location" or "measure of central tendency") and a width of a range of plausible scattering around it (the "spread"). The most common measure of location is the expected value, denoted by $\mathcal{E}(X)$, equal to the integral

$$\mathcal{E}(X) = \int_{-\infty}^{\infty} xf(x)\mathrm{d}x.$$

For discrete random variables, the expected value is

$$\mathcal{E}(X) = \sum\nolimits_k x_k P(X = x_k)$$

which results in $\mathcal{E}(X) = \sum_{k=0}^{\infty} kP(X = k)$ for counts.

b **Spread, variance, standard deviation**. For the spread, the most common measure is the standard deviation, $\mathrm{sd}(X) = \sqrt{\mathrm{var}(X)}$, where

$$\mathrm{var}(X) = \int (x - \mathcal{E}(X))^2 f(x)\mathrm{d}x \qquad \text{or} \qquad \mathrm{var}(X) = \sum\nolimits_k (x_k - \mathcal{E}(X))^2 P(X = x_k)$$

is the variance of the distribution. It plays a very important role in probability theory.

c **Mean value, empirical variance and standard deviation**. For characterizing the location of an empirical distribution, i.e., a sample of values x_i, the number that is most commonly used is the mean, denoted as

$$\overline{x} = \frac{1}{n}\sum\nolimits_{i=1}^{n} x_i.$$

The variance of a sample or the "empirical variance" is defined as

$$\widehat{\mathrm{var}}(x_1, x_2, \ldots, x_n) = \frac{1}{n-1}\sum\nolimits_{i=1}^{n} (x_i - \overline{x})^2,$$

and the empirical standard deviation is its square root, $\widehat{\mathrm{sd}} = \sqrt{\widehat{\mathrm{var}}}$ $(x_1, x_2, \ldots, x_n)$. The argument for using $n - 1$ in the denominator instead of n will be given in Sect. 6.2.5 g. (The "hat" symbol $\widehat{\ }$ is used to denote empirical counterparts of characteristics of theoretical distributions.)

In analogy to the wording for the variance, the mean could be called the empirical expectation. Having two clearly distinct words in this case is helpful for keeping the theoretical concepts distinct from the characteristics of data. The relation between the notions for models or distributions and the respective ones for data will be shown to be important in Sect. 6.3.1 a.

Technical Remark: Combining the foregoing definitions, the mean can be identified with the expectation over the empirical distribution (Sect. 6.2.1 e), and the empirical variance is the variance of the empirical distribution, up to the slight change in the denominator.

d The reason for the dominant role of these measures is that they lead to mathematically simple rules when deriving distributions of sums and averages of random variables (see Sects. 6.2.5 e, f) or describing joint distributions. In practice, they have the serious flaw of being highly influenced by rare extreme possible values leading to the "outliers" which are so familiar to all practitioners of measurements. Characteristics of data that are only influenced by outliers to a limited extent are called **robust**.

e **Median**. An alternative measure of location of a distribution is the median or central value , which divides the possible values such that the probabilities of obtaining values smaller or larger than are $1/2$ each (or less than $1/2$ each, if there is a nonzero probability of obtaining exactly).

f The empirical median, or median of the sample, $\widehat{\text{med}}$ is defined analogously as the value for which half the sample falls on each side. If the number n of observations is odd, it must be one of them, the "middle one". If n is even, there are 2 observations in the middle, and the convention is to define their mean as the median of the sample.

g **Quantiles**. A generalization of the median is the notion of a quantile. The quantile q_p is the value for which the probability of obtaining values below it is p

$$P(X \leq q_p) = F(q_p) = p.$$

The quantile q_p can be regarded as a function of p. It is the inverse function of the cdf F. Thus, a quantile can be obtained from a Fig. 6.3 (right hand panel) in reverse direction, starting from p on the vertical axis and finding the corresponding value on the horizontal axis.

The **empirical quantile** is obtained from using the empirical distribution function $\widehat{F}$ instead of the theoretical one (same figure). Where F or $\widehat{F}$ is horizontal, the quantile function is not well defined. In analogy to the rule for the empirical median of a sample with an even size n, the quantile q_p can then be defined as the midpoint of the interval over which $\widehat{F}(x) = p$.

h **Interquartile Range**. The probabilities for the possible values of X are split into 4 equal parts by the so-called quartiles $q_{0.25}, q_{0.5}, q_{0.75}$, the middle quartile being the median. The quartiles are used to define an alternative measure of spread, the "interquartile range"

$$\mathrm{iqr}(X) = q_{0.75} - q_{0.25}.$$

The median and the interquartile range are measures of location and spread that are not "harmed" by outliers, that is: The shape of the density curve below the first and above the third quartile do not matter for these measures, these "tails" may or may not give extreme outliers a considerable probability. This robustness may be desirable or unfortunate depending on the purpose or the taste of the user. Of course, there are other, less widely used, measures of location and spread.

6.2.3 *The Normal Distribution and Other Families*

a **The normal distribution**. The most widely used distribution is the normal or Gaussian distribution. For given expected value μ and standard deviation σ, its density function is (cf. Fig. 6.4)

$$f_{\mu,\sigma}(x) = \frac{1}{\sqrt{2\pi}\sigma} \exp\left(-\frac{1}{2}\left(\frac{x-\mu}{\sigma}\right)^2\right).$$

We will often use the notation $X \sim \mathcal{N}(\mu, \sigma^2)$ to express that X has a normal distribution with expectation μ and variance σ^2.
The simplest case, $\mu = 0$ with $\sigma = 1$ and plays a central role in statistics and is called **standard normal distribution**, with density

$$\varphi(z) = \frac{1}{\sqrt{2\pi}}\, e^{-\frac{1}{2}z^2}.$$

The cdf does not allow for a simple formula, but is the indefinite integral of the density (as it always is). For the standard normal distribution, we write $\Phi(z) = \int_{-\infty}^{z} \varphi(t)\, \mathrm{d}t$. Some quantiles of the standard normal distribution will be

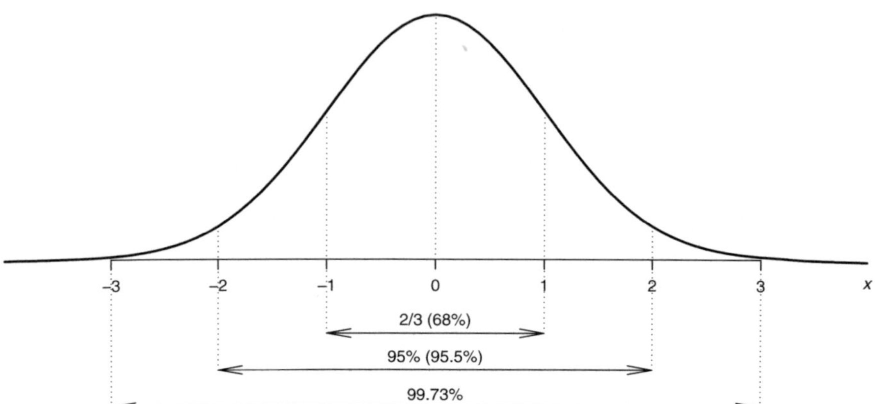

Fig. 6.4 Density of a (standard) normal distribution. Probabilities of the ranges of $\mu \pm k\sigma$ with $k = 1, 2, 3$ are illustrated

Table 6.2 Values of the cumulative distribution function of the standard normal distribution for selected arguments z

z	$\Phi(z)$	z	$\Phi(z)$	z	$\Phi(z)$	z	$\Phi(z)$
0.0	0.500	0.7	0.758	1.4	0.919	0.674	0.750
0.1	0.540	0.8	0.788	1.5	0.933	0.967	0.833
0.2	0.579	0.9	0.816	1.6	0.945	1.282	0.900
0.3	0.618	1.0	0.841	1.7	0.955	1.645	0.950
0.4	0.655	1.1	0.864	1.8	0.964	1.960	0.975
0.5	0.691	1.2	0.885	1.9	0.971	2.326	0.990
0.6	0.726	1.3	0.903	2.0	0.977	2.576	0.995

For negative arguments $-z$, use $\Phi(-z) = 1 - \Phi(z)$. Thus, $\Phi(-1.3) = 1 - \Phi(1.3) = 1 - 0.903 = 0.097$. Quantiles q_p for the values in the columns $p = \Phi(z)$ are the z values to their left. The rightmost column contains quantiles for useful p values, e.g., $q_{0.975} = 1.960$

used repeatedly later in this chapter. Therefore, we collect some values of $\Phi(z)$ in Table 6.2.

For practical purposes, it is useful to know that the interval $[\mu - \sigma, \mu + \sigma]$ contains about 2/3, and the interval $[\mu - 2\sigma, \mu + 2\sigma]$ covers 95 % of probability, see Fig. 6.4.

b **Parameters, families of distributions**. "The normal distribution" only becomes a definite distribution if the values of μ and σ are fixed. So, we should call it "the family of normal distributions". The numbers μ and σ that can vary and thereby constitute the family are called the **parameters** of the distribution (family). Thus, a **parametric family of distributions** is given by a formula $F_{\underline{\theta}}(x)$ for the cumulative distribution function, which contains, apart from the argument x, a parameter $\underline{\theta}$—the notation shows that it may be a whole vector of parameters, indeed. We will see other families shortly. The notation $X \sim F_{\underline{\theta}}$ will be generally used to say that X has distribution (function) $F_{\underline{\theta}}$.

c *Technical Remark*: Distributions may be generated by theoretical insight into a process—and fail to belong to one of the known parametric families. This is the case for **probability based durability design**. The model assumes a distribution over the diffusion coefficient in a diffusion model, e.g., for chloride, and derives a distribution over the penetration depth after any given time t. A second distribution characterizes the depth cover, and finally, this leads to a distribution of the time when the chloride concentration at the rebar first exceeds the critical chloride content (itself a variable with a probability distribution), resulting in a probability of depassivation and corrosion initiation. See Sect. 5.6.3 of this volume for more detail.

d **Importance of the normal distribution**. Again, the popularity of the normal distribution is due to the fact that it has very nice mathematical properties as we will mention shortly. Real data rarely follow this distribution exactly. Often, they show skewed distributions and more extreme values than the normal distribution suggests.

e **Conformity**. The idea of a durable construction can be formalized as follows. Assume that the natural variability of the relevant measure of durability is described by a normally distributed random variable X, and that a construction is durable if no more than $p = 2\,\%$ of the concrete exceeds a threshold c. This translates to requiring that the expected value μ does not fall below $c + 2\sigma$ (see Sect. 6.2.3 a). (For other values of p, the factor 2 is replaced by the quantile q_{1-p} of the standard normal distribution, see Sect. 6.2.3 a) This allows for testing the conformity by examining the expected value μ instead of assessing the probability of exceeding the threshold c directly—if σ is known at least approximately. We come back to this idea in Sect. 6.2.3 a.

6.2.4 *Transformation of Random Variables*

a The permeability data shown in Fig. 6.1 clearly adheres to a skewed distribution, and, to a lesser extent, so does the compressive strength. Thus, the normal distribution is not an adequate model for describing these measurements. This is commonly true for data which cannot be negative and has a larger coefficient of variation $\mathrm{sd}(X)/\mathcal{E}(X)$, above 0.5, say.

b **Log transform**. If we like to apply the strong methodology connected to the normal distribution, we may do so after transforming the data to logarithmic scale, that is, treating the $\log_{10}$(permeability) values as the data rather than the raw values. (Instead of the logarithm with base 10, the natural logarithm (base e) is often used. We prefer base 10 since such logarithms are more easily back-transformed roughly to the original scale in one's mind: $\log_{10}(x) = 3$ corresponds to $x = 10^3 = 1000$.) Figure 6.5 shows a histogram of the log-transformed values with a nicely fitting normal density curve.

c Wide experience shows that quantitative data that cannot be negative by its nature should be log-transformed before being analyzed as well as for graphical displays. Data that are originally observed or measured most often are of this nature. They have been called "amounts" by John Tukey, the father of "exploratory data analysis", and he called the log transformation the "first aid" for such data.

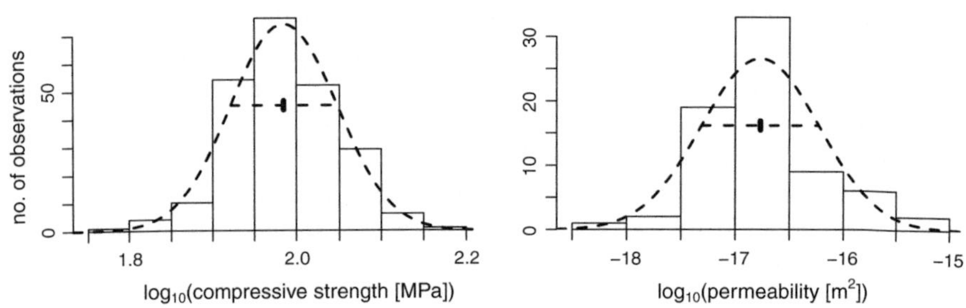

Fig. 6.5 Histograms of the data shown in the first two panels of Fig. 6.1 after $\log_{10}$ transformation. (The discrepancy between the central bar and the corresponding density curve in the *right hand panel* is a non-significant random effect.)

d **Distribution of the transformed random variable**. Assume that X is a random variable described by the cdf $F_X(x)$. Let us apply a transformation g to the values of X. The result is a random variable Y, which we denote by $Y = g(X)$, for which the cdf F_Y can be determined from F_X. If g is a monotonically increasing function, then $F_Y(y) = P(Y \le y) = P(X \le x) = F_X(x)$, where $y = g(x)$.
The link between the densities is more difficult to get. Since the densities are the derivatives of the cdfs, we get

$$f_Y(y) = f_X\left(g^{-1}(y)\right) \Big/ g'\left(g^{-1}(y)\right)$$

e **The log-normal distribution**. We have argued above that quite often, the normal distribution is a reasonable model if applied to log transformed data. Thus, if $Y = \log(X)$ and $Y \sim \mathcal{N}(\mu, \sigma^2)$, what is the distribution of the untransformed variable X? Applying the last formula yields

$$f_X(x) = \frac{1}{\sqrt{2\pi}\sigma}\frac{1}{x}\exp\left(-\frac{1}{2}\left(\frac{\log(x)-\mu}{\sigma}\right)^2\right).$$

(The natural logarithm is used here in accordance with the literature.) Sensible parameters for this family of distributions are $\mu^* = \exp(\mu)$, which is the median and also controls the scale, and $\sigma^* = \exp(\sigma)$, which determines the shape, see Fig. 6.6.

f **Linear transformation**. The most simple transformations are changes of origin and scale of measurements, such as converting degrees Fahrenheit to Celsius. This leads to the linear transformations that play an important role in probability and statistics. In this case, the density of the transformed variable is easy to get.

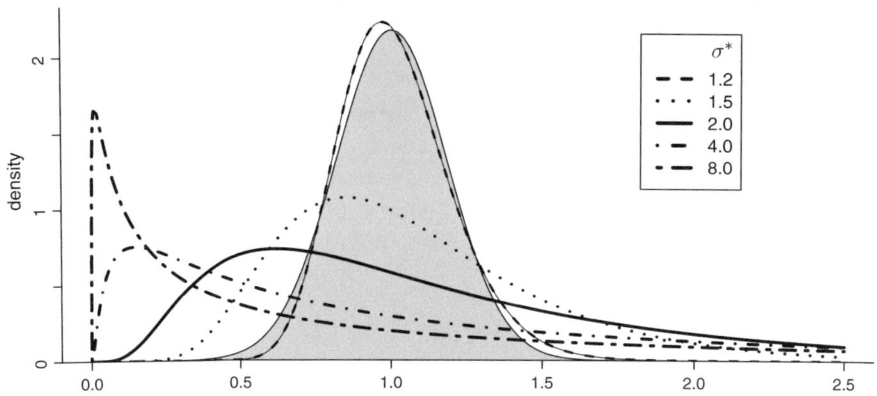

Fig. 6.6 Densities of log-normal distributions with $\mu^* = 100$ and different multiplicative standard deviations σ^*. A normal density with the same median and variance as the log-normal with $\sigma^* = 1.2$ is shown, shaded in grey

If $Y = a + bX$, then $f_Y(y) = f_X(x)/b$, where $y = a + bx$, and there are simple results for the expectation and standard deviation,

$$\mathcal{E}(Y) = a + b\,\mathcal{E}(X)\,, \quad \mathrm{var}(Y) = b^2\mathrm{var}(X)\,, \quad \mathrm{sd}(Y) = b\,\mathrm{sd}(X).$$

6.2.5 *Functions of the Sample Values*

a Means of measurements scatter less than the individual values. This fact is used in many aspects of daily life. Even pupils know that they are promoted only if the *average* of their grades is sufficient. Measurements of properties of concrete are often repeated and then averaged to increase the precision of the assessment. If we consider the data as random, relating to a random variable, we also need to treat the sample mean as a random variable. Probability theory will tell us how the distribution of this derived random variable is obtained from the distribution that we choose for the original (or, alternatively, transformed) data.

b **Random sample**. The basic difficulty in this step is that the assumption just stated must be formulated more precisely and leads to a rather abstract model, the notion of a random sample. The probability model describes what we expect will result if we obtain data, before we have got it. For the first observation, we expect a value with the probabilities given by the model distribution. We denote these ideas about the possible outcomes by a random variable, X_1, characterized by $F_X(x)$. For the second observation, we have the same expectations, even though the result will be different from the first one. We denote this by X_2, a random variable with the same distribution as X_1, and add that the result of the first observation will not change our expectations for the second one. This is expressed by the assumption of (stochastic) independence. We proceed in the same way for the third and following observations. The model of a random sample therefore consists of n independent random variables with the same distribution. The jargon says that they are independent and identically distributed, i.i.d. for short.

Notice that while we have started by using a density as an idealized histogram coming from a sample of n values, we now model the sample in turn by n random variables. This is an abstract concept that is basic for understanding statistics.

c Starting from this basic model, it is a matter of probability calculations to derive the distribution of a function T of the sample values. We now write random variables as the arguments, $T(X_1, X_2, \ldots, X_n)$, expressing thereby that the result is itself a random variable.

d **Samples with dependencies**. Probability calculations are of course also available if the observations X_i are not independent. For controlling the quality of a continuous production process, measurements are taken in regular time intervals. They then form a time series, and a model is needed for describing how the first value, X_1, influences the distribution of the second measurement, X_2—

usually giving values similar to X_1 with higher probability—and how X_3 depends on X_1 and X_2, and so on. Such a model, too, allows for deriving the distribution of $T(X_1, X_2, \ldots, X_n)$.
Measurements of properties of concrete in a construction are located (usually) on a surface, and those that are close together should be expected to give more similar values than those that are far apart. This is a typical case of a spatial correlation and violates the assumption of independence.
Here, we will nevertheless proceed with the assumption of an independent random sample, and postpone remarks about dependent observations to the outlook Sect. (6.4.4 i).

e **Sums of random variables**. A generic problem of probability theory is to determine the distribution of a sum of two independent random variables X_1 and X_2, $S = X_1 + X_2$. Expected values and variances are

$$\begin{aligned} \mathcal{E}(S) &= \mathcal{E}(X_1) + \mathcal{E}(X_2) \\ \text{var}(S) &= \text{var}(X_1) + \text{var}(X_2)\,, \quad \text{if } X_1 \text{ and } X_2 \text{ are independent.} \end{aligned}$$

For the difference $D = X_2 - X_1$, we get $\mathcal{E}(D) = \mathcal{E}(X_2) - \mathcal{E}(X_1)$ and $\text{var}(D) = \text{var}(X_1) + \text{var}(X_2)$. (*Technical Remark*: This is shown by writing $D = X_2 + (-X_1)$ and using Sect. 6.2.4 f and the result for the sum.)
If both variables are normally distributed, so is the sum.
For a derivation of these and many results to follow, we refer the reader to more extensive texts and books on probability.

f **Means**. Calculating means of repeated measurements is often used when characterizing properties of concrete in order to reduce variability. We can now quantify the improvement of precision obtained by such averaging.
Applying these results repeatedly and combining it with the rule on a change in scale (Sect. 6.2.4 f), one gets the following fundamental results for the mean $\overline{X}$ of a sample $X_1, X_2, \ldots, X_n$ of observations X_i with expected value μ and standard deviation σ:

$$\mathcal{E}\left(\overline{X}\right) = \mu\,, \qquad \text{var}\left(\overline{X}\right) = \sigma^2/n.$$

This quantifies the intuitive insight that taking means of several values reduces the variability of the result. A mean over n values has a lower standard deviation than the individual values have, and the ratio is $\sqrt{n}$—**if the "values" are independent** random variables.
Again, if the distribution of the observations X_i is normal, $X_i \sim \mathcal{N}(\mu, \sigma^2)$, then so is the mean. According to the last formulae, specifically,

$$\overline{X} \sim \mathcal{N}(\mu, \sigma^2/n).$$

g **Expectation of the empirical variance**. Let us apply the basic result about the expectation also to the empirical variance. Note first that re-location of the x_i

does not change the variance, that is, the $z_i = x_i - a$ values have the same variance, independent of a. A straightforward calculation shows that $\sum_i (z_i - \bar{z})^2 = \sum_i z_i^2 - n\bar{z}^2$. Now choose $a = \mathcal{E}(X_i)$ to achieve $\mathcal{E}(Z_i) = 0$,

$$\mathcal{E}\left(\sum_i (Z_i - \overline{Z})^2\right) = \mathcal{E}\left(\sum_i Z_i^2\right) - n\,\mathcal{E}\left(\overline{Z}^2\right) = n\sigma^2 - \sigma^2 = (n-1)\sigma^2.$$

(Note that independence is not necessary for the first line in Sect. 6.2.5 e to hold.) This shows that the sum of squares of the deviations $X_i - \overline{X}$ needs to be divided by $n - 1$ rather than n to obtain an estimator for the variance with the desired expected value, cf. Sects. 6.2.2 c and 6.3.1 e.

It is instructive to realize that we have just calculated the "expectation of the variance", and we could also get a "variance of the variance". To sort things out, we need to be aware that the "variance" in the second place is the estimated variance, which is a random quantity with a probability distribution, whereas the first terms are properties of that distribution.

h **Simulation**. Obtaining the distribution of a function T of a random sample $X_1, X_2, \ldots, X_n$ is the fundamental task of probability theory. In some cases, this is straightforward, in others, it may be very difficult. It is helpful, for practical purposes and also for understanding the task better, to notice that there is a very general way of fulfilling the task by "brute-force" computing. It is based on the possibility to generate random numbers that correspond to the assumed model for the sample. For the simple random sample case that we assume here, this is an easy task, see textbooks for details.

The procedure runs as follows: Generate n such random numbers x_i^* and calculate $t^* = T(x_1^*, x_2^*, \ldots, x_n^*)$,. Repeat this r times to get r values $t_1^*, t_2^*, \ldots, t_r^*$. Then take the empirical distribution of these values as an approximation of the distribution of T. This empirical distribution is called the **simulated distribution**, and the procedure is called **statistical simulation**.

i **Simulating outlier rejection**. As an example, we come back to the idea that outliers should not have an undue influence on the estimated location (Sect. 6.2.2 d). A well-known principle to achieve this robustness is called outlier rejection. It consists of first flagging observations that appear too extreme as compared to the others and then to calculate the mean of the unflagged values. A popular rule flags the observations which have a standardized value $Z_i = (X_i - \overline{X})/\widehat{\mathrm{sd}}$ whose absolute value exceeds a threshold γ. As one might expect, if outliers need to be found, it is better to do the standardization using robust measures of location and spread, i.e. $\widetilde{Z}_i = (X_i - \widehat{\mathrm{med}})/(c \cdot \widehat{\mathrm{iqr}})$, where the median $\widehat{\mathrm{med}}$ and the interquartile range iqr are defined in Sect. 6.2.2 e, h and $c = 1.349$ is introduced to adjust the iqr to the standard deviation.

We are now interested in the distribution of the mean after outlier rejection. It is impossible to obtain a nice formula for it by means of probability calculations. A practical means for studying the distribution is therefore simulation. Figure 6.7 shows the results obtained for the mean after outlier rejection for sample size 10.

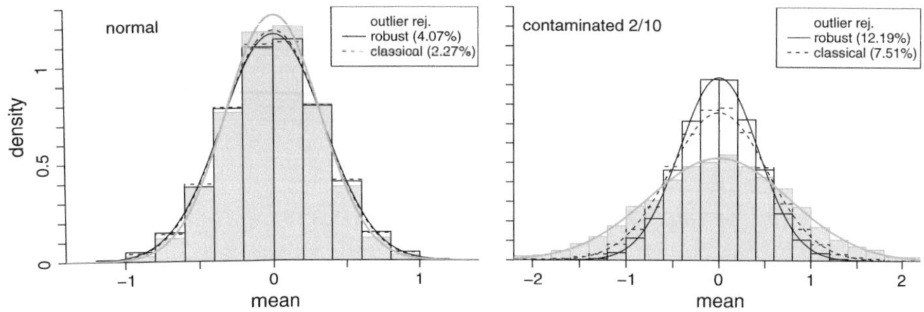

Fig. 6.7 Simulated distributions for means after outlier rejection. The sample of 10 observations follows a standard normal distribution for the *left panel*. On the *right*, 2 of the 10 observations are multiplied by 5. The shaded histogram displays the distribution of the ordinary mean. The distributions for the mean after outlier rejection based on classical and robust standardizations (using $\gamma = 2$ for both) are shown by *solid* and *dashed lines*, respectively. The mean percentage of rejected observations is indicated in the legends. Approximations of the distributions by normal density curves are also shown for all three histograms on each side. The number of simulation replicates is $r = 10,000$

The left panel reflects the situation where outlier rejection is not needed, whereas a model generating outliers is used for the right panel. Clearly, the outlier rejection rules show much narrower distributions for the latter situation than the plain mean. This illustrates a basic idea of robust statistics: By paying a small "premium" in the "ideal" case for the classical method, one achieves a high "protection" against outliers or other deviations from assumptions.

6.2.6 *Approximations by the Central Limit Theorem*

a In Sect. 6.2.5 we have given the results on the distribution of a mean of a random sample. The distribution was only determined if the distribution of the observations X_i was assumed to be normal—then, the result was also a normal distribution. If this was not assumed, we could at least give results for the expected value and the variance (or the standard deviation).

b **Law of Large Numbers**. The result on the variance of the mean, $\text{var}(\overline{X}) = \text{var}(X_i)/n$, tells us that the distribution of $\overline{X}$ becomes narrower and narrower, and in the limit $n \to \infty$, a single point remains, which is the expected value $\mathcal{E}(X_i)$. This result, which is intuitively clear, is called the "Law of Large Numbers". It is also valid for all other sensible "summaries" of samples, like the median, the standard deviation and the interquartile range.

c **Central Limit Theorem**. A more precise result about what happens if n gets large is called the Central Limit Theorem. It also covers more general functions of a random sample—and even cases of samples with dependencies. It says, loosely speaking:

Under suitable conditions, which are usually met in statistical applications, the distribution of any "decent" function $T(X_1, X_2, \ldots, X_n)$ is approximately normal when n is large enough. In a formula:

$$T(X_1, X_2, \ldots, X_n) \approx \sim \mathcal{N}\left(\mu_\infty, \sigma_\infty^2/n\right) ,$$

where μ_∞ is the value of T that would be obtained if one had an infinite number of observations and σ_∞^2 is called the **asymptotic variance** of T. Both values depend on the assumed distribution of the observations. We cannot discuss how to derive the asymptotic variance here. The idea is to linearize the function T around μ_∞ and apply the classical Central Limit Theorem to the linearized version. See [6, Ch. 5.3] for a short text on this topic.

d For practice, it is important to know that, when n is large enough, the approximation is good enough. There is no general answer to the question when n is large enough, even if "good enough" were specified in some way. It is necessary to study this issue in each application of the theorem. In practice, this can be done by using simulation (Sect. 6.2.5 h). For the mean of non-normal observations and the outlier rejection procedures studied in Sect. 6.2.5 i, the figure shows very good approximations already for a sample size of 10.

6.2.7 *Discrete Distributions*

a **Bernoulli trials**. The simplest possible distribution emerges when considering a binary random variable, the result of a "yes/no" type observation. Since only two values are possible, 0 and 1, say, the distribution is given by a single number, $P(X = 1)$. It is usually denoted by p or π. We prefer the latter notation, even though π is often reserved to be 3.14..., since Greek letters are generally used for parameters. Thus, the simplest distribution, called the Bernoulli distribution Bern(π) in honor of Jacob Bernoulli (1655–1705), is given by

$$P(X = 1) = \pi , \qquad P(X = 0) = 1 - \pi.$$

b **Binomial distribution**. If interested in such a probability of obtaining "yes," one needs to get a sample $X_1, X_2, \ldots, X_n$ of observations and then count the number of "successes" $S = \sum_{i=1}^n X_i$. If a simple random sample is obtained, the distribution of S, which we call X again for later reference, is given by the probabilities

$$P(X = k) = \binom{n}{k} \pi^k (1 - \pi)^{n-k} ,$$

where $\binom{n}{k} = n!/(k!(n-k)!)$ is called the **binomial coefficient**, and $n! = n \cdot (n-1) \cdot \ldots \cdot 2 \cdot 1$ is the n **factorial**. The distribution is called the binomial

distribution for its relation to the expansion of a binomial power, $(a+b)^n$. It is characterized by the "number of trials," n, and the "probability of success," π and is denoted as $X \sim \mathcal{B}(n, \pi)$.
Expected value and variance are

$$\varepsilon(X) = n\pi \,, \quad \text{var}(X) = n\pi(1-\pi).$$

Since X is a sum of independent X_i, with Bernoulli distribution, the central limit theorem shows that the binomial distribution can be approximated by the normal distribution $\mathcal{N}(n\pi, n\pi(1-\pi))$. A decent approximation is obtained if $n\pi > 9$ and $n(1-\pi) > 9$.

c **Simple quality control and conformity testing**. An important application throughout industry appears in quality control: A sample of n units from a production lot is examined, and the number of units which fail to meet a certain standard is counted. If this number exceeds a threshold (often 0 or 1), the whole lot is rejected.
This procedure, however, only helps in cases where failing units are allowed to have a non-negligible probability π: If the limit for the tolerable π is 2 %, then $n = 50$ is needed to make sure that one can expect one insufficient unit in the sample and thus has a substantial probability to get the desired indication of the problem.
Let us examine such rules in more detail. Let π_0 be the highest probability of failing units for which the lot is still of sufficient quality. Then, if n units are examined and the number of failing units is X, $X \sim \mathcal{B}(n, \pi_0)$ in the critical case. If the rule says that the lot is rejected as soon as any failing unit occurs in the sample, then the probability of rejection is $1-(1-\pi_0)^n$. Usually, this probability is required to be rather high, like 95 % or more. This means that $0.95 \geq 1-(1-\pi_0)^n$ and thus, taking logs and isolating n on one side, we obtain that n must exceed $\log(1-0.95)/\log(1-\pi_0)$. For $\pi_0 = 2\,\%$, we get $n > \log(0.05)/\log(0.98) = 148$—a very large sample for most practical applications. Therefore, the requirements are usually much less strict. If we assume π_0 of 10 % and $n = 80$ and set the rule that at most 4 failing units (5 %) are acceptable, the probability of accepting the lot is 8.8 %. See [5], Procedure B, for a more detailed description.
These two examples show how a balance can be reached between the risks of the user to accept an insufficient quality, the risk of the constructor to repair the construction even though the true π is below the limit π_0, and the effort needed for deciding acceptance. It is intuitively clear that the use of quantitative measurements instead of simple "yes/no" observations is more efficient and will therefore reduce the required number of samples. We will describe conformity tests based on quantitative data in Sects. 6.3.2 a and 6.3.3 i.

d **Example cover depth**. In the example introduced in Sect. 6.1 f, 80 measurements of the cover depth are made. The required depth shall be 40 mm. If the rule formulated last, with a limit of 4 failed units out of 80, is applied, then the construction will be accepted, since only 3 cover depth readings are below 40 mm.

6.3 Basic Statistical Inference

6.3.1 Estimation for Normal Samples

a **Parametric statistics**. In the last section, we have introduced distributions to describe our expectations about measurements or observations that are not yet made. The distributions came from a specified parametric family, that is, they are determined by a formula which contains one or two (or potentially more) constants, called parameters, that are left free and should be used to adjust the distribution to a given situation. Usually, the "situation" is specified by obtaining data under circumstances that will again be of interest in the future. The task of building the bridge between the parametric models and data, that is, using data to select parameters, is called "parametric statistics". It consists of choosing parameter values that appear "plausible" in the light of the data. We will first assume a normal distribution and will add general considerations in Sect. 6.3.5.

b **Estimator**. What is the most plausible value for the parameter μ of a normal distribution if we have a sample of n observations $x_1, x_2, \ldots, x_n$? The obvious choice is the arithmetic mean of the sample. We say that the mean is the **estimator** of the parameter μ, or the **estimate** to name the resulting value.
Apart from being the "obvious" choice, there are mathematical reasons for using the mean as the estimator of μ: The law of large numbers says that as we get more and more observations, that is, the sample size $n \to \infty$, the mean will tend to the expected value of the distribution, which is the desired parameter value μ. Note, however, that we could also use the (empirical) median $\widehat{\text{med}}(x_1, x_2, \ldots, x_n)$ to estimate μ, since μ is also the (theoretical) median of the normal distribution, and the law of large numbers also says that the empirical median converges to the theoretical one.
Is there a rationale to choose between the median and the mean? To answer this question, we need to recall that any estimator is a function of the random observations and is therefore itself a random variable.

c **Distribution of an estimator**. Since the estimator is a function of the observations, we can apply the results of Sect. 6.2.5 to derive its distribution—if we fix assumptions about the distribution of the observations.
Let us first assume that the random sample follows a normal distribution $\mathcal{N}(\mu, \sigma^2)$. Then the result Sect. 6.2.5 f tells us that the mean $\overline{X}$ has a normal distribution with expectation μ and variance σ^2/n, $\overline{X} \sim \mathcal{N}(\mu, \sigma^2/n)$.

d **Statistical Efficiency**. For the median, the distribution is less easily determined. A symmetry argument shows that its expected value is also μ. More advanced techniques show that its variance is approximately $\mathrm{var}\left(\widehat{\mathrm{med}}\right) \approx 1.57\ \sigma^2/n$. This shows that the median is less precise than the mean in the sense of having a larger variance. The inverse ratio of variances is called the (relative) **statistical efficiency** of the median versus the mean. It is approximately $1/1.57 = 0.637$. This calculation relies on the assumption that the observations are a random sample of a normal distribution. It can be shown theoretically that in this case, the arithmetic mean is the most efficient estimator of the expected value μ. Thus, the mean, which is the omnipresent summary of a batch of numbers, has a sound theoretical justification if the normal distribution is taken to be a good model for the random fluctuations. If outliers occur, this ideal is unrealistic, and outlier rejection (Sect. 6.2.5 i) is a pragmatic way out.

e **Estimation of the variance**. If asked to invent an estimator of the theoretical variance $\mathrm{var}(X)$ on the basis of a sample, a plausible suggestion would be $T = (1/n)\sum_i (X_i - \overline{X})^2$, i.e., the average of the squared deviations $(X_i - \overline{X})^2$. As has been shown in Sect. 6.2.5 g, the expected value of $\sum_i (X_i - X)^2$ is $(n-1)\mathrm{var}(X)$. Therefore, the expected value of T would be $\mathcal{E}(T) = \frac{n-1}{n}\mathrm{var}(X)$ and thus not equal to the quantity $\mathrm{var}(X)$ that it should estimate. (For large n, the difference is negligible.)

f **Bias, unbiasedness**. The difference between the expected value of an estimator T and the quantity θ that it should estimate is called the bias of the estimator,

$$\mathrm{bias}(T) = \mathcal{E}(T) - \theta.$$

For T defined above, we have a bias of $-\mathrm{var}(X)/n$. The empirical variance, with the denominator $n-1$ corrects this bias. It is unbiased—a desirable property for any estimator. Coming back to the mean and the median as estimators for μ, we find that both are unbiased.

g **Distribution of the empirical variance, chi-squared distribution**. The distribution of the empirical variance $\widehat{\mathrm{var}}$ shall not be derived here. The distribution of $(n-1)\widehat{\mathrm{var}}/\mathrm{var}(X)$ is called the chi-squared or, in symbols, χ^2 distribution. This family of distributions (Sect. 6.2.3 b) has a parameter called degrees of freedom ν. It takes natural numbers 1, 2, 3, … as values and equals its expected value. For $(n-1)\widehat{\mathrm{var}}$, the degrees of freedom are $n-1$.
Technical Remark: The chi-squared distribution with m degrees of freedom is defined as the distribution of the sum of m independent squares Z_i^2 of standard normal variables Z_i.

h **Robust estimation**. Mean and empirical standard deviation are the best estimators of the expected value and the theoretical standard deviation of normally distributed observations—but they characterize the distribution poorly if outliers are present. Figure 6.8 shows the data of the first team in Example 6.1.d for wall 2 with two normal densities. The solid line is fitted through mean and standard deviation. The dashed density is based on robust estimators of location and scale and is meant to summarize the ``bulk of the data" better by giving less weight to outliers.

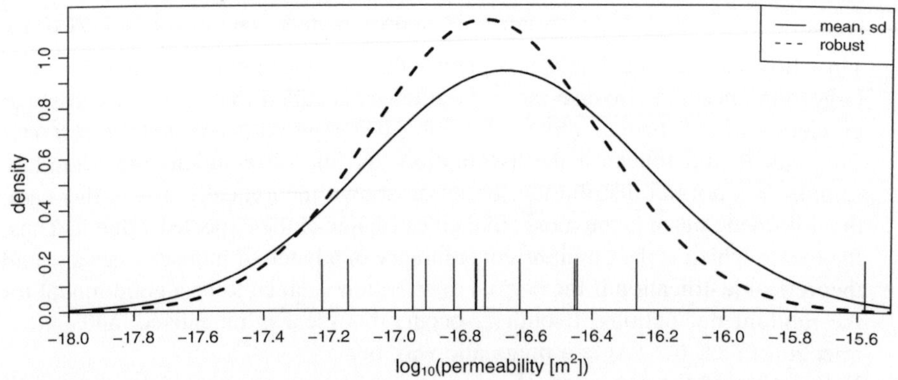

Fig. 6.8 Two normal densities fitted to 13 measurements of permeability. Since a histogram gives a crude picture of so few observations, the data is shown by the *vertical lines*

Technical Remark: The robust estimators are so-called M estimators of location and scale, as implemented in the function 'lmrob' of package 'robustbase' in the statistical software R.

6.3.2 Statistical Tests

a **Conformity testing**. Assume that the true expected value of the quantity X to be checked for conformity—like compressive strength or cover depth—is μ_0. The sample mean $\overline{X}$ of n measurements will never coincide exactly with μ_0, but will deviate from it as described in Sect. 6.2.5 f, $\overline{X} \sim \mathcal{N}(\mu_0, \sigma^2/n)$. Thus, the respective standardized variable $T = (\overline{X} - \mu_0)/(\sigma/\sqrt{n}) \sim \mathcal{N}(0, 1)$. (The standardization is only strictly possible if σ is known. Usually, it has to be estimated from the data. We come back to this point below, Sect. 6.3.2 k.) If we get a value of T that exceeds 2 (or is below -2), we will doubt that the assumed expected value μ_0 is the true expected value underlying the observations, since such extreme values would have little probability if this was the case.

Assume that the expected value μ of the measured quantity must be larger than μ_0 to meet conformity. (This threshold differs from the safety limit c by a suitable multiple of the standard deviation, $\mu_0 = c + q_{1-p}\sigma$, see Sect. 6.2.3 e.) If $T > 2$ or, equivalently, $\overline{X} > \mu_0 + 2\sigma/\sqrt{n}$, we will conclude that indeed $\mu > \mu_0$, and the lot can be accepted. In the opposite case, $\overline{X} < \mu_0 - 2\sigma/\sqrt{n}$, the lot is "certainly" bad. If $\overline{X}$ falls between these limits, it is common to reject the lot to be on the safe side.

Technical Remark: Alternatively, one might decide to get more measurements in the last case, possibly in as many steps as to get a clear indication in one direction or the other. Note, however, that the limits for such a sequential procedure should be adjusted.

b In order to fix a rule for deciding about conformity, we choose a probability α, usually 0.05, and determine the critical value γ for the value of T such that $P(T > \gamma) = \alpha$ if μ_0 is the expected value. The critical value is the $1-\alpha$ quantile (Sect. 6.2.2 g) of the standard normal distribution (Sect. 6.2.3 a), which is $\gamma = \Phi^{-1}(1-\alpha) = 1.645$. If $T < 1.645$, the "null hypothesis" $\mu \leq \mu_0$ appears **"plausible" in the light of the data**, otherwise, it should be **rejected**—and the lot is accepted, with a high confidence that the specification $\mu > \mu_0$ is fulfilled.

c **Example cover depth**. Assume a known value of $\sigma = 12$ mm and a minimum cover depth of $c = 40$ mm, which should be exceeded with $1-p = 90\,\%$ probability. This leads to a threshold $\mu_0 = 40 + 1.28 \cdot 12 = 55.4$ mm, using $\Phi^{-1}(0.9) = 1.28$. Since $\overline{X} = 59.3$, we obtain $T = (59.3 - 55.4)/(12/\sqrt{80}) = 2.91 > 1.64$ and conclude that the concrete conforms to the requirement. For $1-p = 95\,\%$, we get $\mu_0 = 59.7$, and this stronger requirement is not satisfied.

d **Identity testing, comparison of two samples**. For an **identity test**, a sample of size n_1 from the prequalified mix of concrete is compared with a sample of size n_2 from the concrete used on the construction site. We would like to prove that their expected values μ_1 and μ_2 are identical, or $\Delta = \mu_2 - \mu_1 = 0$. If this is true, the difference of means, $U = \overline{X}_2 - \overline{X}_1$ has the normal distribution $\mathcal{N}(0, \sigma_1^2/n_1 + \sigma_2^2/n_2)$ according to Sect. 6.2.5 e, f (where σ_1 and σ_2 are the theoretical standard deviations of the measurements in the two samples). Standardization leads to $T = U/\sqrt{\sigma_1^2/n_1 + \sigma_2^2/n_2} \sim \mathcal{N}(0,1)$.
In this situation, the null hypothesis $\Delta = 0$ arguably should be rejected if T is either extremely large or extremely small (strongly negative). This can be expressed as $|T| > \gamma$, and γ is again selected such that $P(|T| > \gamma) = \alpha = 0.05$. This leads to $\Phi(\gamma) = 1 - \alpha/2$ and $\gamma = 1.96$, see Sect. 6.2.3 a.
This type of rule is called a **two-sided test** in contrast to the previous, **one-sided** rule that the null hypothesis was only rejected for large positive deviations between the estimator and the hypothesized parameter value. (Admittedly, one can argue that there is no problem accepting concrete on the construction site that has better quality than the prequalified mix. Then, the situation is one-sided as in the conformity situation.)

e **Test "recipe" and notions**. Formalizing and generalizing these ideas leads to the notion of a statistical test. To construct a test, we follow these steps:

1. Choose a suitable **model** for the observations that contains the **parameter of interest**. (In the two cases above, we have assumed normal distributions for the measurements, and the parameter of interest was the expected value μ and the difference Δ, respectively.)
2. Choose the value(s) of the parameter of interest that shall be tested. It is called the **null hypothesis** ($\mu \geq \mu_0$ and $\Delta = 0$). All other values form the **alternative hypothesis**.

3. Choose a **test statistic** U that will typically take different values under the null than under the alternative hypothesis. Usually, this is an estimator of the parameter of interest ($\overline{X}$ and $\overline{X}_2 - \overline{X}_1$).
4. Derive the **distribution of the test statistic**, assuming that the null hypothesis is true. Often, this is done after a suitable standardization of the test statistic in order to obtain a distribution that does not depend on the parameters of the model. (Standardization leads to $T \sim \mathcal{N}(0, 1)$ in both cases.)
5. On the basis of this distribution, determine a critical value which is exceeded only with a predetermined probability of α ($\gamma = 1.645$ for T and $\gamma = 1.96$ for $|T|$). This probability is called the **level of the test**.

Note that all these steps do not involve any actual data. They lead to a rule. When data is available, the (standardized) test statistic T can be calculated and compared to the critical value. If it is exceeded, **the null hypothesis is rejected**, and this is called a **statistically significant** test result. Otherwise, it is **compatible with the data**, the **null hypothesis is not rejected**. The values for which this is the case are called the acceptance region for the test statistic, the others form the **rejection region.**

f **P value**. Rather than just ending with a yes/no alternative—as is the formal result of a hypothesis test—it is more informative to give a kind of measure of (non-) significance of the test result. Figure 6.9 shows the idea. For a given value t of the test statistic, the more densely shaded area is the probability that this value is exceeded,

$$p(t) = P(T > t) = 1 - F_T(t)$$

for a one-sided test, and the double of this number if both high and low values of T form the rejection region. The figure makes it clear that the p value is $\leq \alpha = 5\,\%$ if and only if the observed value of the test statistic is in the rejection region. Therefore, if the p value is known, the result of the test becomes clear by comparing it to the "universal critical value" of α (5 %)—but note that low p values lead to a significant result rather than high ones, as is usual for other test statistics.

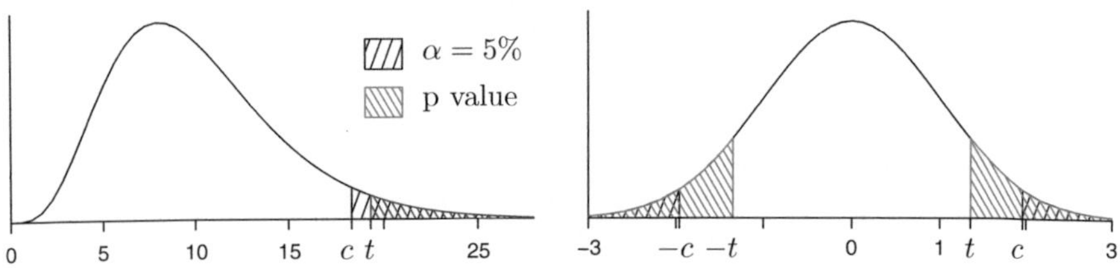

Fig. 6.9 Rejection region and p value for a one-sided (*left*) and a two-sided (*right*) test. In the *left panel*, the observed value t of the test statistic is larger than the critical value c, and the p value is $<\alpha$. The null hypothesis is rejected. On the *right*, it is the other way around, and the null hypothesis is not rejected

g **Remarks**. The statistical test is a rather delicate concept.

1. It is possible that we reject the null hypothesis even if it is true. This wrong conclusion is called the "**error of the first kind**." The construction of the rule is such that this happens with the chosen probability of α.
2. If the null hypothesis is not rejected, we cannot conclude that it is "true". It is easy to see that a small, but non-zero difference of expected values would also lead to small values of the test statistic and therefore, with high probability, to failing to reject the (wrong) null hypothesis. This "wrong" conclusion is called the "**error of the second kind**". Note that saying that the null hypothesis is "**accepted**" instead of "not rejected" is too easily misinterpreted as "proven," and should therefore be avoided.
3. Because we cannot "prove" a specific null hypothesis like $\Delta = 0$, the statistical test is mainly used for a **contradiction argument**: We believe that a difference between the groups indeed exists, and apply the test for zero difference with a hope to reject it; this would "prove" our belief.
   For conformity testing, we have also used this wording in Sect. 6.2.3 a: Since we want to prove that security is warrantied, we call the case that the sampled concrete is bad the null hypothesis, which we hope to reject with the sample values.

h **Power**. Instead of choosing the difference of means as the test statistic, we could take the difference of medians of the two samples. Intuition says that using the more precise estimator is advantageous. This is formalized by considering the "error of the second kind" introduced in remark 2 above. Its probability can be calculated if a specific alternative hypothesis (a specific value for the difference of expectations Δ) is fixed. The probability of not committing this error, i.e., of correctly rejecting the null hypothesis, is called the power of the test. If it is calculated as a function of the "effect" Δ, we obtain the power curve. Figure 6.10 shows the power curves for the four tests to be discussed in the following paragraphs.

i **The testing paradox**. When the sample size n increases, the distribution of the parameter's estimator is concentrated more and more around the true value. This makes clear that the power of any reasonable test and any fixed alternative hypothesis tends to one—the test will eventually reject the null hypothesis certainly.

This is good news. On the other hand, it leads to a paradox for tests of null hypotheses that specify the parameter to a single value, typically 0, as done for the two-sample comparison above. When we have reason to examine such a null hypothesis, it would be very unlikely that the respective null hypothesis be *exactly* true. Therefore, **the test result will only tell us if we have used a large enough sample and/or a powerful enough test** to obtain a statistically significant effect.

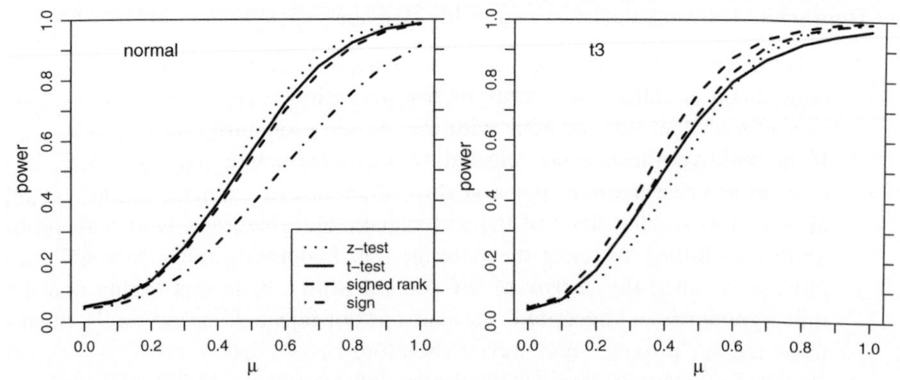

Fig. 6.10 Power of 4 tests for $\mu = 0$ as a function of the true parameter μ, for sample size $n = 20$ and standard normal observations (*left*) and t_3 distributed ones (scaled to standard deviation 1 before shifting; *right*)

Thus, the concept of a statistical test **should not be applied in practice if it can be avoided! Its important role is that it provides the basis for defining the notion of confidence interval** to be discussed in the next subsection.
The paradox does not occur for composite null hypotheses, which are relevant for conformity testing. Nevertheless, confidence intervals should also be preferred to tests in this context, as they provide more complete information about the parameter than significance tests do.
For this reason, we postpone practical examples to the next subsection.

j **One-sample z-test**. The following test is fundamental, as it applies to the simplest situation in some sense and is used as an approximation in very general contexts. We have already met it above in Sect. 6.2.3 a. We follow the recipe Sect. 6.2.3 e to describe it.

1. Assume that we have a single sample of n observations, which we assume to follow a normal distribution, $X_i \sim \mathcal{N}(\mu, \sigma^2)$.
2. Null hypothesis $\mu = \mu_0$ (often $\mu_0 = 0$).
3. The test statistic is the deviation of the estimated parameter from the hypothesized expected value, $U = \overline{X} - \mu_0$.
4. Standardization: Assume that σ is known. This corresponds to the situation of measurements in routine setups: We may be interested if the "location" μ has shifted in a production line, assuming that the precision, expressed by σ, is the "usual one". Then, we can standardize U as $T = U/\mathrm{sd}_{\overline{X}}$, where $\mathrm{sd}_{\overline{X}} = \sigma/\sqrt{n}$. We have $T \sim \mathcal{N}(0, 1)$.
5. The critical value is, as before, $c = 1.96$. The null hypothesis is rejected if $|T| > c$.

k **One-sample t-test**. Usually, σ is not known in advance. Then, the obvious modification of the z-test is to use the estimated value $\widehat{\sigma}$ instead of σ in the standardization step, obtaining

$$T = \overline{X}/\mathrm{se}_{\overline{X}}\,, \quad \text{where} \quad \mathrm{se}_{\overline{X}} = \widehat{\sigma}/\sqrt{n}$$

is called the **standard error** of $\overline{X}$. This changes, of course, the distribution of T by making it a bit "more random," that is, increasing its variability. The extent of the change depends on the sample size n that governs the precision of the estimated standard deviation, $\widehat{\sigma}$—more precisely, the degrees of freedom of the distribution of $\widehat{\sigma}^2$, see Sect. 6.3.1 g. The result is called Student's t distribution with parameter $\nu = n - 1$. The critical value is, as in the normal case, the $1 - \alpha/2 = 0.975$ quantile of this distribution, denoted as $q(0.975; t_{n-1})$. The test is called the t-test.

l **Signed rank test**. The derivation of the test statistics' distributions roots in the assumptions on the distribution of the observations $X_1, X_2, \ldots, X_n$ (the normal distribution). It is of course desirable that this assumption is as weak as possible. In fact, there is a trick to make it very weak, if the test statistic is based on ranks of the observations. Here is the specification of "Wilcoxon's signed rank test:"

1. Model: A simple random sample $X_1, X_2, \ldots, X_n$ of observations from any symmetric distribution, with symmetry center μ.
2. Null hypothesis: $\mu = \mu_0$ (often $\mu_0 = 0$).
3. Test statistic: Drop the X_i that are $= \mu_0$. Let R_i be the rank of $|X_i - \mu_0|$. Then, $U = \sum_{i|X_i > \mu} R_i$.
4. Standardization: The expected value of U is half the sum of all ranks, $\mathcal{E}(U) = n(n+1)/4$ for symmetry reasons. The variance $\mathrm{var}(U)$ is more difficult to obtain. It only depends on n and the tied ranks which appear in the sample. (* That is, the conditional distribution, given the pattern of ties, is considered.) If there are no ties, $\mathrm{var}(U) = n(n+1)(2n+1)/6$. Otherwise, see textbooks. The standardized test statistic is, as usual, $T = (U - \mathcal{E}(U))/\sqrt{\mathrm{var}(U)}$.
5. The distribution of T under H_0 is approximately standard normal, and therefore, rejection occurs if $|T| > 1.96$. Exact calculations of the distribution are possible and should be used for $n < 10$.

m **Sign test**. A simple test, that does not even need the assumption of symmetry, is the sign test. It tests a hypothesis about the median of the model distribution and relies simply on counting the number of observations that exceeds the hypothesized median:

1. Model: The X_i's have a distribution with median μ.
2. Null hypothesis: $\mu = \mu_0$.
3. Test statistic: Exclude observations with $X_i = \mu_0$. Then $U = \#\{i|X_i > \mu_0\}$.
4. Standardization: Clearly, $U \sim \mathcal{B}(n, \pi = 1/2)$ under H_0. Thus, standardization is only needed if the approximation of the binomial distribution by the normal is to be used. According to Sect. 6.2.7 b, $T = (U - n/2)/\sqrt{n/4} = \sqrt{n}(2U/n - 1)$.

5. The critical value for $|T|$ is approximated by 1.96 as usual, for $|U - n/2|$, it is obtained from the binomial distribution. Note that it is usually impossible to achieve that $P(|U - n/2| > c) = 0.05$ precisely because U is discrete, and therefore, $P(|U - n/2| > c)$ jumps at the possible values of U.

n **Power of one-sample tests**. The power (Sect. 6.3.2 h) of the four tests just discussed is shown in Fig. 6.10 for $n = 20$. In the left panel, a normal (standard) distribution is assumed. On the right, we use a t-distribution with 3 degrees of freedom. Although this distribution has been introduced as characterizing a test statistic, it may also serve to model original observations. It has "fatter" or "longer tails" than the normal, thus giving extreme observations a higher probability, which often appears adequate in practical applications.

The curves show that under the normal distribution, the z-test is the winner. This is not surprising, as it is adapted to the normal model and in addition uses the assumption that the variance is known. Estimation of the parameter σ costs some power. The t-test is only slightly better than the signed rank test. The sign test, based on minimal assupmtions, is clearly worse. For the t_3 distribution, the signed rank test outperforms the others. For sample size 10, the curves (not displayed) show that the signed rank test becomes almost equivalent to the t-test for both distributions. For $n<6$, neither the signed rank nor the sign test can become significant, since the most extreme case—all observations on one side of μ_0—has probability > 5 % under the null hypothesis.

Summarizing, since extreme observations, perceived as outliers, are usually a realistic feature, these arguments suggest that **the signed rank test should be applied** for testing a hypothesis about the "location" μ for sample sizes larger than a dozen. For smaller samples, the t-test is appropriate.

o **Two samples**. Let us turn to the situation that two conditions are compared, that is, we are interested in the **difference** of a property of concrete between two compositions, concentrations of an additive, locations in the construction, or the like.

For later generalizations, we change the wording and notation slightly, saying that we compare a **target variable** Y between two groups, $g = 0$ and $g = 1$. Assume that a simple random sample has been obtained for each group, denoted as $Y_{01}, Y_{02}, \ldots Y_{0n_0}$ and $Y_{11}, Y_{12}, \ldots Y_{1n_1}$. The simplest model assumes that the Y's have a normal distribution with possibly different expected values, but the same variance,

$$Y_{gi} \sim \mathcal{N}(\mu_g, \sigma^2) , \quad g = 0, 1 , \quad i = 0, 1, \ldots, n_g .$$

The obvious testing problem checks the null hypothesis that the difference $\Delta = \mu_1 - \mu_0$ equals a given value Δ_0. Usually $\Delta_0 = 0$, expressing that there is no difference, and the two groups have exactly the same distribution.

p **Two sample z- and t-test**. The two sample test for known (possibly different) variance(s) of the two samples (z test) has been developed in Sect. 6.3.2 d. Let us present the case for unknown, but equal variances σ^2 in the present notation here.

The raw test statistic is $U = \overline{Y}_{1.} - \overline{Y}_{0.}$. We also need to estimate the variance σ^2. This is done by

$$\widehat{\sigma}^2 = \frac{1}{n_0 + n_1 - 2} \sum_g \sum_i (Y_{gi} - \overline{Y}_{g.})^2.$$

The estimated variance of U is then

$$\widehat{\mathrm{var}}(U) = \mathrm{se}_\Delta^2 = \widehat{\sigma}^2 (1/n_0 + 1/n_1).$$

Then, $T = U/\mathrm{se}_\Delta$ has a t distribution with $n_0 + n_1 - 2$ degrees of freedom under the null hypothesis. A modification is needed if equal variances for the two groups are not assumed.

q **Mann-Whitney test**. There is also a test based on ranks, called **rank sum test** and named after Mann and Whitney, and sometimes also after Wilcoxon. It works with a general distribution for the observations and simply tests if both groups have the same distribution. The test statistic is obtained from ranking all values Y_{gi} together, obtaining ranks R_{gi}. If group 1 tends to show higher values of Y, then the large ranks will belong to observations of group 1. Therefore, the test statistic sums the ranks of group 1, $U_1 = \sum_i R_{1i}$. The distribution of U_1 can be calculated precisely for small n_0 and n_1 and can be standardized and approximated by the standard normal distribution for larger samples.

r **Goodness of fit**. The basic idea of a statistical test can also be used for null hypotheses that are of a more general nature than fixing a parameter value in a parametric family of distributions. For example, the null hypothesis can be that the data follows any normal distribution (or another fixed shape of distribution). This formalizes the comparison of an empirical distribution with the supposed theoretical one that we mentioned in Sect. 6.2.1 d.

As a test statistic, we need a measure of discrepancy between the empirical and theoretical cumulative distribution functions. Alternatively, the test statistic measures the difference between a kind of histogram with sensibly selected bins and the density curve. This is done by the popular "Chi-squared goodness of fit test".

We avoid details here, not only due to the lack of space, but also because these tests are not very useful. The idea behind applying them is to prove that assumptions are fulfilled. We have made it clear above (Remark 2 in Sect. 6.3.2 g) that a null hypothesis can never be proved. Failure to reject it may simply be a consequence of having a small sample or applying a test with little power (cf. Sect. 6.3.2 i). Assumptions should therefore be avoided if possible without too big losses in precision, or checked informally by graphical means.

6.3.3 Confidence Intervals

a **Definition**. When we compare two groups, it is rarely plausible to assume that they do not differ *at all* in their expected values. Therefore, even if the test does not reject the null hypothesis, it is not reasonable to believe in a zero difference —and it is not justified to declare the null hypothesis to be "proven," as we have seen before (Remark 2 in Sect. 6.3.2 g).

Rather than examining if a *specific* value of the parameter is plausible in the sense that it is not rejected by the test, we can ask for **all values of the parameter that are plausible**. They form the confidence interval.

By this definition, every test for a parameter leads to a confidence interval for this parameter.

b **z- and t-interval**. The generic example of a confidence interval is based on the one-sample z-test (Sect. 6.3.2 j). The testing rule says that μ is accepted as a plausible expected value if $|\overline{X} - \mu| < 1.96\,\mathrm{sd}_{\overline{X}}$, where $\mathrm{sd}_{\overline{X}} = \sigma/\sqrt{n}$. Solving this inequality for μ leads to an interval of plausible values with the bounds $\overline{X} - 1.96\mathrm{sd}_{\overline{X}}$ and $\overline{X} + 1.96\,\mathrm{sd}_{\overline{X}}$. This determines the confidence interval, often denoted as $\overline{X} \pm 1.96\,\mathrm{sd}_{\overline{X}}$. If σ is unknown and therefore estimated from the data, the t-test yields in exactly the same way the "t confidence interval"

$$\overline{X} \pm q(0.975; t_{n-1})\,\mathrm{se}_{\overline{X}}\,, \qquad \mathrm{se}_{\overline{X}} = \widehat{\sigma}/\sqrt{n}.$$

c **Confidence interval for the signed rank test**. Based on the principle of collecting all parameter values that are "compatible" with the data in the sense of a specified test, it is possible to derive a confidence interval from the signed rank test. Since the latter applies under a very general assumption and is still excellent for the cherished normal model, the use of this confidence interval is highly recommended. The details are omitted here. Sound programs for the signed rank test also provide this confidence interval.

d **Example**. Figure 6.11 shows, in the upper part, the permeability measurements (Sect. 6.1 e) of one team for the two walls, with confidence intervals corresponding to the t- and the signed rank test for each sample. For wall 2, there is an outlier, which causes the two intervals to differ somewhat.

e **Interpretation**. The idea of a confidence interval is that we can be "almost sure" that the true value of the parameter is within its limits. By its construction, the probability for this to happen is $1 - \alpha = 95\,\%$, which is called the **confidence level**.

It is recommended to digest this statement thoroughly. The statement "The parameter is contained in the interval $[a, b]$" may suggest that the parameter should be a random quantity. In the concepts presented in this article, the parameter always was a fixed number, even though it was taken as unknown in this section. The solution to this "paradox" is that the bounds of the confidence interval are random variables as they are obtained from the random data. Thus, the statement should be more clearly formulated as "The confidence interval

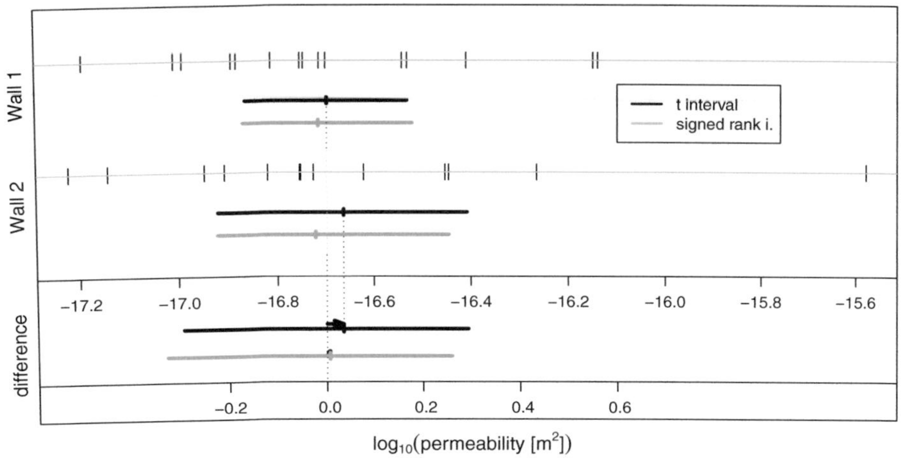

Fig. 6.11 Measurements of permeability for two sites with t (*black*) and signed rank (*gray*) confidence intervals. For the difference between the two walls, the estimates and the two confidence intervals are displayed in the lower part

contains the parameter with probability of 95 %." This statement is true since fixing any value θ_0 for the parameter (which in turn determines probabilities) we have: The confidence interval contains the parameter exactly if the null hypothesis $\theta = \theta_0$ is not rejected. By construction of the confidence interval, this has the probability $1 - \alpha$.

f **Length of the confidence interval**. When we want a confidence interval for the expected value μ of normally distributed data with unknown σ, we may use the t-interval (Sect. 6.3.3 b) or the interval based on the signed rank test (Sect. 6.3.3 c). It is intuitively clear that we should choose the interval based on the test with the largest power (Sect. 6.3.2 h). This will lead to intervals that are generally shorter—more precisely, they may be shorter or longer for a given data set, but their expected length will be smaller for the interval based on the more powerful test.

g **Two groups and confidence intervals**. When comparing two groups, it is tempting to use a confidence interval for the expected value of each group, and then conclude that the two groups are significantly different if the two intervals do not overlap, and not significantly different otherwise.

Note that this is not a valid test. While the first conclusion is correct, the second is not: The two confidence intervals can overlap even if the null hypothesis of no difference should be rejected. Correct procedures for testing whether two groups differ have been discussed above (Sect. 6.3.2 o and following paragraphs). They lead to confidence intervals for the difference Δ of the expected values (or other location parameters) of the two groups, which have a clear interpretation.

The lower part of Fig. 6.11 shows the two confidence intervals for the difference between the two walls in the example. The estimated difference is tiny, indeed.

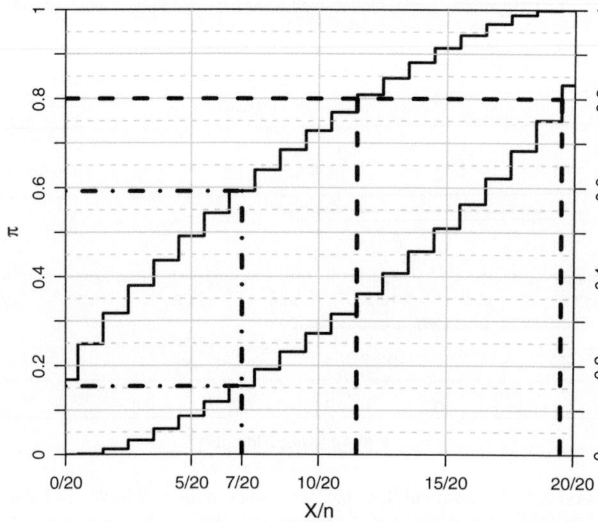

Fig. 6.12 Nomogram for tests and confidence intervals for the binomial distribution with $n = 20$. The *dashed lines* illustrate how to get the acceptance region for $\pi = 0.8$. The *dashed-dotted lines* lead to the confidence interval for $X/n = 7/20$. The stepping nature of the limits of "compatible values" of π and proportions X/n reflects the fact that X can only take 21 different values

h **Identity testing**. Even though we have not shown an example in which a sample of concrete characteristics on the construction site was compared with a sample of prequalified concrete, it is clear how a two-sample inference applies to this generic case. As argued above, for any two-sample problem, identity testing should be presented by giving a confidence interval for the difference in expected values (or other location parameters), and possibly a more detailed comparison of the distributions.

i **Conformity testing with unknown** σ. In the introduction to testing in Sect. 6.3.2 a, conformity testing was discussed for the case of known standard deviation σ. The one-sample t-test provides the version for testing the conformity with a required minimal expected value μ_0 of the relevant measure X for the case of unknown σ. However, μ_0 is usually itself determined by a lower threshold c and the standard deviation as $c + q_{1-p}\sigma$, see Sect. 6.2.3 e. Writing this requirement as $\theta = \mu - q_{1-p}\sigma > c$ suggests that θ should be estimated and the respective lower confidence limit should be obtained. The latter has the form $\widehat{\mu} - k\widehat{\sigma}$, where k is a complicated function of n, p and α, see Table 9.1 and Appendix A 9.2 of [5]. For $p = 5\,\%$ and $\alpha = 5\,\%$, $k = 2.91,\ 2.40,\ 2.06,\ 1.98$ and 1.73 for $n = 10,\ 20,\ 50,\ 80$ and 1000, respectively.

j **Example cover depth**. With the data of the cover depth example, we obtain $\widehat{\mu} = 59.3$, $\widehat{\sigma} = 11.1$, an estimated θ of $59.3 - 1.28 \cdot 11.1 = 45.1$ mm and a lower tolerance limit of $59.3 - 1.98 \cdot 11.1 = 37.3$ mm. Whereas the estimator suggests that the requirement of 40 mm cover depth is fulfilled, the tolerance limit shows that this cannot be concluded with the desired confidence level.

6.3.4 *Inference for the Binomial Distribution*

a The rules for conformity discussed in Sect. 6.2.7 c are in fact a form of statistical test for a hypothesis about the parameter π of a binomial distribution. Such rules can be formulated also if a qualitative event defines failure, like the occurrence of spallings. The sign test (Sect. 6.3.2 m) examines if $\pi = 0.5$ for a binomial distribution.
In these situations, we want to draw inference about a proportion of "trials" in which a certain "event" occurs. Let us call the event a "success"—even though it often means the contrary in practice.

b **Model**. The model in these situations assumes that the "trials" i have a common probability π of a success, and that this occurs independently. Then, the number X of successes among a number n of trials has a binomial distribution $\mathcal{B}(n, \pi)$, see Sect. 6.2.7 b.
The situation is much simpler than with the sample of normally distributed observations since there is only one number X on which inference about the parameter π can be based.

c **Estimator**. The only sensible way to estimate the probability π of a success is the proportion of successes $\widehat{\pi} = X/n$.

d **Test**. The test can only concern the parameter π. For testing the null hypothesis $\pi = \pi_0$, the natural test statistic is $\widehat{\pi}$. Choosing X itself is somewhat more convenient in this case, and of course leads to the same result. The distribution under the null hypothesis is the binomial $\mathcal{B}(n, \pi_0)$. We need to determine a range with probability $1 - \alpha = 95\,\%$ of most plausible values of X under this distribution.
Technical Remark: There is a complication that applies to the binomial and other discrete distributions. Since there are only $n + 1$ possible outcomes, the desired probability cannot be achieved precisely for most combinations of n and π. The convention says that one should then choose a range that has probability $> 1 - \alpha$. In addition, if unplausible values on both sides should be flagged, the range is selected such that the probabilities on both sides are as equal as possible. The resulting ranges for $n = 20$ and all possible values π_0 are shown graphically in the "nomogram" shown in Fig. 6.12. The proportion $\widehat{\pi}$ is used on the horizontal axis instead of X to make the correspondence with the parameter even more direct.

e **Confidence interval**. The confidence interval for a given number X collects the π values that are not rejected by the test if used as a null hypothesis π_0. The nomogram allows it to be read off immediately for any given X/n.
It is fundamental to note that in this case, the confidence intervals are **not symmetric** around the estimated value—and neither are the acceptance regions, the ranges of plausible values mentioned above, symmetric around the parameter π_0.

f **Example Cover Depth**. In the Example 6.1f, with 3 out of 80 measurements below the specified threshold of 40 mm of cover depth, a two-sided confidence interval for the true probability of insufficient depth turns out to reach from 0.781 % to 10.6 %. This can be verified by calculating $P(X \geq 3)$ for $X \sim \mathcal{B}(80, 0.00781)$ and $P(X \leq 3)$ for $X \sim \mathcal{B}(80, 0.106)$, which indeed both result in 0.025. In connection with conformity testing (Sect. 6.2.7 d), a "one-sided confidence interval" may be calculated, ranging up to 9.41 % for a confidence level of 95 %.

6.3.5 Inference for General Parameters

a In this subsection, we summarize the foregoing arguments and introduce some more general considerations and concepts.

b **The three basic questions of parametric statistics**. In the last three subsections, we have considered the situation where the observations are assumed to follow a parametric model with a parameter θ on which we want to draw inference. We have asked three questions leading to three fundamental notions.

1. Q: What is the most plausible value of the parameter θ in the light of the data?

   A: Estimator $\widehat{\theta}$.
2. Q: Is a certain value θ_0 plausible in the light of the data?

   A: Test, based on an estimator $\widehat{\theta}$ as a test statistic.
3. Q: Which values of θ are plausible in the light of the data?

   A: Confidence interval, based on the test, collecting all θ that are plausible in the sense for Question 2.

c **Maximum likelihood**. We have discussed methods for the most simple situations where inference is requested for the expected value of a sample with a symmetric distribution, or the difference between two such samples. For other situations, specific methods can be found in textbooks, and some will be given in Sect. 6.4.

There is a general principle that produces useful methods for almost any parametric model, called Maximum Likelihood. Assume that the model has a density $f_\theta(x)$. High densities are an indication that an observation x and a parameter value θ "fit well". The estimator is therefore based on considering the (joint) density for given observations as a function of θ and maximizing it. Based on a general version of the Central Limit Theorem (Sect. 6.2.6 c), probability theory finds approximate (normal) distributions for this estimator. Based on this result, a version of the z-test (Sect. 6.3.2 j) and the respective confidence interval (Sect. 6.3.3 b) can be used.

d **Nonparametric and semi-parametric methods**. The validity of the methods, i.e., the correctness of the probability statements obtained for tests and confidence intervals, such as the level α, the confidence level $1 - \alpha$, as well as quality measures, such as the efficiency of estimators, the power of tests or the expected length of confidence intervals, depend on the adequacy of the model for the data. It is therefore recommended to choose as general a model as possible. This has been achieved for the one-sample situation by the signed rank test (Sect. 6.3.2 l) and the respective confidence interval (Sect. 6.3.3 c), since a normal distribution was not required there, but the symmetry of the distribution was a sufficient assumption. The idea of leaving the distribution of the observations unspecified, except possibly for a symmetry assumption needed to determine the parameter of interest, leads to so-called nonparametric procedures, or, in the context of regression models, to semi-parametric procedures. Note that in spite of these names, the idea is to draw inference about a parameter, e.g., the location parameter characterizing a sample.

e **Outliers**. Extreme observations that are more distant from the bulk of the data than would be expected under the normal distribution are quite frequent in many practical applications. Sometimes, these outliers can be attributed to gross errors, i.e., malfunctions of a measurement device or wrongly written numbers, sometimes they may be attributed to somewhat different circumstances, and sometimes there is no explanation. They may go unnoticed when the data is analyzed by automated procedures without graphical inspection. Unless gross errors can be identified and corrected or dropped, they will have an undesired, large effect on the result of the methods based on the assumption of a normal distribution.

f **Outlier rejection**. One way to deal with outliers is to apply rules that decide when they should be "rejected" in the sense of being dropped as if they had never occurred, before applying the methods based on the normal distribution. Such rules are usually based on the idea of a statistical test for the null hypothesis that even the extreme observation(s) come(s) from a normal distribution (cf. Sect. 6.3.2 r). Two simple rules have been introduced in Sect. 6.2.5 i. The threshold γ for such outlier rejection rules can be determined as the appropriate critical value from the distribution of the maximum of a sample of size n of normally distributed random variables.

A problem with these rules is that the methods applied after their application are affected by the "cleansing" of the data. Most prominently, the standard deviation estimated from the remaining data is smaller (its distribution has a smaller expected value). Unless this bias is corrected, this leads to confidence intervals that are generally too short and cover the true parameter with a smaller than the assumed probability.

g **Robust statistics**. Such considerations have called for modifying the classical procedures in more refined ways that still lead to valid procedures. A fruitful idea to create such methods is to replace the "hard rejection" of outliers by a gradual downweighting, combined with the necessary bias corrections for the estimation of the scale parameter. Another way of thinking about the problem is to assume a distribution that describes the occurrence of outliers—a so-called **long-tailed** or **heavy-tailed distribution,** and applying the principle of

maximum likelihood.(Sect. 6.3.5 c). Both ideas lead to the same class of methods, called **M-estimators**.
The general paradigm of considering the effect of outliers and other deviations from model assumptions and developing methods that are adequate under such situations is called robust statistics.

h **Approximations**. Justifications for statistical methods are based on determining the distribution of estimators or test statistics, given a distribution of the observations. In the case of the normal distribution, this can be achieved by basic rules of probability theory. In most other cases, the precise distribution of the estimator or test statistic would be difficult to obtain in any practically useful way. Nevertheless, there are rather general principles that lead to approximations that are good enough in most cases.

- **Asymptotic approximation**. The first and usually simplest approximation relies on the Central Limit Theorem, which says that $\widehat{\theta} \approx \sim \mathcal{N}(\mu_\infty, \sigma_\infty^2/n)$, see Sect. 6.2.6 c.
- **Simulation**. Simulation has been discussed in Sect. 6.2.5 h. It needs known parameters for the distribution of the observations. When parameters are replaced by their estimated values, the procedure is often called the **parametric bootstrap**.
- **Bootstrap**. These two methods are based on an assumption about the distribution of the observations X_i. What if we "estimate" not only the parameters, but also the true distribution from the data? We have seen that the empirical distribution approaches the true distribution for a growing number n of observations. This leads to the idea of plugging the empirical distribution (Sect. 6.2.1 e) into the place of the model distribution and then applying simulation. Since this leaves the distribution open like the "nonparametric" methods do, the procedure is called the **nonparametric bootstrap**. If unspecified, the notion "bootstrap" means its nonparametric version.
  Due to lack of space, we cannot go deeper into this topic. In some software packages, procedures are available for practical application.

6.4 Analysis of Variance and Regression

6.4.1 Analysis of Variance

a **Several groups**. The comparison of two groups of observations discussed in Sect. 6.3.2 o calls for an extension to more than two. For example, we may want to compare 3 different composite cements with varying supplementary cementing material (SCM) types and amounts and a control with a pure ordinary portland cement (OPC) with respect to their effect on the chloride resistance.

A natural generalization of the simplest model for two groups of observations is to assume k normal distributions with different expected values μ_g but equal variances σ^2, $Y_{gi} \sim \mathcal{N}(\mu_g, \sigma^2)$, $g = 1, 2, \ldots, k$; $i = 1, 2, \ldots, n_g$. A different way of writing the same model splits the observations into a "structural part" $\mu_g = \mu + \alpha_g$ and a "**random deviation**" or "random error" E_{gi} which has expected value 0 and the same distribution for all observations,

$$Y_{gi} = \mu + \alpha_g + E_{gi} \ , \qquad E_{gi} \sim \mathcal{N}(0, \sigma^2) \ ,$$

where the parameters α_g are conceived as deviations of the expected values μ_g from a kind of "overall expectation" μ. While this notation appears unnecessary for describing k groups, it is suitable for generalizing the model later.

b **Identifyability**. Note that this model is "**over-parametrized**": If any number c is added to μ and subtracted from each α_g, then the distribution of the observations Y_{gi} remains exactly the same. Therefore, the parameters are not **identifiable** on the basis of even an infinite number of observations. In order to retrieve identifiability, we need to add a side condition that defines the meaning of the "overall expectation" μ, the most natural one being

$$\sum\nolimits_g \alpha_g = 0$$

c **Pairwise tests and contrasts**. A straightforward way of dealing with k groups is to compare any suitable pair of groups by a two-sample test—or better, to calculate an estimator with confidence interval for the difference within each pair.
There is a nice graphical way to display this kind of analysis, called the **notched box plot**, see Fig. 6.13. The notches are constructed in such a way that the following rule applies: If the ranges defined by two notches do not overlap, there is a statistically significant difference between the respective groups. (Note that the notches do not delimit confidence intervals, since such intervals can slightly overlap even if the difference is significant, compare Sect. 6.3.3 g.)
Technical Remark: The elements of the box plot are defined as follows: The box covers the inner 50 % of the data, ranging from the first quartile to the third (Sect. 6.2.2 h). The line cutting it into two parts represents the median. The dashed lines on both sides are delimited by the "whiskers," which are defined as follows: Find the point that is 1.5 interquartile ranges (IQR) away from the upper quartile, and go back to the largest observation smaller than this value. This is the upper whisker. The lower whisker is found in the same way.

d **Multiple tests**. The problem with such an analysis is a version of a general issue when many tests are evaluated: Assume that we simulate 7 groups of observations, using the same distribution for all of them. This ensures that the null hypothesis is fulfilled for all pairwise comparisons. When testing all of them, we calculate $7 \cdot 6/2 = 21$ test results. If we test with the usual level $\alpha = 0.05$, we

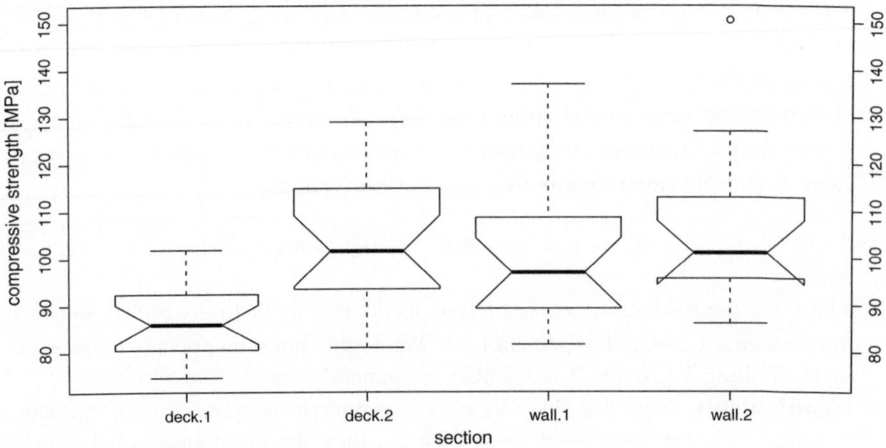

Fig. 6.13 Box plots of the compressive strength in layer 3 for the various "sections" in the tunnel example (Sect. 6.1 d). The "notches" allow for an approximate pairwise test according to the rule: Differences are statistically significant if notches do not overlap

should expect $21 \cdot 0.05 = 1.05$ formally significant results among the 21—even though we have made sure that there is no "true" difference between any groups. The problem always occurs when many tests are calculated for the same dataset. It also lurks when not all the tests are explicitly calculated, but the data is informally scrutinized for salient patterns, which are subsequently formally tested in a suitable way.

A formal approach to the problem is to ask for procedures that guarantee a bound on the "global type 1 error" probability, the probability that at least one of the m tests becomes significant even though all null hypotheses are valid. A general procedure that guarantees this probability to be $\leq \alpha$ is to use a level of α/m for each test. The rule is due to **Bonferroni**. For more specific situations, as the k groups model, there are more powerful procedures than Bonferroni's, as follows.

e **Test for no effect**. One way to avoid the multiplicity problem in the k groups model is to collect the null hypotheses $\mu_g = \mu_{g'}$ into a single one, $\mu_g = \mu$ or $\alpha_g = 0$ for all g. Then, a single test statistic is chosen according to the recipe Sect. 6.3.2 e: It should attain large values if the null hypothesis fails, that is, if the α_g are different from zero. A natural choice is the "mean square of the means,"

$$\text{MS}_\alpha = \sum\nolimits_g n_g(\overline{Y}_g - \overline{Y})^2/(k-1),$$

where n_g is the number of observations in group g and $\overline{Y}$ is the mean of all observations. It must be compared to the variance σ^2 of the observations within the groups, which is best estimated by the "mean square of residuals,"

Table 6.3 Pairwise comparisons and analysis of variance for logarithms (base 10) of compressive strength [MPa] for 4 sections in the tunnel example, restricted to layer. The p-values for the pairwise tests indicate clear differences between the first and the other three sections, but no significant differences within the last three. The test for no effect is highly significant with a p-value below 10^{-4}. The standard deviation within the groups is estimated as 0.0576 = 3

			Difference			p value	
Section	Mean	Deck 1	Deck 2	Wall 1	Deck 1	Deck 2	Wall 1
Deck 1	1.938						
Deck 2	2.013	0.075			0.000		
Wall 1	2.004	0.066	−0.009		0.001	0.670	
Wall 2	2.02	0.082	0.007	0.017	0.000	0.721	0.475

F-statistic: 8.31 on 3 and 69 d.f., p.value: 8.56e-05
St.dev.error: 0.0576 on 69 degrees of freedom

$$\mathrm{MS}_E = \sum\nolimits_g \sum\nolimits_i (Y_{gi} - \overline{Y}_g)^2/(n-k).$$

The test statistic is then $T = \mathrm{MS}_\alpha/\mathrm{MS}_E$. Its distribution under the null hypothesis only depends on the number of groups k and the number of observations n and is called the **F distribution** with $k-1$ and $n-k$ degrees of freedom. (This is true regardless of whether the group sizes are equal or unbalanced.)

f **Example**. Let us collect results for the data shown in Fig. 6.13. We apply a logarithmic transform to the data because the boxes in the box plot indicate skewed distributions and the spread (box height) tends to increase with the median. Such patterns are typical for technical measurements, and a logarithmic transform is a general recommendation for measurements of concentrations and other "amounts," i.e., continuous variables that are restricted to positive values (cf. Sect. 6.2.4 b).
Table 6.3 shows various results of the analysis of these four groups.

g **Analysis of variance**. The terms MS_α and MS_E have a simple relation to the overall variability of the observations Y_{gi},

$$\sum\nolimits_g \sum\nolimits_i (\overline{Y}_{gi} - \overline{Y})^2 = (k-1)\,\mathrm{MS}_\alpha + (n-k)\,\mathrm{MS}_E = \sum\nolimits_g n_g(\overline{Y}_g - \overline{Y})^2 + \sum\nolimits_g \sum\nolimits_i (\overline{Y}_{gi} - \overline{Y}_g)^2.$$

In words, the "total" sum of squares splits into a sum of squares due to the variability between the groups and a sum of squares stemming from the deviations of the observations from their groups' means, called the "error sum of squares". This decomposition of the variability is the root of the name "analysis of variance" of the model and the inference methods attached to it. The simplest case discussed here, with a single grouping variable or "**factor**" is called "one way analysis of variance".

h **Two-way analysis of variance**. The model for the k groups problem can be easily extended to include the effect of a second factor on the response variable Y. The simplest version is

$$Y_{ghi} = \mu + \alpha_g + \beta_h + E_{ghi} \,, \qquad E_{ghi} \sim \mathcal{N}(0, \sigma^2).$$

This model describes the effect of the factors as additive: Each factor adds its contribution, α_g or β_h, to the response, independent of the value of the other factor.

i **Interactions**. If this assumption fails, the model should be extended to include an interaction term,

$$Y_{ghi} = \mu + \alpha_g + \beta_h + \gamma_{gh} + E_{ghi}.$$

Interaction means that **the effect of one factor depends on the value of the other factor**. This is, of course, an important extension of the flexibility, but also a complication of the interpretation of the model.

j **Generalization**. Generalizations of these models to include more factors are straightforward. They play an important role in research for optimizing a property by varying the composition of input materials or the procedure of an industrial production process.

The models discussed here all study the relationship between a **response variable** Y and one or several "**explanatory**" or **input variables**, which here are factors. Continuous input variables will be considered in Sect. 6.4.3.

6.4.2 *Interlaboratory Studies*

a Measurements of permeability involve a procedure of several steps, which must follow a protocol and need experience to produce a valid result. There is reason to venture that such measurements depend to some extent on a laboratory's or technician's interpretation of the guidelines for the measurement. In order to assess such an influence, two or more samples of a homogeneous batch are sent to different laboratories or measured by different teams on site, and the resulting measurements are collected, cf. Sect. 6.1 e.

b **Model**. Such an interlaboratory study leads to data that can be modelled as a k groups or one-way analysis of variance problem, with a special twist: The group effects are not considered as fixed, unknown numbers α_g, but as random variables that, as the simplest case, again follow a normal distribution. This results in

$$Y_{gi} = \mu + A_g + E_{gi} \,, \quad A_g \sim \mathcal{N}(0, \sigma_A^2) \,, \quad E_{gi} \sim \mathcal{N}(0, \sigma^2).$$

All random quantities are assumed to be statistically independent. The random group effect A_g models the variability between labs.

c **Variance components**. Any single measurement Y_{gi} then deviates from the "true" value μ by an effect A_g of the laboratory and a remaining error E_{gi} of individual measurements within lab g. The combined deviation $Y_{gi} - \mu = A_g + E_{gi}$ has variance

$$\text{var}(Y_{gi}) = \sigma_Y^2 = \sigma_A^2 + \sigma^2.$$

The two variances σ_A^2 and σ^2 that contribute to the total variance σ_Y^2 are called variance components.

d **Estimation**. Estimation of the parameters μ, σ_A^2 and σ^2 can be based on the means $\overline{Y}_g$ and the mean squares introduced in Sect. 6.4.1 e for the fixed effects model. However, there are more precise methods that follow a modified principle of maximum likelihood (Sect. 6.3.5 c) call **REML**, interpreted as "reduced" or "restricted maximum likelihood," see [6] and other textbooks.

e **Example**. The second example briefly described in the Introduction (Sect. 6.1 e) is such an interlaboratory study. The 15 measurements of each of 5 teams (wall 1) are shown in Fig. 6.14 together with the results of the analysis to be addressed in the following.

f **Repeatability and reproducibility**. The model reflects the fact that two measurements coming from different labs will have a tendency to differ more than two measurements coming from the same lab. The latter has variance $\text{var}(Y_{gi} - Y_{gi'}) = \text{var}(E_{gi} - E_{gi'}) = 2\sigma^2$. An interval $[-\gamma, \gamma]$ covering the difference with approximately 95 % probability will therefore have half width $\gamma = 2\sqrt{2}\sigma$. This quantity is called the **repeatability**. The difference between

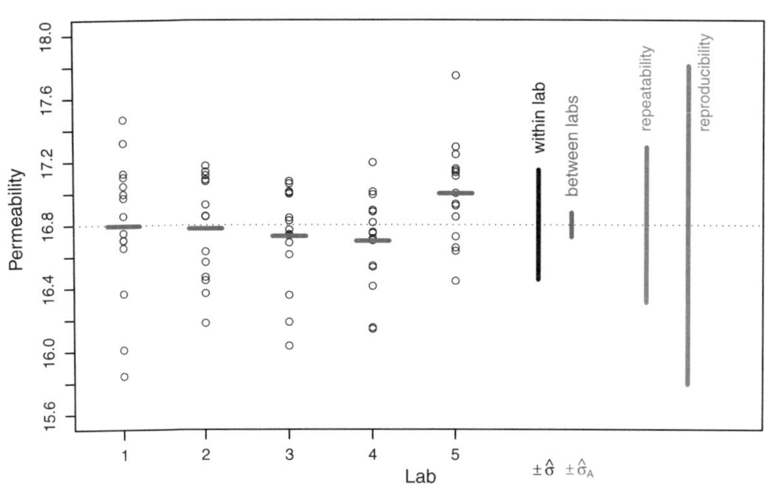

Fig. 6.14 Measurements from an interlaboratory assessment, with estimated standard deviations of the errors, $\widehat{\sigma} = 0.308$, and of the lab effects, $\widehat{\sigma}_A = 0.097$, as well as repeatability and reproducibility

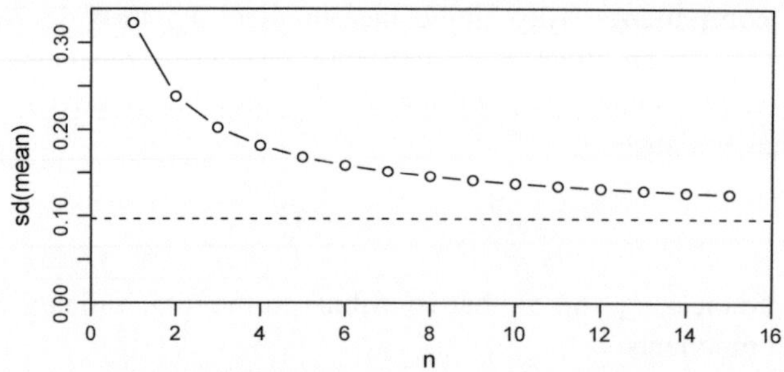

Fig. 6.15 Precision of a mean of n measurements from the same lab (or team, in the example), as a function of n, for the estimated parameters $\widehat{\sigma} = 0 : 308$ and $\widehat{\sigma}_A = 0.097$

two measurements from different labs has variance $\text{var}(Y_{gi} - Y_{g'i'}) = \text{var}(A_g + E_{gi} - A_{g'} - E_{g'i'}) = 2(\sigma_A^2 + \sigma^2)$. Thus, the interval covering it with probability 95 % has half width $2\sqrt{2}\sigma_Y$, called the **reproducibility**.

g **Precision of measurements**. The confidence interval for the true value, based on a single measurement Y, is approximately $Y \pm 2\sqrt{\sigma_A^2 + \sigma^2}$ according to Sect. 6.4.2 c. If this is not precise enough, more samples may be measured and the results averaged. If the ℓ samples are measured by the same lab, the precision of the result is given by the variance $\sigma_A^2 + \sigma^2/\ell$, which never gets smaller than σ_A^2. Thus, the increase in precision obtained by averaging over multiple measurements from the same laboratory is limited. Figure 6.15 shows the standard deviation of a mean of n such measurements as a function of n, for the parameter values estimated in the example (Sect. 6.4.2 e).

Reflecting this argument, one might judge that taking averages is worthwhile only up to around 6 repeated measurements. Any noticeable further increase in precision would need measurements done by different labs or teams.

h **Generalization**. The model used in this subsection is the simplest instance of a **random effects model**. Generalization to more than one grouping variable with random coefficients is straightforward. If they are hierarchically nested, like samples within production lots within production periods, this will lead to a variance component for each of them.

Other factors may have fixed coefficients (as in Sect. 6.4.1 a), in which case a **mixed effects model** appears.

6.4.3 *Simple Linear Regression*

a In the last subsections, we have studied the relationship between a quantitative response or target variable Y and one or several grouping variables or factors. Let us turn to the case where the "**input variable**" is also a quantitative or

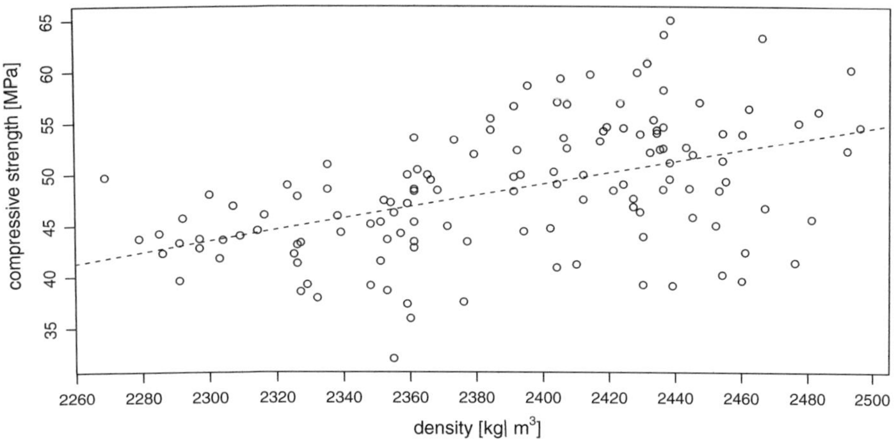

Fig. 6.16 Relation between compressive strength and density in the tunnel example 6.1d for layers ≥4. The line shows the estimated linear regression function

continuous variable, X. Figure 6.16 shows the dependence of the compressive strength of concrete on its density (Sect. 6.1 d) with a straight line determined by the following methods.

b **Model**. The simplest model for such data describes a linear dependence of Y on X and allows for random deviations of the observations from it. In other words, the expectation of an observation Y_i, given the value of X as x_i, is given by the linear **regression function** $\alpha + \beta x_i$. This is the equation of a straight line with **intercept** α and **slope** β. As in Sect. 6.4.1 a, we introduce a random deviation or **error term** E_i with a normal distribution. This results in the **simple linear regression model**

$$Y_i = \alpha + \beta x_i + E_i \ , \qquad E_i \sim \mathcal{N}(0, \sigma^2).$$

We also assume that the random errors E_i are statistically independent. In most applications, the slope β is the quantity of interest, since it describes the change in the response variable when the input variable X is increased by one unit.

c Note that the input values x_i are modelled as fixed values, not as random. This is the obvious choice if the input values x_i can be set by the experimenter, as in our example. In our application, the density X is also random, as we study the relationship between the observed variations of density and compressive strength. Nevertheless, we choose to ask for the (conditional) distribution of the response Y_i, given the observed value x_i of X, also in this situation. This turns out to be appropriate for examining most questions for which a model is needed, especially prediction of Y, given X (see Sect. 6.4.3 j).

d **Estimation of the coefficients**. The parameters of the model are the **coefficients** α and β of the straight line regression function and the standard deviation σ of the random error. Applying the principle of maximum likelihood for normally distributed errors leads to the more accessible criterion of **Least Squares**: Consider, for any choice of the coefficients α and β, the deviations of the observations from the respective straight line, $r_i(\alpha, \beta) = Y_i - (\alpha + \beta x_i)$. The coefficients shall be chosen such that the sum of squared residuals,

$$Q(\alpha, \beta) = \sum\nolimits_i r_i(\alpha, \beta)^2$$

is minimal. This simple optimization problem leads to the solution $[\widehat{\alpha}, \widehat{\beta}]$ given by

$$\widehat{\beta} = \frac{\sum_i (x_i - \overline{x})(Y_i - \overline{Y})}{\sum_i (x_i - \overline{x})^2}, \quad \widehat{\alpha} = \overline{Y} - \widehat{\beta}\overline{x}.$$

e **Correlation**. It is interesting to note the relationship of $\widehat{\beta}$ to the correlation coefficient

$$\widehat{\rho} = \frac{\sum_i (x_i - \overline{x})(Y_i - \overline{Y})}{\sqrt{\sum_i (x_i - \overline{x})^2 \sum_i (Y_i - \overline{Y})^2}} = \widehat{\beta}\,\frac{\sqrt{\sum_i (x_i - \overline{x})^2}}{\sqrt{\sum_i (Y_i - \overline{Y})^2}}.$$

If the variables X and Y are standardized to have variance 1, then $\widehat{\beta}$ equals the correlation coefficient.

f **Estimation of** σ. An estimate of the standard deviation σ must be based on the **residuals**

$$R_i = r_i\left(\widehat{\alpha}, \widehat{\beta}\right) = Y_i - (\widehat{\alpha} + \widehat{\beta} x_i).$$

The best estimator is given by

$$\widehat{\sigma}^2 = \frac{1}{n-2} \sum\nolimits_i R_i^2.$$

The somewhat surprising denominator $n - 2$ makes the estimator $\widehat{\sigma}^2$ unbiased for the error variance σ^2 (analogously to Sect. 6.3.1 e).

g **Test and confidence interval for** β. The estimated slope $\widehat{\beta}$ can be written as a linear function of the observations, $\widehat{\beta} = \sum_i c_i Y_i$. Therefore, it has a normal distribution, and it is straightforward to calculate its parameters using Sects. 6.2.4 f and 6.2.5 e,

$$\widehat{\beta} \sim \mathcal{N}\left(\beta, \mathrm{sd}_{\beta}^2\right), \quad \mathrm{sd}_{\beta}^2 = \sigma^2 / \mathrm{SS}_X,$$

where $\mathrm{SS}_X = \sum_i (x_i - \overline{x})^2$ is the **sum of squares** of the x_i values. This leads, in the same way as in Sect. 6.3.2 k, to a **t-test** of the null hypothesis $\beta = \beta_0$ based on the test statistic

$$T = (\widehat{\beta} - \beta_0)/\mathrm{se}_\beta \ ,$$

which has a t distribution with $n - 2$ degrees of freedom. The respective confidence interval is, in analogy to Sect. 6.3.3 b,

$$\widehat{\beta} \pm q(0.975;\, t_{n-2})\ \mathrm{se}_\beta .$$

A test and confidence interval for the intercept α can be obtained in the same way once the variance of $\widehat{\alpha}$ is known, see Sect. 6.4.3 k below, with $x = 0$.

h **Checking assumptions**. The probability calculations used for deriving the distribution of the test statistic relies on the assumptions of the model, which were:

1. The regression function is a straight line.
2. The variances of the random deviations E_i are all equal.
3. The distribution of the E_i is normal.
4. The random deviations E_i are independent.

It is important to make sure that the data supports these assumptions. If the regression function is incorrect, the model is of little use, and so is inference for its coefficients. The independence assumption is the most critical for the correctness of tests and confidence intervals. It may be violated by intra-group correlations as those modelled in Sect. 6.4.2 c, or by (auto-) correlations of subsequent observations in time order.

i **Example**. In the example shown in Fig. 6.16, the dependence of Y on X does not seem to be linear, but somewhat convex. Furthermore, the variance of the deviations increases with increasing compressive strength. These two fallacies often come together like the syndrome of an illness. As in Sect. 6.4.1 f, a good cure relies on transforming the response variable by the logarithmic transformation, cf. Sect. 6.2.4 b. Figure 6.17 displays the transformed data. The input variable compressive strength has also been transformed in order to remain consistent in some way—with little effect, since the range of values is such that the transformation is almost linear in this case. The estimated straight line is displayed, together with two lines through the "center of gravity" $[\overline{x}, \overline{Y}]$, with slopes given by the limits of the confidence interval for β.

j **Prediction**. Regression models are typically applied for "predicting" a value of the response Y that will be obtained if a new observation is made for a given input value x_0 of X. Note that this is different from a prediction in the sense of extrapolation of the past into the future.
The best prediction according to the model is obviously the value of the estimated regression function, $\widehat{y}_0 = \widehat{\alpha} + \widehat{\beta} x_0$. The more difficult problem is to give an adequate indication of precision of such a value through an interval. In fact, there are two basically different ways of posing this problem: Should the interval cover the expected value of Y, given x_0, or should the new observation Y_0 itself be contained in it?

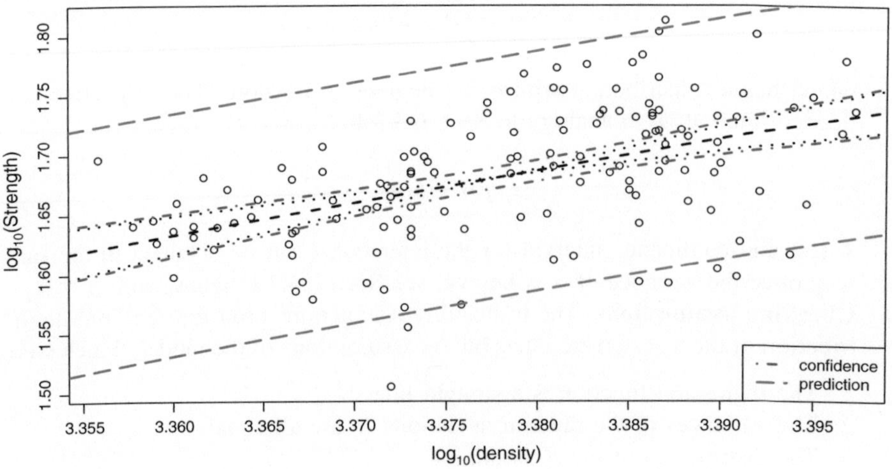

Fig. 6.17 Regression between the log transformed data, with estimated straight line. The *fine dotted lines* have the slopes corresponding to the endpoints of the confidence interval for β. The two curved bands display the confidence band for the values of the regression function and the prediction band

k **Confidence interval for the regression function.** The first problem asks for a confidence interval for $\gamma = \alpha + \beta x_0$. It is obtained in the usual way by deriving the variance of the estimator $\widehat{\gamma} = \widehat{\alpha} + \widehat{\beta} x_0$,

$$\mathrm{var}(\widehat{\gamma}) = \mathrm{sd}_{\gamma}^2(x_0) = \sigma^2 \left(\frac{1}{n} + \frac{(x_0 - \overline{x})^2}{SS_X} \right).$$

This leads to the confidence interval

$$\widehat{\alpha} + \widehat{\beta} x_0 \pm q(0.975;\, t_{n-2}) \mathrm{se}_{\gamma}(x_0)\,,$$

where se_{γ} is sd_{γ}, with σ replaced by $\widehat{\sigma}$.

This interval can be visualized for varying x_0 through the **confidence band** shown in Fig. 6.17. It expresses the insecurity of the true value of the regression function due to the randomness of the sample to which the straight line was fitted.

l. **Prediction interval.** A new observation of Y_0 of Y contains, according to the model, a new random error E_0. This adds an insecurity expressed by the standard deviation σ to the uncertainty just discussed. Since the new random deviation E_0 is independent of the sample used to fit the straight line, the two variances, σ^2 and $\mathrm{sd}_{\gamma}^2(x_0)$, add. Therefore, replacing se_{γ} in Sect. 6.4.3 k by se_Y with $\mathrm{se}_Y^2(x_0) = \widehat{\sigma}^2 + \mathrm{se}_{\gamma}^2(x_0)$ produces an interval that contains the new observation with 95 % probability. Note that this is not a confidence interval, since it does not characterize the precision of an estimated parameter, but instead should cover a random quantity, Y_0. It is called a prediction interval instead. The intervals for all values of x_0 can again be shown graphically as a **prediction band**, see Fig. 6.17.

6.4.4 *Outlook*

a The models of analysis of variance and simple regression, introduced in this section, are basic building blocks for a multitude of more general regression models. We mention those which might be fruitfully applied in the concrete industry.

b **Multiple linear regression**. The simple regression model easily generalizes to include the influence of more input variables on the response, simply by adding more terms,

$$Y_i = \alpha + \beta_1 x_i^{(1)} + \beta_2 x_i^{(2)} + \cdots + E_i, \quad E_i \sim \mathcal{N}(0, \sigma^2).$$

It is important to note that the input variables $X^{(j)}$ are not modelled as random variables, but fixed values $x_i^{(j)}$ are used as above. Therefore, they do not need to show any distribution, but can be of any nature, including discrete or even binary. They do not need to be independent, even though they are sometimes called the "independent variables" (and Y, the "dependent" variable). They may even be deterministic functions of each other. Choosing $X^{(2)} = (X^{(1)})^2$ leads to a **quadratic regression** function, and letting $X^{(3)} = X^{(1)}X^{(2)}$ provides a generic form of **interaction**. Finally, grouping variables (factors, see Sect. 6.4.1 g) can be included as input variables, too. Combining this with the possibility of transformations (Sect. 6.4.3 i), these ingredients make the model of multiple linear regression very versatile, allowing for much more general relations between input variables and response than just linear ones.

The methods of inference for coefficients as well as prediction follow the same principles as for the analysis of variance and the simple regression.

c **Response sufaces**. Finding the mix that optimizes a certain property of concrete is of course an important task. It calls for fitting a model to data from pertinent experiments. Since multiple linear models may include quadratic and other functions with a localized maximum or minimum, they are suitable for finding conditions which lead to an optimum of the response variable. The basic model uses a quadratic function in all the input variables to be examined as the regression function and is called the "response surface model".

d **Nonlinear models**. The notion of (multiple) *linear* models supposes that the regression function is linear in the coefficients β_j. For some phenomena there may be theoretical knowledge entailing a regression function which involves parameters in a more complicated way (and a linear form cannot be achieved by transforming variables). Even though computation then gets more difficult, Least Squares is still the principle underlying estimation, and the theory of inference for linear regression can be generalized to such models.

e **Other models for a quantitative response**. All these models specify a normal distribution for the random deviations E_i. For example, the response may be a time until a device or a structure breaks, known as **failure time**. A normal distribution is inappropriate for such variables. A popular alternative is the **Weibull distribution** family. In addition, for units which have not broken down at the end of a study, one knows that the response variable is larger than the time in service at the end of the study. This leads to **censored** values, and there are corresponding models.

f **Logistic regression and generalized linear models**. The response variable may also be binary, distinguishing between presence or absence of a characteristic or between any kind of "success" and "failure," like the situations leading to the binomial distribution. A corresponding model is called **logistic regression**. It relates the probability of a success to a **linear predictor** that has the form of the right hand side of the model Sect. 6.4.4 b. It belongs to the more general class of **generalized linear models**.

g **Random effects**. Random coefficients have been introduced in Sect. 6.4.2. They can be combined with fixed effects, leading to **mixed models**.

h For all the models mentioned here, there is well established methodology for estimating parameters, testing, getting confidence intervals and generating predictions. They are generally based on the principle of maximum likelihood and use asymptotic approximations for obtaining the necessary distributions of estimators and test statistics.

i **Spatial correlation**. We have briefly mentioned spatial correlation in Sect. 6.2.5 d. Apart from requiring correction terms for the variability of a mean or other estimators or test statistics, this leads to a fundamental problem of definiting conformity: There is an obvious difference of risk if the fraction p of "true" values of the criterion X that falls short of the treshold c (see Sect. 6.2.3 e) is a patch of bad concrete or if it consists of isolated, very local bad spots. A discussion and treatment of this problem clearly goes beyond the scope of this text.

References

1. Montgomery D, Runger G. Applied statistics and probability for engineers. 4th ed. Hoboken, NJ: Wiley; 2007.
2. Rice JA. Mathematical statistics and data analysis. 3rd ed. Belmont, California: Duxbury Press; 2007.
3. Leemann A, Hoffmann C, Malioka V, Faber M. Variability of concrete properties in structures. Technical Report 611, Bundesamt für Strassen (ASTRA), Bern, Switzerland, 2006.
4. Jacobs F, Leemann A, Denarié E, Teruzzi T. Recommendations for the quality control of concrete with air permeability measurements. Technical Report 641, Bundesamt für Strassen (ASTRA), Bern, Switzerland, 2009.
5. Monteiro AV, Gonçalves A, Gulikers J, Jacobs F. Basis for the statistical evaluation of measured cover depths in reinforced concrete structure. In *Performance-Based Specifications and Control of Concrete Durability*, RILEM TC 230-PSC. RILEM, 2014.
6. Pinheiro JC, Bates DM. Mixed-effects models in S and S-plus. Statistics and computing. New York: Springer, 2000.

Chapter 7
Responsibilities

R.D. Hooton, M. Khrapko, M. Otieno and M.A. Ismail

7.1 Introduction

While there is widespread interest in moving from prescriptive to performance-based specifications for concrete, very few true performance-based specifications exist [1] and many engineers are more comfortable with the traditional (conservative) prescriptive approach than the performance-based one. Some of the barriers to the wide acceptance of performance-based specifications include perceptions of increased costs associated with extra testing, extra time and increased quality control or quality assurance measures. Prescriptive specifications typically evaluate durability indirectly using measures such as limits on strength, water-to-binder ratio (w/b), cover depth, grade of concrete and in some cases minimum binder content and binder type for a given exposure environment. Pertinent issues such as resistance to chloride ingress, sulphate ingress and cracking are often ignored. These specifications often inhibit innovation. Even though there is enough impetus to shift to performance-based concrete specifications, there are currently no purely performance-based specification codes. For example, EN 206 [2] specifies minimum binder contents as well as maximum w/b ratio, AS 3600 [3] and NZS 3101 [4] specify minimum strength and cover to reinforcement, ACI 318-11 [5] and CSA A23.1 [6] have limits on maximum w/b and minimum strength, although CSA A23.1 also includes limits on fluid penetration using an

R.D. Hooton (✉)
University of Toronto, Toronto, Canada

M. Khrapko
CBE Consultancy Ltd, Auckland, New Zealand

M. Otieno
University of the Witwatersrand, Johannesburg, South Africa

M.A. Ismail
Universiti Teknologi Malaysia (UTM), Johor Bahru, Malaysia

H. Beushausen and L. Fernandez Luco (eds.), *Performance-Based Specifications and Control of Concrete Durability*,
RILEM State-of-the-Art Reports 18, DOI 10.1007/978-94-017-7309-6_7

index test (ASTM C1202 [7]) and allows concrete to be specified based on performance. Recently formed ACI Committee 329 on Performance Criteria for Ready Mixed Concrete is developing a performance specification as a potential alternative to the current ACI 301 Specifications for Structural Concrete [8].

For performance-based specifications to be successful (fully implemented), concrete needs to be specified in terms of the required physical and durability performance rather than prescriptive limits on ingredients and mix designs. In principle, a performance-based specification should state the minimum essential performance requirements of the hardened concrete which can be measured by accepted industry standards and test methods. The fresh concrete properties should not be stated in a performance-based specification because they do not link directly to durability and/or mechanical performance of the hardened concrete. Fresh concrete properties (characteristics) relate to the means of delivering and assuring required durability and mechanical properties/performance and hence are to be agreed between concrete supplier/producer and contractor. The processes, materials, or activities used by the contractors, and suppliers should be left to their discretion but subject to satisfying an initial approval (pre-qualification) process to ensure that what is being proposed meets the performance requirements, as well as ongoing quality assurance during construction. Therefore, the performance-based specification should provide a system for the owner/specifier, contractor and supplier/producer to assess and maintain a quality concrete. To achieve this, the responsibilities of all parties need to be clearly defined in the contract document [9]. In addition, proper communication and improved partnership (co-operation) between the parties (owner/specifier/engineer (design professional), contractor and supplier/producer) must be ensured to address any problems and deficiencies quickly in order to achieve the desired concrete performance.

Some of the key components of a workable performance-based specification system should include [1, 10]:

i. A qualification/certification system that establishes the requirements for a concrete production facility, the facility's quality control management system, and the facility's personnel
ii. Suppliers/producers and contractors that partner to ensure that the right concrete mixture is developed, delivered, placed and finished
iii. Sufficient flexibility to allow the supplier to provide a concrete mixture that meets the performance criteria (including pre-qualification test results) while satisfying the contractor's requirements for placing and finishing
iv. Requirements for field acceptance tests needed to verify that the in-place concrete meets the performance criteria, as well as a clear set of instructions defining the actions required if those test requirements are not met i.e. non-conformance.

This chapter gives a general view of the responsibilities of the main parties in a performance-based specification construction project i.e. owner/engineer/design professional, contractor and concrete supplier/producer. This also includes the identification of industry-accepted, reliable and repeatable standard tests to assess relevant fresh and hardened concrete properties, as well as acceptance test limits corresponding to the desired durability performance. The objective is to encourage a more clear, target-oriented and co-operative relationship between these parties to ensure that the desired durability performance of the structure is achieved by clearly defining both the acceptance criteria and the responsibilities of the various parties.

7.2 Performance-Based Durability Limits

The application of a prescriptive specification approach implies that, if the specified limits (e.g. minimum cement contents and maximum w/b ratios for all durability classes, specific binder types, acceptable range of air contents and limitations on the types and quantity of chemical and mineral admixtures) and good construction practices are strictly followed, durable concrete will be produced. However, previous experience has shown that this is not always the case.

Contrary to a prescriptive approach, in a performance-based approach, limits are set based on test results for specific standard and reliable test methods. These limits, should also allow for test variability by use of both average values and allowances for individual values to exceed those average limits due to variability inherent in the test method (this is similar to what is currently allowed in most specifications for occasional understrength test results). Furthermore, it is important to note that since test results are variable and it is more difficult to consistently meet the set performance limits based on field cast samples than laboratory batched concrete, it is conceivable that on-site job acceptance limits will be less stringent than the limits required to pre-qualify concrete mixtures. It is encouraged that all specification documents move towards adopting the use of statistical quality control to assure consistent conformity with the desired performance at the lowest cost [11].

7.3 Verification of Durability

Adoption of performance-based specifications assumes:

i. That there are appropriate performance test methods in place to evaluate all the essential properties of concrete
ii. That performance can either be measured in time to affect the construction outcome, and/or can be used to pre-qualify concrete mixtures.

While new and better test methods will become available with time, nearly all the relevant properties of concrete can be measured to a practical acceptable level provided the implementation is governed by appropriate procedures (such as location of sampling points, sampling frequency and handling/storage of test specimens). These procedures need to be incorporated in contract specifications, and mutually agreed on by all the parties. However, while most parties to construction are familiar with testing for fresh and hardened concrete properties, the biggest challenges in this regard relate to evaluation of durability performance.

The verification of the performance-based functional requirements requires that standard established test methods and acceptance criteria are clearly defined a priori, with some testing required for pre-qualification and some for in situ acceptance both before and after the concrete is placed. A standard acceptance test to measure rates of ingress of relevant aggressive fluids, or a related rapid index test, is therefore fundamental to the development of performance-based durability specifications. The tests must not only be shown to be useful and reliable, but must also be standardized and have precision data based on inter-laboratory evaluations (i.e. repeatability) in order to develop confidence in the results and to be able to set realistic specification limits that take into account the inherent variability of the test results. For large infrastructure projects, especially those with stated long service life requirements, often during pre-qualification, for example, meeting chloride diffusion or permeability limits has been required. However, these tests are often too slow to be useful for acceptance purposes, so during prequalification, the results need to be correlated to rapid index test results [9, 12–14]. Table 7.1 summarizes a number of established test methods that are applicable to performance specifications and list the time frames required to obtain test results.

Some of the challenges facing concrete testing with respect to meeting performance requirements include the following

i. There are no methods for checking how well a (pre-qualified) concrete that meets the desired specifications based on the results of current test methods relates to concrete performance in the field
ii. Lack of reliable, consistent (reproducible) and standardized test procedures for evaluating all the relevant concrete properties related to the desired performance
iii. Some tests do not adequately represent any or all of the in situ exposure conditions
iv. Some of the available tests are expensive and complex, and results may not be as precise as desired
v. Some tests take long to perform and cannot therefore determine the essential concrete properties soon enough to affect the construction outcome (i.e. some tests used during prequalification may not be suitable for construction acceptance)
vi. Short bid times and quick construction can create a difficult situation for a concrete supplier faced with the need to develop a performance-based mixture and perform pre-qualification testing. Furthermore, due to time constraints during construction, the durability tests available for pre-qualification of concrete mixtures will not typically be appropriate for use for quality assurance/quality control purposes during construction
vii. Identification and use of “identity tests” to confirm on site that the concrete mixture being delivered is the one that was pre-qualified. Of necessity, these (rapid) tests need to be done at the point of discharge and provide immediate

Table 7.1 Examples of established test methods applicable to performance specifications (based mostly on ASTM tests, adapted from [11, 15])

Property	Standard	Required lead time (after casting except as noted)
Compressive strength	ASTM C31 and C39	35 days to obtain materials and make and test concrete mixtures at ages up to 28 days
Compressive strength in place	ASTM C900 and C1074	
Density (unit weight), yield, and air content of fresh concrete	ASTM C138	
Density of fresh and hardened structural lightweight concrete	ASTM C567	
Early-age strength	ASTM C39	
Flexural strength	ASTM C78	
Density, absorption, and permeable voids in hardened concrete	ASTM C642	
Splitting tensile strength	ASTM C496	
Modulus of elasticity	ASTM C469	
Drying shrinkage	ASTM C157	180 days. Or less, based on requirements
	CSA A23.2-21C	35 days
	AS 1012.13-1992	56 days
Chloride bulk diffusion	ASTM C1556	60 days
Rapid chloride resistance	ASTM C1202	30 or 58 days
Sulphate resistance	ASTM C1012	6, 12, or 18 months depending on level of resistance required
Resistance to freezing and thawing	ASTM C666	90 days
Modulus of elasticity	ASTM C469	35 days
Creep	ASTM C512	1 to 2 years
Splitting tensile strength	ASTM C496	35 days
De-icer salt scaling	ASTM C672	98 days after casting
	CSA A23.2-22A	105 days after casting
Alkali-silica reaction, to evaluate aggregates	ASTM C1260 ASTM C1293	16 days 1 year
Alkali-silica reaction, to evaluate mixture	ASTM C227 and C1293 CSA A23.2-2-28A	3 to 6 months 2 years
Alkali-silica reaction, to evaluate job combinations except when low-alkali cement is used	ASTM C1567	16 days

(continued)

Table 7.1 (continued)

Property	Standard	Required lead time (after casting except as noted)
Rapid chloride resistance	ASTM C1202	28 to 56 days
	Chloride conductivity index test [16]	10 days after sampling
Air void system	ASTM C 457	Age plus 14 days
Resistivity	Bulk or Wenner probe AASHTO TP 95-11 [12, 17]	One hour plus age of concrete (assuming in saturated condition)
Gas permeability	Oxygen permeability index test [18]	10 days after sampling
Uniformity of water content in fresh concrete	AASHTO T-318	On-site as delivered
Sorptivity	ASTM C1585	28 to 56 days
	Water sorptivity index test [19]	10 days after sampling
Rapid chloride migration test	Nordtest NTBuild 492 AASHTO TP 64	28 to 56 days
Chloride bulk diffusion	ASTM C 1556, Nordtest NT Build 443	42 days after sampling

For precision statements, see text of the cited test standards

confirmation that the mix is essentially the same as the pre-qualified one. However, the use of rapid index tests for acceptance during construction should not be construed as ignoring other more rigorous test methods or use of service life modelling

viii. There is also a school of thought of the opinion that current testing technology has not yet caught up with the performance-based specification philosophy (however, waiting for perfect test methods is seen as an excuse for preventing any progress) [9].

These challenges will need to be adequately addressed by the owner/specifier, contractor and supplier/producer to ensure a performance-based specification is implemented. Advances have been made towards addressing these challenges to facilitate the evolution and adoption of performance specifications and the development of more rapid and reliable test methods. Some of these are summarized in Table 7.1. Due to the different time frames required to obtain test results, some test methods listed in Table 7.1 maybe suitable for pre-qualification testing but not for quality control testing.

7.3.1 *Types of Performance Testing*

Performance tests carried out at various stages before or during construction can be of three types:

i. Pre-qualification testing: to provide a concrete mixture that when placed under defined conditions can meet the specification requirements. Required one cubic meter or larger monolith or mock-up trials have been used successfully as part of pre-qualification, with tests performed on cores removed from these trials [9]. For this at least 2.5 m^3 of concrete would have to be batched using plant equipment to ensure uniform distribution of concrete constituents in the mix.

ii. Quality control testing: to document that (a) the concrete supplied meets the desired specification(s), (b) the concrete supplied is equivalent to that which was pre-qualified (sometimes called identity testing), and (c) pre-qualified placing practices are being followed (i.e. test(s) at each change of ownership, such as the point of discharge from the concrete delivery truck). In some cases the owner/specifier may prefer to evaluate concrete performance at the point of placement. However, suitable methods must be put in place to ensure proper sampling. For example, AS 1012.1 [20] and NZS 3112.1 [21] provide guidance for on-site sampling of fresh concrete—this is referred to as 'snatch'/ 'individual' sampling in New Zealand and Australia respectively, or 'grab' sampling in North America. While this is logically sound, it is logistically difficult for several reasons. First, while some guidance exists in various documents, no standard method exists for sampling concrete at the point of placement, and methods in use vary widely. Secondly, questions remain about how to obtain a representative sample, how to do so in a safe manner (e.g. individual samples from the same batch of concrete can be combined to provide representative samples which can be used to assess the nature and condition of a defined volume of concrete), and whether to attempt to sample immediately prior to impact at the point of placement (e.g. individual sampling 5 min before discharge would indicate if concrete is of an acceptable quality), after impact, before or after consolidation, and whether to retrieve and then re-compact a sample that has been placed, consolidated, and perhaps finished. Note that while limits on the w/b ratio of concrete are prescriptive in nature, the variability of water content of different loads of concrete is one of the major problems facing the industry in terms of consistently meeting both strength and durability requirements. Therefore, an acceptance test providing a measure of the uniformity of water content, as a measure of the uniformity of as delivered concrete is quite useful. For this purpose, the AASHTO T318 [22] microwave water content test has been used successfully on several projects [23].

iii. In-place testing: using non- invasive testing and/or tests on statistically sampled cores extracted from the structure to ensure that the concrete supplied and the placement methods meet owner-defined performance levels.

For a successful performance testing to be achieved, the owner/specifier should discern the performance characteristics appropriate for the intended use of the concrete unambiguously and quantitatively together with the test procedures to be used for acceptance so that the desired performance can be quantitatively evaluated [11]. This will help shift from conventional (default) testing which often concentrates on properties such as slump, air content and 28-day strength even though one or more of these properties may not be relevant to meeting the owner's desired performance, while more relevant performance requirements may not be tested at all. As an example, one way to look at a broader range of concrete properties that could be specified for testing in a performance-based specification as shown in Table 7.2, provided that industry-accepted test methods and limits exist.

Nevertheless, while various performance tests can be used for pre-qualification, quality assurance, or in situ testing, there are many more issues which need to be addressed to obtain the desired performance in aggressive environments. This type of information should be detailed in performance-based specifications as it is in, for example, an annex to CSA A23.1 [6]—some of these are listed as follows:

i. Require all contract bidders to attend a pre-bid meeting to know the special performance requirements (so they cannot complain afterwards)
ii. Require contractors, including sub-contractors, to detail in their bid how they intend to meet the special performance requirements part of the bid submittal,

Table 7.2 Concrete performance properties of interest [10]

Fresh concrete	Transition	Hardened concrete
• Slump • Plastic viscosity • Response to vibrator • Pumpability • Finishability • Segregation • Bleeding • Air content • Stability of air bubbles • Uniformity of mixing • Consistency of properties • Temperature • Yield	• Rate of slump loss • Time to initial set • Time to final set • Rate of strength gain (compression) • Rate of strength gain (tension) • Rate of stiffness gain • Time to frost resistance • Tolerable rate of evaporation • Plastic shrinkage • Drying shrinkage • Temperature	• Strength (compressive, tensile, flexural, shear, fatigue) • Fracture toughness • Elastic properties • Shrinkage • Creep • Porosity • Pore size distribution • Permeability and chloride resistance • Air void system • Frost resistance • Abrasion resistance • Sulphate resistance • Acid resistance • Alkali-aggregate resistance • Thermal volume change • Heat capacity • Thermal conductivity • Electrical conductivity • Density • Radiation absorption • Colour • Texture • Cost

for example, concrete placement methods, protection, curing, hot/cold weather provisions, etc

iii. Do not accept low-price bids that are not fully addressing the special concrete performance requirements

iv. Once construction has commenced, require pre-placement meetings for important concrete pours. The contractors, suppliers, sub-contractors, including finishers need to be aware of what needs to be done, and what equipment and supplies need to be in place to ensure that the concrete can be delivered, placed, compacted, protected, finished, and cured to achieve the desired performance. Everyone needs to be present to understand the importance of their role(s).

7.3.2 *Durability Tests and (Typical) Limiting Values*

To address the need to limit fluid penetration into hardened concrete (hardened cement paste), most (hybrid) concrete codes e.g. AS 3600 [3], EN 206 [2], CSA A23.1 [6] and ACI 318 [5] set limits on the maximum w/b ratio permitted for various exposure classes. However, while it is clear that in general the permeability of concrete increases exponentially with w/b ratio, the specific value of permeability at any given w/b ratio varies significantly in response to materials characteristics such as total water content and total paste content, aggregate content and grading, and the type and proportions of supplementary cementitious materials. Time-temperature history (maturity) as well as the duration and type of curing also critically influence permeability and related properties, as does the age of the concrete or specimen at the time of test. Therefore, the limiting values of w/b ratio required by, for example, ACI 318 [5] are prescriptive in nature, and do not necessarily correlate to specific values of measurable concrete performance.

In a performance-based specification approach, the contractor and producer should demonstrate the acceptability of a proposed concrete mixture on the basis of measured performance of physical and durability properties of hardened concrete by means of standard industry-accepted tests specified by the owner. Depending on the exposure class specified by the owner, limits on fluid penetration resistance properties of hardened concrete could be adopted for pre-qualification purposes based on standard tests (e.g. ASTM C1202 [7] or other ASTM tests [13] or diffusion or sorptivity tests (see Table 7.1) [13], Resistivity tests [12, 14, 24, 25], South African chloride conductivity test [16, 19] that assess chloride penetrability). For example, in the CSA A23.1 standard [6], where resistance to chloride penetration is quantified using the ASTM C 1202 test method [7], for exposure class C-1 or A-1 (35 MPa air-entrained concrete intended for exposure chlorides and freezing (C-1) or aggressive chemicals (A-1), 'a single test value is allowed to be up to 1750 coulombs as long as the average value remains below 1500 coulombs at 56 days'.

Other examples of such test limits will be discussed in Chap. 8. However, it is important to note that the limits should allow for the variability of the test results if used for acceptance testing during construction. This can be accommodated by incorporating statistical limits for the test results.

7.4 Quality Management Using Performance-Based Specifications

The verification of concrete quality to ensure performance is the responsibility of the owner/specifier. Quality plans must take into account that there are quality management elements both internal and external to the owner's concrete acceptance requirements, and that these elements must be tailored to each specific project and the concrete performance that is being sought. This includes ensuring that the contractor has in place an industry-recognized quality control plan (e.g. [26]) to prevent or correct defects and non-conformity in the concrete. Care must be taken by the owner during the contractor selection and award stages of a project to ensure that contractors and suppliers are provided with the necessary incentives for the added effort and cost of maintaining such a quality control process.

The external quality control effort (e.g. inspection and testing for verification and acceptance) made by the owner must complement and balance the internal quality control effort made by the contractor, ensuring that the contractor's quality control systems are in place, operating effectively, and preventing or correcting non-conformance.

In a performance-based specification environment, a high level of responsibility is placed on the contractor and all of his/her suppliers (ready-mix, hardware, reinforcing steel, etc.) and sub-contractors (formwork, reinforcing steel, pumping, placing, finishing, etc.) for the internal quality control effort. The owner must in turn balance this effort by reviewing the quality control plans and records of primary contractors, sub-contractors, suppliers, and secondary suppliers, and by conducting independent quality assurance, testing and verification of concrete and other material properties to validate the results of the contractor's processes. The owner should also undertake an independent audit of the quality management system.

A contractor's quality plan should define the contractor's responsibilities and actions required to meet the owner's performance specifications. The management of the plan, including compliance with the quality plan and any modifications remain the responsibility of the contractor. The quality plan should include:

a. Organization charts, roles and responsibilities, identification of the person in charge of the quality management for the project (this can include personnel for the supplier and sub-contractor as well as the contractor)
b. Document management and retention processes

c. Concrete construction processes, including placing, protection, finishing and curing
d. Verification of concrete mixtures and submittal processes
e. A non-conformance management process including identification, reporting and procedure to correct non-conformance
f. Quality control testing and inspection plan complete with reporting of test results
g. Change management process. This should include a procedure for informing all parties of changes to the construction process or concrete mix design affecting performance and if required, indicate how the quality control will be adjusted to assess how performance criteria will still be met.

The quality plan may be implemented wholly or partially by a contractor, sub-contractor, supplier or an independent organization. Changes to the plan should be in writing and accepted in kind by the owner. It is important to note that acceptance of the contractor's initial quality plan does not exclude that changes may be requested by the owner at any time, following observations from audits.

In addition, in relation to concrete quality management, when ordering concrete, the following items must be selected by the owner [6]:

a. Intended application including exposure class
b. Quantity of concrete required
c. Compressive strength at age
d. Nominal maximum size of aggregate
e. Air content of air-entrained concrete
f. Finish requirements
g. Other characteristics as required e.g. volume stability including shrinkage limits.

7.5 Responsibilities of the Owner/Specifier, Contractor and Supplier/Producer

The advent of performance-based specifications has significantly changed the distribution and sharing of responsibility among the owner/specifier, contractor and supplier/producer. The first step towards a successful performance-based specification is to ensure that each party clearly understands their responsibilities in a "performance specification". Hooton et al. [11] state that that the term means many things to different people—both local and international. This is not necessarily because of any misinterpretation but because there is such a wide array of options and interpretations, making it imperative that the roles and responsibilities be carefully defined in a specification. Without this, parties could agree in principle to execute work under the 'performance specification' umbrella and yet have divergent views about mutual expectations, leading to problems. Performance specifications have prompted the need (i) for improved partnership (communication and networking) between all parties, as well as inspection, auditing and end-result (in-place) verification to achieve the desired in-place performance, and (ii) as already mentioned, to clearly define the responsibilities of each party involved in the construction process.

In most cases, although many owners/specifiers want to place performance limits on concrete durability performance, they are often unwilling to give up prescriptive requirements on mix design and materials but are reluctant to take on the associated responsibility for their prescriptive requirements on the resulting performance of the fresh and hardened concrete. However, because both the concrete mixture and construction practices have an impact on concrete durability, achieving the owner's performance requirements demands co-operation (improved communication and networking) between the concrete suppliers, contractors, and finishers e.g. the contractor (not the owner/specifier) should set the target slump to allow for proper placement and compaction for the situation, and the producer needs to provide this without compromising the owner's desired performance [1, 11, 27, 28]. Furthermore, even though both concrete producers and contractors are often primarily interested in meeting pre-qualification requirements and passing as-delivered quality control tests, the onus for meeting the actual performance clearly rests with the supplier/producer up to point of placement, but the contractor is responsible for placement methods and practices that affect in-place performance. Therefore, as stated earlier, the contractor needs to develop a quality plan to demonstrate that proper equipment, personnel and resources are available to place, compact, finish, and cure the concrete to attain the specified durability over the service life of the structure [29].

It is also imperative that because owners/specifiers are usually interested in quality/performance of the hardened concrete in the structure, they need to develop, review, audit and monitor the execution of the quality plan. To meet this end, some owners who require the concrete supplier to assume responsibility for the performance of the concrete as delivered and the contractor to assume responsibility for the concrete in place [6] (e.g. a number of highway agencies in North America) have adopted, or are currently considering the use of "in-place" or "end-result" specifications (ERS) where contractors are paid bonuses based on consistently meeting or exceeding specified performance requirements using in-place testing of the structure. Some of these ERS incorporate well-defined financial penalties for failure to meet the in-place requirements, some of which exceed the cost of the concrete and if performance is lower than a certain threshold, removal is required. This approach has also been successfully applied in some major infrastructure projects in South Africa using the durability index tests.

In summary, in a performance-based specification, the responsibilities of all the parties (owner/specifier, contractor and supplier/producer) should be clearly defined, including detailed guidance on how to fulfil the responsibilities, and commentary on communication of non-conformity and for making changes to rectify non-conformity. Some of these (based largely on the performance option in Canadian standard [6]) are summarized in the following sections.

7.5.1 *Responsibilities of the Owner/Specifier/Design Authority (Engineer)*

Prior to endorsing the use of a performance-based specification, the owner/specifier must have confidence that the approach will meet his/her objectives. This requires reliance on the design team to prepare an effective performance-based specification and on the implementation of a reliable quality assurance process that will verify that the pre-defined performance criteria will be met. The owner/specifier is therefore responsible for:

1. Appointing a competent design authority and implementing an appropriate quality assurance process and management system (In most cases, responsibility for quality assurance will be delegated to the specifier.). The owner should demand that successful bids be clearly responsive to special requirements for achieving the desired concrete performance
2. Establishing the performance criteria depending on the expected concrete exposure conditions during placement and in service
3. Preparing the technical specification that states the performance criteria in clear terms i.e. discern the performance characteristics appropriate to the owner's intended use of the concrete unambiguously and quantitatively together with the test procedures used for acceptance so that the performance can be evaluated
4. Pre-qualification or verification criteria quality management requirements
5. Conducting quality assurance, reviewing quality assurance reports, or both, to ascertain that the performance criteria have been met
6. Defining the relevant exposure class for the concrete. These include the EN 206 [2] and ACI 318-11 [5] exposure classes and conditions (and any modifications thereof)—see Table 7.3 for a summary of the exposure classes. It is important to note that for any durability exposure class, there are no default requirements with respect to the type and proportion of concrete ingredients, and the production process, and the contractor/supplier is not expected to assume such
7. Stating any other concrete properties that may be required to meet the desired performance.

Table 7.3 Main exposure classes for EN 206 [2], ACI 318-8 [5], CSA A23.1 [6] and AS 3600 [3]

EN 206 exposure classes[a]:	No risk of corrosion or attack	X0
	Carbonation-induced attack	XC
	Chlorides not from sea water	XD
	Chlorides from sea water	XS
	Freeze-thaw with or without de-icing salts	XF
	Chemical attack	XA

(continued)

Table 7.3 (continued)

ACI 318-08 exposure classes[a]:	Freezing-thawing	F
	Sulphates	S
	Requirements for low permeability	P
	Corrosion protection of reinforcement	C
CSA A23.1 exposure classes[a]:	Chloride exposures	C
	Freezing and thawing	F
	Not exposed to exterior influences	N
	Exposed to chemical attack	A
	Exposed to sulphate attack	S
AS 3600 exposure classes:	Surfaces in contact with the ground	
	Members protected by damp-proof membrane	A1
	Residential footings in a non-aggressive soils	A1
	Other members in non-aggressive soils	A2
	Members in non-aggressive soils	U
	Surfaces in interior environments	
	Fully enclosed within a building (except for a brief period of weather exposure during construction)	A1
	In industrial buildings (subjected to repeated wetting and drying)	B1
	External surfaces above ground	
	within 1 km of coastline	B2
	within 1 to 50 km of coastline	B1
	Further than 50 km from coastline and	
	within 3 km of industrial polluting area	B1
	in non-industrial and tropical zone	B1
	in non-industrial and temperate zone	A2
	in non-industrial and arid zone	A1
	Surfaces in contact with water	
	in soft or running water	U
	in fresh water	B1
	in sea water and	
	permanently submerged	B2
	in tidal or splash zone	C
	Surfaces of members in other environments	U

[a]The exposure classes shown here have sub-classes (more detail is provided in [28])

7.5.2 *Responsibilities of the Contractor*

The contractor (and sub-contractor):

1. Is responsible for procuring concrete and related materials and incorporating them into the structure in a manner that meets the performance requirements
2. Is responsible for conducting appropriate and sufficient quality control to demonstrate and document that the performance requirements have been met. The quality control documents must be communicated to the owner/engineer (design authority) in a manner, and according to a schedule, that will accommodate the quality assurance process
3. Must be aware of and share the responsibility for handling, constructability, curing concrete and scheduling issues that influence the in-place concrete properties
4. Should detail in their bid how they intend to meet the special performance requirements part of the bid e.g. placement methods, protection, curing, etc
5. Should understand that any errors or deficiencies (non-conformance) must be corrected immediately, the owner notified of the incident, and the corrective action taken and all data transmitted to all the parties involved without delay
6. Needs to be aware of the performance test programme prior to bidding in order to allow for associated costs
7. Should work with the supplier to establish the concrete mixture properties to meet the performance criteria for plastic and hardened concrete, considering the contractor's criteria for construction and placement and the owner's performance criteria
8. Should submit documentation demonstrating the owner's pre-qualification performance requirements have been met
9. Should prepare and implement a quality control plan to ensure that the owner's performance criteria will be met and submit documentation demonstrating the owner's performance requirements have been met.

7.5.3 *Responsibilities of the Supplier/Producer*

The concrete supplier is responsible for procuring materials and producing concrete that will, in its plastic and hardened states, meet the owner's performance requirements. This includes responsibility for implementing a quality control programme to demonstrate and document that the product as delivered is of appropriate quality and will meet the performance requirements. In summary, the supplier should:

1. Certify that the concrete production plant, equipment, and all materials to be used in the concrete comply with the requirements of the performance standard

2. Certify that the concrete mix design satisfies the requirements of the performance standard i.e. the supplier should approve the proposed concrete mix in advance of the construction operations (i.e. pre-qualification). The specified properties should be verified by the owner (using standard industry-accepted test methods) for acceptance at the point of discharge
3. Certify that production and delivery of concrete will meet the requirements of the performance standard
4. Prepare and implement a quality control plan to ensure that the owner's and contractor's performance requirements will be met
5. Provide documentation verifying that they meet industry certification requirements, if required
6. At the request of the owner, submit documentation to the satisfaction of the owner demonstrating that the proposed mixture design will achieve the required strength and durability performance requirements.

Finally, it is important to note that aspects such as method and rate of placement, required slump/slump flow at point of discharge and finishability should be agreed on between the contractor and supplier.

7.6 Conclusion

In a performance-based specification, 'performance' means more than acceptance of plastic concrete at the end of the truck chute. It also means in-place performance of the hardened concrete. Therefore, the contractor and concrete supplier/producer have to work as a team to meet 'in-place' or 'end-result' concrete specifications. Just as it is with prescriptive specifications, it is important that the owner is clear when specifying concrete performance as to their own roles and responsibilities as well as those of the contractor and supplier. Further, since in a typical construction project the custody of the concrete transfers from the supplier to the contractor while in its plastic state, a high degree of co-ordination is required between the supplier and contractor to ensure that the final product meets the performance criteria and that the quality control processes are compatible and demonstrate compliance. The supplier-contractor team should be flexible enough to choose suitable combinations of materials, concrete mixtures and construction techniques to meet the desired performance criteria so that projects can be planned and bid, risks and costs can be assessed, and materials and construction operations adjusted to comply with performance requirements.

References

1. Bickley JA, Hooton RD, Hover KC. Preparation of a performance-bases specification for cast-in-place concrete, Report to the Ready-mixed Concrete Research Foundation. Maryland, USA: Silver Springs; 2006. p. 168.
2. EN-206, Concrete: specification, performance, production and conformity. European Standard; 2013.
3. AS-3600, Concrete Structures, Incorporating amendments No. 1 and 2. Sydney, New South Wales: Standards Australia; 2001. p. 430.
4. NZS-3101. Concrete structures standard. Part 1—the design of concrete structures. Section 3 —Design for Durability. Wellington: Standards New Zealand; 2006.
5. ACI-318-11. Building code requirements for structural concrete. Farmington Hills, Michigan: American Concrete Institute; 2011.
6. CSA-A23.1-09. Concrete materials and methods of concrete construction. Mississauga, Ontario, Canada, L4 W 5N6: Canadian Standards Association; 2009.
7. ASTM-C1202-12. Standard test method for electrical indication of concrete's ability to resist chloride ion penetration. 100 Barr Harbour Dr., P.O. box C-700, West Conshohocken, Pennsylvania USA: ASTM Annual Book of Standards, V.04.02, ASTM International; 2012.
8. ACI-ITG8R-10. Report on performance-based requirements for concrete (adopted in 2013 by ACI Committee 329 on Performance Criteria for Ready Mixed Concrete). Farmington Hills, Michigan: American Concrete Institute; 2010. p. 46.
9. Hooton RD, Bickley JA. Prescriptive versus performance approaches for durability design—the end of innocence? Mater Corros. 2012;63(12):1097–101.
10. Hover KC, Bickley JA, Hooton RD. Guide to specifying concrete performance: phase II Report of preparation of a performance-based specification for cast-in-place concrete. Report, RMC Research & Education Foundation, http://www.nrmca.org/p2p; 2008. p. 39.
11. Hooton RD, Hover KC, Bickley J. Performance standards and specifications for concrete for promotion of sustainable construction. In: Toutlemonde F, et al. Proceedings of the international conference on Concrete under Severe Conditions: Environment and Loading (CONSEC '07); 2007. p. 16.
12. Hooton RD, Karkar E. Evaluating durability of concretes using rapid measurements for fluid penetration resistance. Proceedings, concrete structures for sustainable community. Stockholm: FIB; 2012. p. 315–318,
13. Lane DS, Detwiler RJ, Hooton RD. Testing transport properties in concrete. Concr Int. 2012;32(11):33–8.
14. Stanish KD, Hooton RD, Thomas MDA. Evaluation of four short-term methods for determining chloride penetrability in concrete, water-cement ratio and other durability parameters: techniques for determination. ACI SP-191, 2000. p. 81–98.
15. Hooton RD, Mindess S, Roumain JC, Boyd AJ, Rear KB. Proportioning and testing concrete for durability. Concr Int. 2006;28(8):38–41.
16. SANS 3001-CO3-3. Concrete durability index testing—part 3: chloride conductivity testing (Draft). Pretoria, South Africa: South African Bureau of Standards—Standards Division; 2015.
17. Alexander MG, Ballim Y, Stanish K. A framework for use of durability indexes in performance-based design and specifications for reinforced concrete structures. Mater Struct. 2008;41(5):921–36.
18. SANS 3001-CO3-2, Concrete durability index testing—part 2: oxygen permeability test (Draft). Pretoria, South Africa: South African Bureau of Standards—Standards Division; 2015.
19. Alexander MG, Ballim Y, Mackechnie JR. Durability index testing procedure manual—revised research monograph No. 4 (1999), Department of civil engineering, University of Cape Town. Concrete Materials and Structural Integrity Research Unit, Department of Civil Engineering, University of Cape Town; 2009. p. 30.
20. AS-1012.13. Methods of testing concrete. Sydney, New South Wales: Standards Australia; 1993.

21. NZS-3112:1986, Methods of Test for Concrete. Part 1—Tests Relating to Fresh Concrete. Wellington: Standards New Zealand; 1986.
22. AASHTO-T318, Standard method of test for water content of freshly mixed concrete using microwave oven drying. Washington DC: American Association of State Highway and Transportation Officials; 2007.
23. Bognacki CJ, Pirozzi M, Marsano J, Scriffiano A. Increasing the services lives of concrete pavements. Concr Int. 2012;34(1):27–33.
24. Nokken MR, Hooton RD. Electrical conductivity as a pre-qualification and quality control tool. Concr Int. 2006;28(10):61–6.
25. Polder R, Andrade C, Elsener B, Vennesland Ø, Gulikers J, Weidert R, Raupach M. Test methods for on-site measurement of resistivity of concrete. Mater Struct. 2000;33(10):603–11.
26. ISO-9000. Quality management, International Organization for Standardization, 1, ch. de la Voie-Creuse, CP 56, CH-1211 Geneva 20, Switzerland; 2005.
27. Bickley JA, Hooton RD, Hover KC. Issues related to performance-based specifications for concrete. In: Soutos M, editor. Concrete Durability—A practical guide to the design of durable concrete. Thomas Telford; 2012. p. 481–541.
28. Hooton RD, Mindess S, Roumain JC, Boyd AJ, Rear KB. Proportioning and testing concrete for durability. Concr Int. 2006;28(8):38–41.
29. Geiker MR. On the importance of execution for obtaining the designed durability of reinforced concrete structures. Mater Corros. 2012;63(12):1114–8.

Chapter 8
Application Examples of Performance-Based Specification and Quality Control

H. Beushausen, M.G. Alexander, C. Andrade, M. Basheer, V. Baroghel-Bouny, D. Corbett, R. d'Andréa, A. Gonçalves, J. Gulikers, F. Jacobs, A.V. Monteiro, S.V. Nanukuttan, M. Otieno, R. Polder and R. Torrent

8.1 Introduction

This chapter presents an overview on several performance-based approaches for concrete durability specification and conformity assessment of the as-built structure. Various authors contributed to the following sections and presented approaches that have already been applied in practice for some time, often based on regional or national traditions and experiences. The application of various test methods is covered, including those measuring air permeability, oxygen permeability, water permeability, resistivity, conductivity, and ionic migration. For some of these material properties different test methods and their application are reviewed, and some are discussed with specific reference to particular countries. Where possible, background information with respect to the development of limiting values is presented based on fundamental or empirical relationships between test values and actual deterioration processes such as chloride ingress and carbonation. Based on

H. Beushausen (✉) · M.G. Alexander
University of Cape Town, Cape Town, South Africa

C. Andrade · R. d'Andréa
Eduardo Torroja Institute (IETcc-CSIC), Madrid, Spain

M. Basheer
University of Leeds, Leeds, UK

V. Baroghel-Bouny
Laboratoire Central des Ponts et Chaussées (LCPC), Chelles, France

D. Corbett
Proceq SA, Schwerzenbach, Switzerland

A. Gonçalves
Laboratório Nacional de Engenharia Civil—LNEC, Lisboa, Portugal

H. Beushausen and L. Fernandez Luco (eds.), *Performance-Based Specifications and Control of Concrete Durability*,
RILEM State-of-the-Art Reports 18, DOI 10.1007/978-94-017-7309-6_8

the structure of this chapter, i.e. the separate discussion of various approaches in individual sections, some information may be repeated as sometimes a similar discussion is necessary to explain the background and application of the various approaches.

A number of test methods and performance approaches exist worldwide, which are not covered in this chapter as they are still under development or have so far only been used for research purposes. A comprehensive list of various test methods and their application for research purposes, pre-qualification, or conformity assessment in different countries is presented in Appendix A. For the future, it can be expected that the number of performance-based approaches for concrete durability design increases with growing experience and further development of various test methods.

The information presented in the following sections is necessarily just a condensed summary of all the knowledge and experience available for the various approaches. A comprehensive list of references is presented at the end of the chapter for those readers seeking further details.

8.2 Site Air-Permeability ("Torrent Method")

8.2.1 Introduction

Since the early 90s, ASTRA (Swiss Federal Bureau of Roads) has been supporting research and development (R&D) projects oriented at developing a suitable approach for specifying and controlling the quality of the cover concrete on site [1–5]. This work, complemented by other investigations, led to the standardization in 2003 of a non-invasive test method, originally developed by [6], to measure the

J. Gulikers
Rijkswaterstaat Bouwdienst, Rotterdam, The Netherlands

F. Jacobs
TFB, Zurich, Switzerland

A.V. Monteiro
National Laboratory for Civil Engineering, Lisboa, Portugal

S.V. Nanukuttan
Queen's University Belfast, Belfast, UK

M. Otieno
University of the Witwatersrand, Johannesburg, South Africa

R. Polder
TNO Built Environment and Geosciences, TU Delft, Delft, The Netherlands

R. Torrent
Materials Advanced Services, Buenos Aires, Argentina

coefficient of air-permeability of the cover concrete on site [7]. Hence it is generally known as "Torrent Method" for site air-permeability.

In the same year, a new Swiss Code for Concrete Construction, SIA 262:2003, based on Eurocode 2, was issued [8]. This Code describes the measures to be adopted in order to ensure durability and, acknowledging the importance of the "impermeability" of the cover concrete, specifically states:

- "With regard to durability, the quality of the cover concrete is of particular importance"
- "The impermeability of the cover concrete shall be checked, by means of permeability tests (e.g. air permeability measurements), on the structure or on cores taken from the structure"

However, no limiting values of the coefficient of air-permeability (*kT*) were specified nor conformity rules for compliance given in the Code. The coefficient of air-permeability, determined with instruments based on the principle and formula given in [2], is abbreviated with *kT*.

In a recent report [5] recommendations are given on:

- Limiting values of kT_s for typical exposure classes found in Switzerland
- Sampling measurement points within a structure
- Site measurement of air-permeability of concrete
- Age, Temperature and Moisture conditions for testing *kT*
- Compliance criteria to check conformity with the specified kT_s values

Many of these recommendations have been incorporated in a new version of the Swiss Standard covering the test method [9].

There are two commercial instruments, complying with Standard SIA 262/1-E on the market, the "Torrent Permeability Tester" produced by Proceq SA and the "*PermeaTORR*", produced by Materials Advanced Services Ltd. The former yields higher *kT* values for low permeability concretes, say *kT* below 0.5×10^{-16} m^2 [10].

8.2.2 Specified Limiting KT_s Values

Recommended limiting or specified values of the coefficient of air-permeability (kT_s), as a function of the exposure classes of the Swiss version of Standard EN 206, are shown in Table 8.1. As discussed later, the kT_s values are "characteristic" upper values.

Several researches have shown that the coefficient of air-permeability *kT* correlates quite well with other standardized durability-related tests [5, 10]. For instance, Fig. 8.1 shows data of water sorptivity (SIA 262/1 Annex A) and Fig. 8.2 of carbonation depth (RILEM Recommendation CPC 18) of concretes after 500 days of indoor exposure (20 °C, 50 % RH), in both cases as function of their *kT* values measured at 28 days.

Table 8.1 Recommended limiting values kT_s specified for measurements on-site, as function of the exposure conditions

Exposure	EN 206 classes	kT_s (10^{-16} m^2)
Moderate carbonation	XC1, XC2, XC3	Not required
Severe carbonation Moderate chlorides Moderate frost	XC4 XD1, XD2a[a] XF1, XF2	2.0
Severe chlorides Severe frost	XD2b[a], XD3 XF3, XF4	0.5

[a]Swiss regulation: XD2a: chloride content ≤ 0.50 g/l; XD2b: chloride content > 0.50 g/l

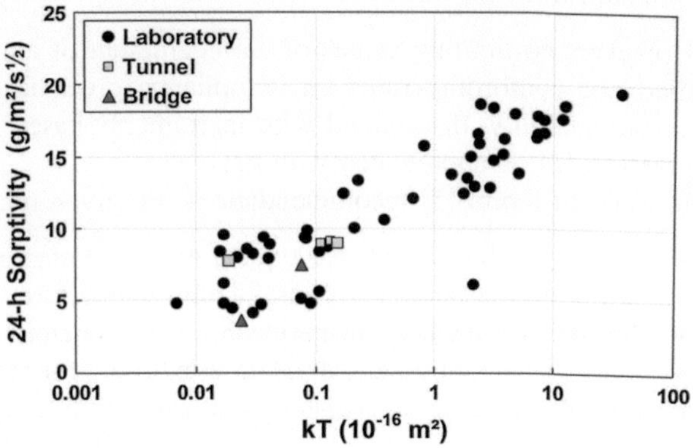

Fig. 8.1 Relation between water sorptivity and air-permeability *kT*, data from [1, 2]

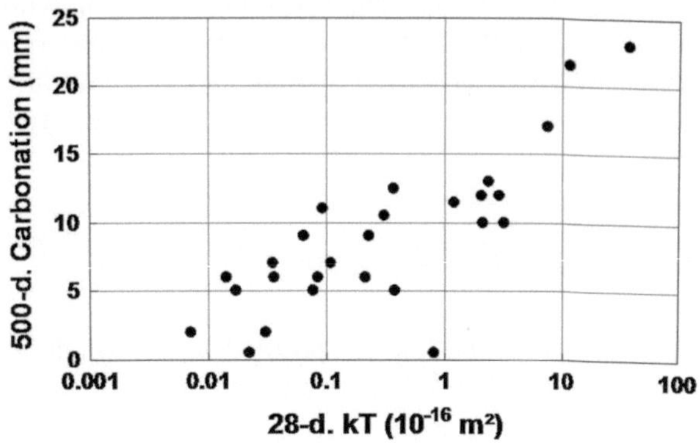

Fig. 8.2 Relation between indoor carbonation and air-permeability *kT* mean, data from [1, 2]

Figure 8.3 shows data from measurements of carbonation depths and air permeability, performed at ages between 35 and 40 years, on several constructions.

Figure 8.4 presents the correlation of *kT* with the mean penetration of water under pressure (EN 12390-8 [12] and DIN 1048 [13]), coming from several sources.

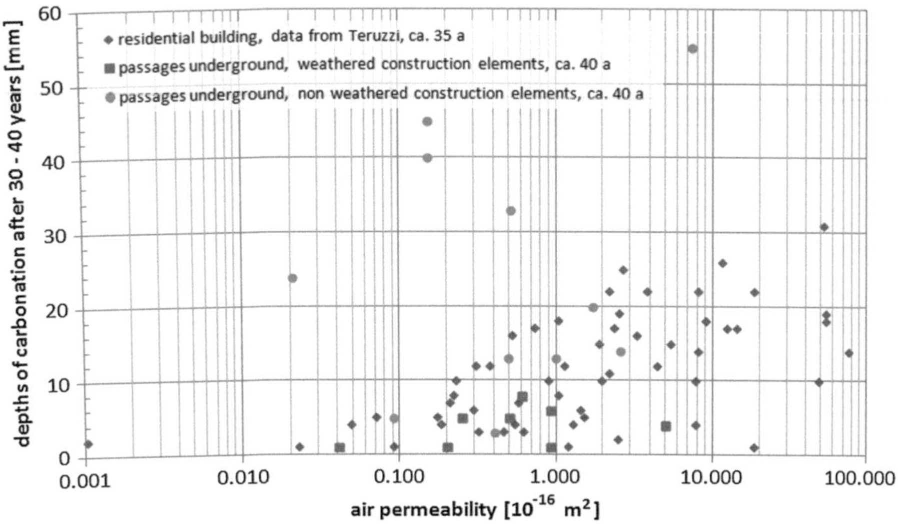

Fig. 8.3 Carbonation depth and air permeability *kT* of several constructions, data from [11]

Fig. 8.4 Relation between mean water penetration under pressure and air-permeability *kT*, data sources reported in [10]

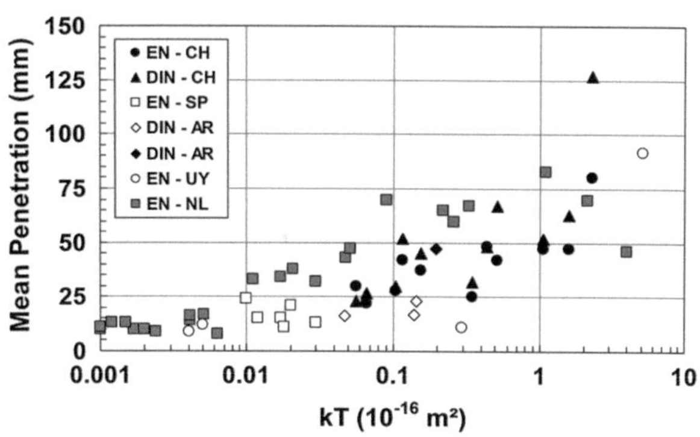

Finally, Fig. 8.5 presents the correlation between *kT* and Coulombs passed in the 'Rapid Chloride Permeability Test' [14]. Interesting to remark is the variety of countries contributing to the results.

Based on investigations in the laboratory and on more than 100 new and old construction elements tested on site [1–4] the specified limiting values kT_s were set up. This means that the air permeability was measured on elements and specimens made of concrete fulfilling the actual requirements of the Swiss concrete standard (SIA 262, SN EN 206) in terms of composition and conformity testing. Based on these results the limiting values of kT_s, for measurements on-site, were recommended [5].

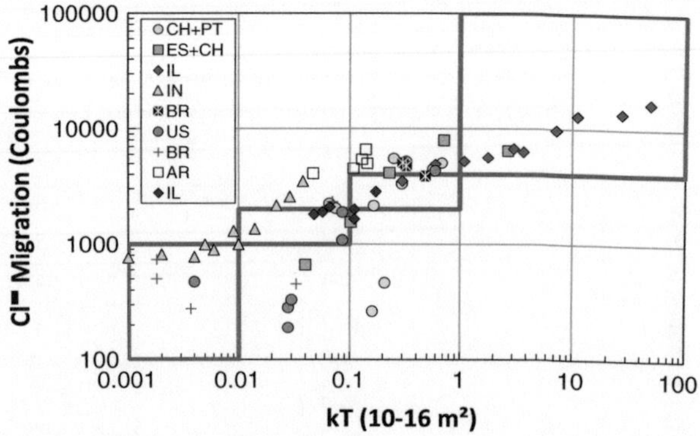

Fig. 8.5 Relation between and air-permeability *kT*, data sources reported in [15]

8.2.3 *Conformity Rules and Reporting*

The background of the conformity rules can be found in [5]. Each Test Area (Lot) must satisfy the following conditions:

Condition 1: Out of 6 air-permeability values kT_i, measured on a Test Area, as described in Sect. 8.2.4 , not more than 1 can exceed the specified Air-permeability limit value kT_S. In case that just 2 of the 6 air-permeability values kT_i, measured on a Test Area, exceed the specified air-permeability limit value kT_s, a further 6 Air-permeability tests should be conducted on 6 new Measurement Points selected from the same Test Area.

Condition 2: Not more than 1 air-permeability value kT_i out of the 6 new determinations can exceed the specified air-permeability limit value kT_s. If neither Condition 1 nor Condition 2 is satisfied, the Test Area is considered as not in conformity with the specifications and complementary/remedial measures have to be taken.

The Operation Characteristic Curve (OC) of a compliance criterion gives the probability of accepting a Lot as function of the proportion of "defectives" in the Lot (in this case, area with *kT* values above the specified kT_s).

Figure 8.6 presents the OC curve of the adopted conformity criterion for *kT*. It means that a Test Area (Lot) composed by just 10 % of non-compliant concrete (i.e. with $kT > kT_s$) has 97 % probability of being accepted. On the other hand, a Test Area composed by 40 % of non-compliant concrete has only 30 % probability of being accepted. This gives a clearer statistical meaning of kT_s as 'characteristic' air-permeability upper limit.

A report form is proposed to present results and relevant information as well as special circumstances that shall be present during the measurements (e.g. cracks, Surface Protection Treatment).

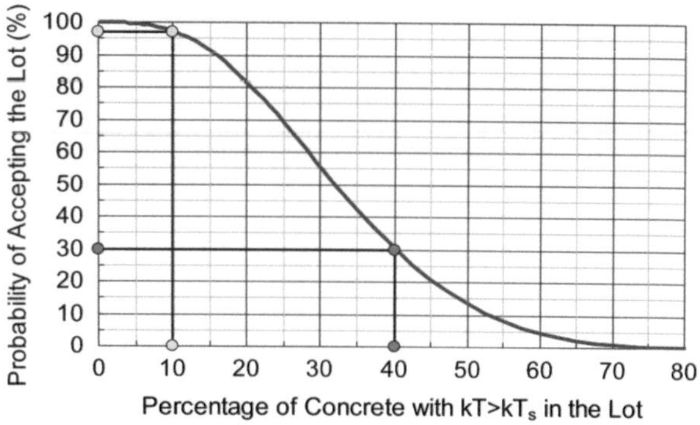

Fig. 8.6 OC curve of the compliance criterion

8.2.4 *Sampling of Test Areas and Measurement Points*

8.2.4.1 Grouping

The structure to be evaluated should be divided into groups of elements that have the following features in common:

- Same specified Air-Permeability value kT_s (see Table 8.1)
- Built with concrete belonging to the same EN 206 class (same strength, aggregate size and exposure class)
- Built applying similar concreting practices (placing, compaction, curing, etc.)

For compliance purposes, all the elements in the structure having the same features described above, will constitute a Group. They should be listed chronologically, within each Group, by date of concreting; in the case of continuous elements (e.g. walls or deck slabs), segments concreted on the same day should be identified.

8.2.4.2 Test Areas

The elements within each Group are divided into Test Areas (Lots) according to the following criteria (the resulting maximum number of Test Areas should be adopted):

- 1 Test Area per each 500 m^2 of exposed surface area or fraction thereof
- 1 Test Area per 3 days of concreting of the elements of the Group

8.2.4.3 Measurement Points

From each resulting Test Area, 6 Measurement Points are sampled at random, avoiding excessive closeness to edges (especially top and bottom) and to each other, as shown in Fig. 8.7.

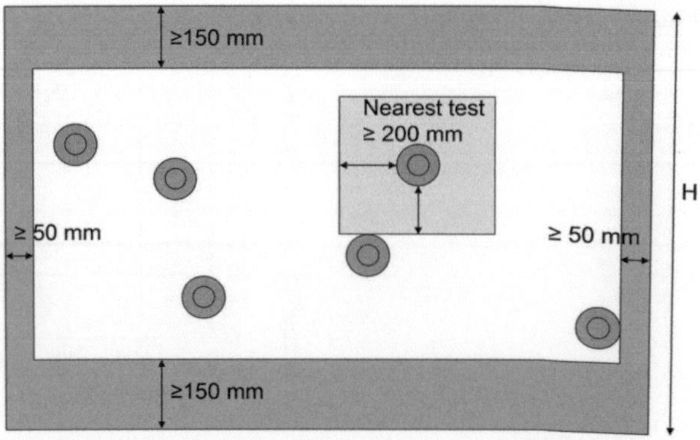

Fig. 8.7 Selection of measurement points within a test area of an element of height *H*

For the selection of the Measurement Points, the following should be observed:

- Measurements can be made on sufficiently smooth surfaces; if too rough, it can be manually polished with care
- Surface Protection Treatment (SPT): If possible it should be locally removed and the circumstance should be mentioned in the test report
- Re-bar cover depth: It should be controlled that there are not re-bars, cable ducts or pipes closer than 20 mm from the concrete surface at the Measurement Point
- De-dusting: Before the test, the surface shall be de-dusted with a brush or hard, dry sponge
- Measurement Points on visible cracks should be avoided, i.e. the Measurement Points should previously be inspected for cracks, e.g. by spraying the zone with an alcohol solution
- The Measurement Points should be marked on the surface (e.g. pencil or chalk), to avoid two measurements being conducted on the same spot and for further investigation, if required.

8.2.5 Age, Temperature and Moisture Conditions of the Concrete

- Age of concrete: the age of concrete when tested should be between 28 and 90 days. In particular, when slow-reacting cements (e.g. CEM III/B) or significant amount of slow-reacting mineral additions such as fly-ash are used, a minimum age of concrete of 60 days should be considered.
- Temperature of concrete: the surface temperature of the construction element, measured for instance with an infrared thermometer, should be above 10 °C. Experienced users can, if necessary, perform the test measure at temperatures between 5 and 10 °C.

- Moisture conditions of concrete: the moisture content should not exceed 5.5 % (by mass) when determined by electrical impedance method (Concrete Moisture Encounter instrument manufactured by Tramex or equivalent)
- The above condition is likely to be met if the curing ended 3–4 weeks prior to the test and more than 2–5 days have passed after the last ingress of water in the concrete by, for instance, rain, spray or thaw.

8.2.6 Application

The proposed approach was tested on two constructions by several laboratories (different measuring devices and users) [5]. The measurements were made prior to the setup of the proposed conformity control. Therefore the number of measurements does not always equal 6 or 12. All measurements were performed on the same construction elements but on different spots.

The first construction was a trough bridge where the measurements were made on two wall segments. The concrete quality specified by the engineer was C30/37, XD3, XF3, D_{max}32, Cl 0.10, C3 (comment: XD3 and XF3 are a contradiction). Therefore the required air permeability kT_s was selected by the most demanding exposure class XD3 as 0.5×10^{-16} m^2. Figures 8.8 and 8.9 illustrate the bridge and

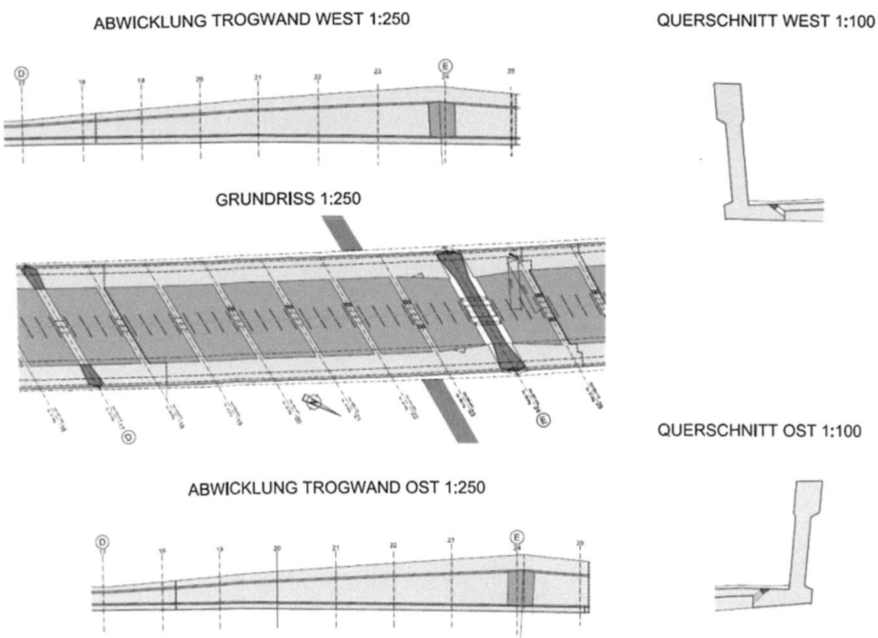

Fig. 8.8 Sketch of trough bridge

Fig. 8.9 Measurement points (*squares*) at one construction element

the comparative measurements. A summary of the results is shown in Fig. 8.10 and Table 8.2. On section XI all five laboratories found that the air permeability fulfilled the requirement (Table 8.2), on section D–E four out of five laboratories found that the air permeability fulfilled the requirement. A visual inspection and measurements on drilled cores confirmed that some parts of the walls have a low quality. In summary, the air permeability measurements gave a realistic picture of the cast concrete quality.

The second construction was a cut-and-cover-tunnel where measurements were made on three wall segments. The ordered concrete quality was C30/37, XD3, XF1, $D_{max}32$, Cl 0.10, CT1.10 (comment: CT: target value for C; XD3 and XF1 are a contradiction). Therefore the required air permeability kT_s was selected by the most demanding exposure class XD3 as 0.5×10^{-16} m^2. A summary of the results is presented in Fig. 8.10 and Table 8.3. On section 41 E three out of four laboratories

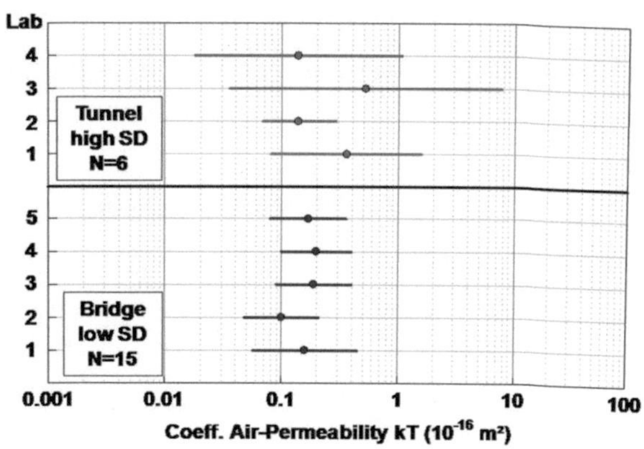

Fig. 8.10 Geometric mean ± standard deviation of logarithms in bridge (section XI) and tunnel, measured by different labs

Table 8.2 Results of air permeability measurements on the trough bridge

Laboratory	1	2	3	4	5
Section XI					
Geometric mean of kT (10^{-16} m^2)	0.20	0.16	0.17	0.19	0.10
sLOG: Standard deviation log(kT)	0.30	0.45	0.33	0.32	0.32
Number > 0.5 × 10^{-16} m^2 [−]	2 of 15	2 of 15	1 of 14	2 of 12	0 of 15
Requirement fulfilled?	Yes	Yes	Yes	Yes	Yes
Section D–E					
Geometric mean of kT [10^{-16} m^2]	0.22	0.25	0.17	0.12	0.13
sLOG: Standard deviation log(kT)	0.42	0.58	0.61	0.32	0.68
Number > 0.5 × 10^{-16} m^2 [−]	2 of 13	5 of 15	3 of 15	0 of 15	2 of 15
Requirement fulfilled?	Yes	No	Yes	Yes	Yes

Table 8.3 Results of air permeability measurements on the cut-and-cover-tunnel

Laboratory	1	2	3	4
Section 41 E				
Geometric mean of kT (10^{-16} m^2)	0.99	0.10	0.19	0.25
sLOG: Standard deviation log(kT)	29	0.38	2.09	0.46
Number > 0.5 × 10^{-16} m^2 [−]	2 of 6	2 of 12	1 of 9	2 of 12
Requirement fulfilled?	No	Yes	Yes	Yes
Section 42 E				
Geometric mean of kT (10^{-16} m^2)	0.11	0.02	0.07	0.08
sLOG: Standard deviation log(kT)	0.41	0.01	0.02	0.04
Number > 0.5 × 10^{-16} m^2 [−]	2 of 12	0 of 6	0 of 6	0 of 6
Requirement fulfilled?	Yes	Yes	Yes	Yes
Section 41 W				
Geometric mean of kT (10^{-16} m^2)	0.14	0.53	0.14	0.36
sLOG: Standard deviation log(kT)	0.10	10	10	2.31
Number > 0.5 × 10^{-16} m^2 [−]	0 of 6	2 of 7	2 of 12	2 of 7
Requirement fulfilled?	Yes	No	Yes	No

found that the air permeability fulfils the requirement. On section 42 E all laboratories confirmed the fulfilment and on section 41 W two confirmed and two denied the fulfilment. A visual inspection and measurements on drilled cores confirmed that parts of the walls have a low quality. In summary, the air permeability measurements gave a realistic picture of the cast concrete quality.

The applications show that contradictory results are more likely to be found in air permeability measurements on site than with tests on specimens (cubes, cores). It is believed that this reflects mainly the higher number of non-invasive measurements compared to the mainly single results of tests on specimens and the higher variability of site concrete compared to lab specimens, together with the less controlled testing conditions.

Figure 8.10 summarizes the results of the investigations on both constructions. The good reproducibility achieved can be seen.

8.2.7 *Estimation of Service Life*

The Application Test dealt with in Chap. 10 required that the participants provide not just the test results but also an approach on how the measured values could be applied for service-life prediction. In particular, the participants had to analyse the adequacy of the panels to stand 100 years of life exposed to a severe de-icing salts chlorides environment (XD3 of EN 206).

8.2.7.1 Carbonation

Recent applications to estimate service life of old and new structures subject to carbonation have been presented [16–18]. Since carbonation-induced corrosion is not the object of the Application Test analysis, the topic will not be discussed in this report.

8.2.7.2 Chlorides

Regarding Chlorides-induced corrosion, the approach presented in [19] will be applied, as described below. Ten assumptions made for Service Life estimation are presented in detail.

Assumption 1: The end of the service life is reached when the concentration of chlorides at the level of the reinforcement equals the critical threshold (Service Life = Corrosion Initiation Time T_i).

Assumption 2: The mechanism of chloride penetration is non-steady state diffusion, following Fick's 2nd law.

Assumption 3: A decay in the coefficient of chloride diffusion D_{Cl} can be introduced in the explicit solution (error function) of Fick's 2nd law.

Assumptions 1–3 lead to Eqs. (8.1) and (8.2) to estimate service life:

$$T_i = \frac{c^2}{4\left(D_0\left(\frac{t_0}{t}\right)^m\right)} A^2 \tag{8.1}$$

$$A = \frac{1}{erf^{-1}\left(1 - \left(\frac{C_{cr}}{C_s}\right)\right)} \tag{8.2}$$

where T_i is the time for initiation of corrosion in years (equalled to the service life), c is the cover depth in mm, D_0 is the coefficient of chloride diffusion considered/measured at age t_O (typically 28 days), t is the hydration time ($t \leq t_{max}$, t_{max} corresponding to the end of hydration), m is the "ageing factor" or "diffusion decay exponent", erf^{-1} is the inverse error function, C_{cr} is the critical concentration of chloride capable of initiating the corrosion process, and C_s is the concentration of chloride at the surface of the element.

The term in brackets in Eq. (8.1) is the coefficient of chloride diffusion at time t.

Any attempt to use values of air-permeability measured on site to predict service life of concrete exposed to chlorides requires a relation between kT and the coefficient of chloride diffusion D_{Cl}.

Results of kT and D_{Cl} (measured under Cl^- ponding/immersion long-term tests) are plotted with black symbols in Fig. 8.11. The empty circles in Fig. 8.11 correspond to kT and Coulomb [14] values found in the literature as detailed in [15]. The Coulomb values were converted into D_{Cl} applying the following formula, established at Purdue University [20]:

$$D_{Cl} = 0.4 + 0.002 \cdot \text{Coloumb} \tag{8.3}$$

where D_{Cl} is the chloride diffusion coefficient in 10^{-12} m²/s.

Since the large majority of values of D_{Cl} and kT reported in Fig. 8.11 are measurements made at early ages, the coefficient of diffusion values correspond to D_O;

Assumption 4: Eq. (8.4), fitted to the results of Fig. 8.11, expresses the relation between kT and D_O.

$$D_0 = 315kT^{\frac{1}{3}} \tag{8.4}$$

where D_O is in mm²/year and kT is in 10^{-16} m².

Now, substituting Eq. (8.4) into Eq. (8.1):

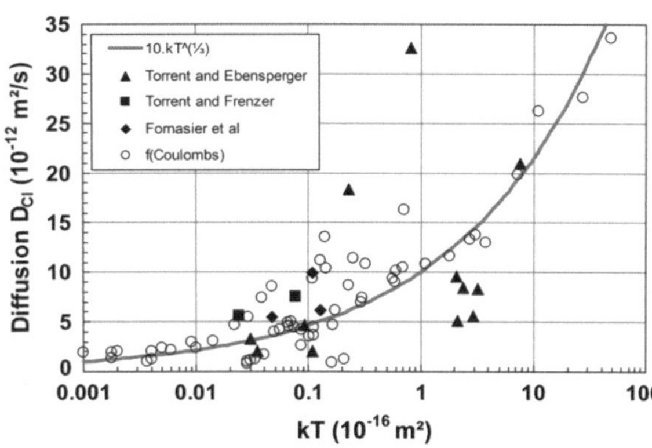

Fig. 8.11 Tentative relation between D_{Cl} and kT

$$T_i = \frac{c^2}{4\left(315kT^{\frac{1}{3}}\left(\frac{t_0}{t}\right)^m\right)}A^2 \tag{8.5}$$

Establishing the correct values of C_s, C_{cr} (and therefore of A) and of m is not easy and is a matter of controversy and discrepancy among specialists and prediction methods.

Assumption 5: To eliminate the influence of those factors on the estimated T_i, a Reference Condition is assumed. This Reference Condition is that a concrete structure designed and built according to the provisions of European Standards EN1992-1-1 (Eurocode 2) [21], EN 206 [22] and EN 13670 [23] will reach a service life of 50 years. This involves the following assumptions:

Assumption 6: The mean w/c ratio of the Reference concrete produced is given by Eq. (8.6).

$$w/c_{ref} = w/c_{max} - 0.02 \tag{8.6}$$

where w/c_{max} is the value specified in EN 206 for the applicable exposure class.

Assumption 7: The mean cover depth of the Reference structural elements is given by Eq. (8.7).

$$c_{ref} = c_{min} + 10 \tag{8.7}$$

where c_{min} is the value specified in EN1992-1-1 (Table 4.4 for Structural Class S4) in mm.

Assumption 8: The quality of execution (placement, compaction, finishing, curing) is according to EN 13670

Assumption 9: The relation between gas-permeability and *w/c* ratio, proposed in the CEB-FIB Model Code 1990, is applicable to estimate the kT of well processed concretes. Hence, the coefficient of air-permeability of the Reference concrete is (Eqs. 2.1–107 of [24]):

$$\log kT_{ref} = -19 + 5 \cdot w/c_{ref} \tag{8.8}$$

where kT_{ref} is in m^2.

Figure 8.12 shows that Eq. (8.8) gives a reasonable estimate of the expected kT_{ref} for a given value of w/c_{ref}, when compared with experimental results presented in [5].

If we apply the values of the Reference Condition to Eq. (8.7), we get:

$$T_{ref} = \frac{c_{ref}{}^2}{4\left(315kT_{ref}{}^{\frac{1}{3}}\left(\frac{t_0}{t}\right)^m\right)}A^2 \tag{8.9}$$

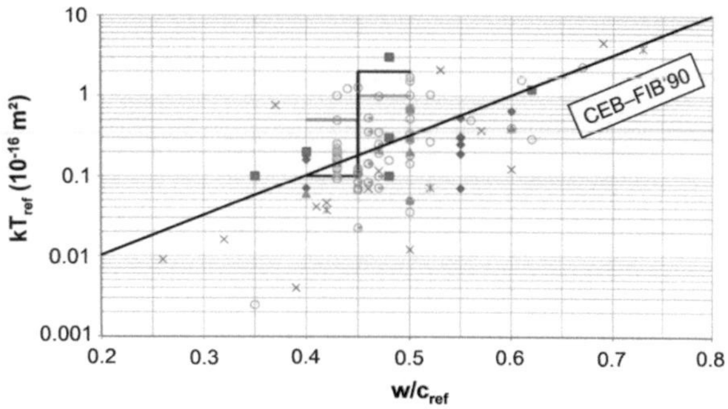

Fig. 8.12 Eq. (8.8) versus *kT* and *w/c* data compiled in Fig. D-8 of [4]

Assumption 10: If the Service Life estimate will correspond to ages ≥25 years, as is usually the case, it can be assumed that m, C_s and C_{cr} (and hence A) will not differ from those of the Reference Condition (50 years of age).

Then, dividing Eq. (8.5) by Eq. (8.9) and reorganizing terms, we get:

$$T_i = T_{ref}\left(\frac{c}{c_{ref}}\right)^2 \cdot \left(\frac{kT_{ref}}{kT}\right)^{\frac{1}{3}} \tag{8.10}$$

In the particular case of severe de-icing salts chloride environment XD3, is:

- $w/c_{max} = 0.45 \rightarrow w/c_{ref} = 0.43$
- applying Eq. (8.8) we get $kT_{ref} = 0.14 \times 10^{-16}\ \text{m}^2$
- $c_{min} = 45$ mm $\rightarrow c_{ref} = 55$ mm
- $T_{ref} = 50$ years

Entering the Reference values in Eq. (8.6) we get:

$$T_i = 50\left(\frac{c}{55}\right)^2 \cdot \left(\frac{0.14}{kT}\right)^{\frac{1}{3}} = 0.0086 \cdot \frac{c^2}{kT^{\frac{1}{3}}} \tag{8.11}$$

Similarly, for other chloride-aggressive environments is:

$$\text{for XD1} \quad T_i = 50\left(\frac{c}{45}\right)^2 \cdot \left(\frac{0.45}{kT}\right)^{\frac{1}{3}} = 0.0189 \cdot \frac{c^2}{kT^{\frac{1}{3}}} \tag{8.12}$$

$$\text{for XD2} \quad T_i = 50\left(\frac{c}{50}\right)^2 \cdot \left(\frac{0.45}{kT}\right)^{\frac{1}{3}} = 0.0153 \cdot \frac{c^2}{kT^{\frac{1}{3}}} \tag{8.13}$$

By measuring on site the cover depth c (by means of a covermeter) and the coefficient of permeability to air kT (supported by checking that the moisture content of concrete is sufficiently low for kT measurements), an estimate of the T_i of a particular zone of the structure can be made applying Eqs. (8.11)–(8.13).

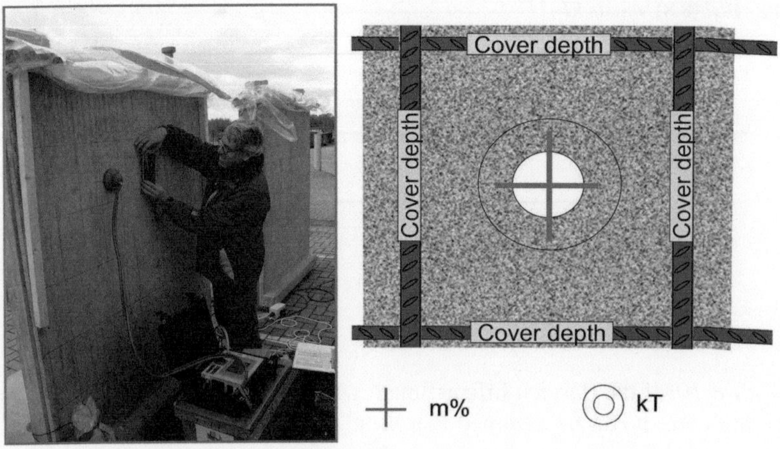

Fig. 8.13 Measurement scheme of *c*, *kT* and moisture content (*m%*)

Figure 8.13 shows the scheme followed in the Application Test. First, the position of the steel was identified and the cover depth measured, recording its minimum value as variable *c* in Eqs. (8.10) and (8.11). Then, it was verified that the moisture content of the concrete delimited by the bars was ≤5.5 % (electrical impedance method) and the coefficient of air-permeability was measured in the reinforcement-free zone. The result of the air-permeability test is the *kT* value to be entered in Eqs. (8.10) and (8.11).

A similar approach has been developed for the case of carbonation-induced corrosion [25].

8.3 Oxygen Permeability Index (South Africa)

The South African Oxygen Permeability Index (OPI) test method is described in Chap. 5. The oxygen permeability index obtained with this method is defined as the negative log of the coefficient of permeability (*K* (m/s)). OPI values for South African concretes range from approximately 8.5–10.5 (equivalent *K* from approximately 3.2×10^{-9} to 3.2×10^{-11} m/s), with a higher OPI value indicating a higher impermeability and thus a concrete of potentially higher quality. Note that oxygen permeability index is measured on a log scale and therefore an apparently small difference in the OPI value may correspond to a large difference in permeability. For example, a concrete with an OPI value of 8.5 is 100 times more permeable than a concrete with an OPI of 10.5.

8.3.1 Prediction of Carbonation Depth Development

Oxygen permeability results may be used to characterize young concretes (typically 28 days age) for influences such as concrete grade, binder type, initial curing and construction effects such as compaction. Based on empirical relationships, the carbonation resistance of concrete was found to be sufficiently related to the early age (28 days) OPI value (Fig. 8.14), so that OPI can be used in a carbonation-type service life model [26]. A typical outcome of the OPI-based carbonation model is shown in Fig. 8.15.

A recently updated version of the South African carbonation prediction model combines the aspects of binder chemistry, mix composition, environmental conditions, and the concrete's diffusivity as characterized by the OPI value, for the prediction of carbonation depth development [27]. Based on this model, the permeability coefficient assessed with the OPI test can be related to the carbonation coefficient as shown in Fig. 8.16.

It is possible to relate theoretically the permeability coefficient k to the diffusion coefficient D, where these two mechanisms occur in the same porous medium [28]. The latter is affected by binding reactions between the diffusing gas (CO_2) and the cement paste. Therefore the diffusion coefficient calculated from the carbonation coefficient represents the *effective* diffusion coefficient—the diffusion coefficient of CO_2 through concrete. By further adjusting D for the amount of carbonatable material (primarily the $Ca(OH)_2$ content—which differs for plain vs. blended binders) and normalising to a uniform humidity condition (65 % RH), a relationship emerges between permeation and diffusion that is essentially independent of binder type. From literature, the relationship will be of the kind $D = m.k^n$, with m and n constants. Data used in Fig. 8.16 and Fig. 8.17 show the normalised, where it is

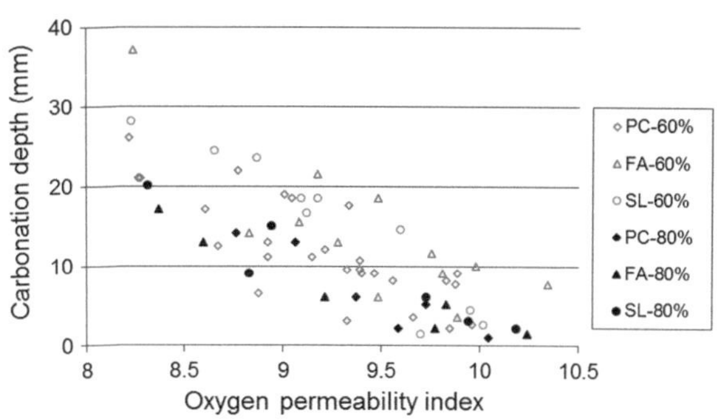

Fig. 8.14 Carbonation depth in various concretes (*PC* Portland cement, *FA* Fly ash (30 %), *SL* GGBS (50 %)) versus oxygen permeability index (measured at 28 days) for 4 years exposure at an average relative humidity of 60 or 80 % [26]

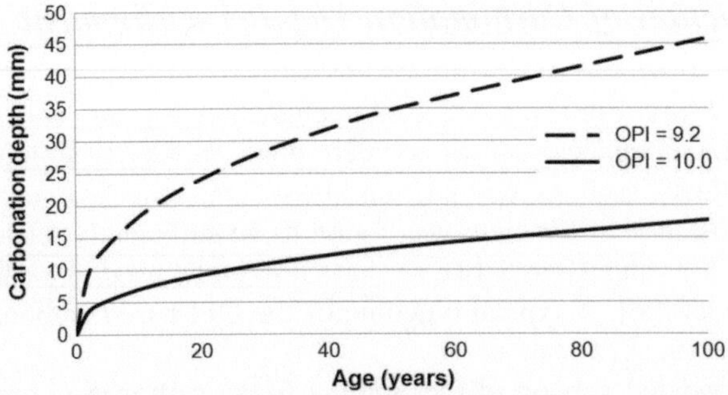

Fig. 8.15 Example for the prediction of carbonation depth development using the South African OPI approach (the data are based on concrete containing portland cement (70 %) and fly ash (30 %) situated in an environment with average RH = 80 %)

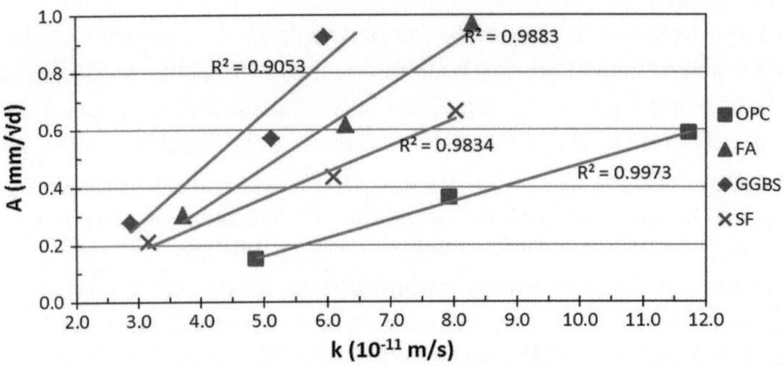

Fig. 8.16 Permeability *k* versus carbonation coefficient *A* for concretes with various binder types (100 % OPC, 70/30 OPC/FS, 50/50 OPC/GGBS, 90/10 OPC/SF), based on experimental results [27]

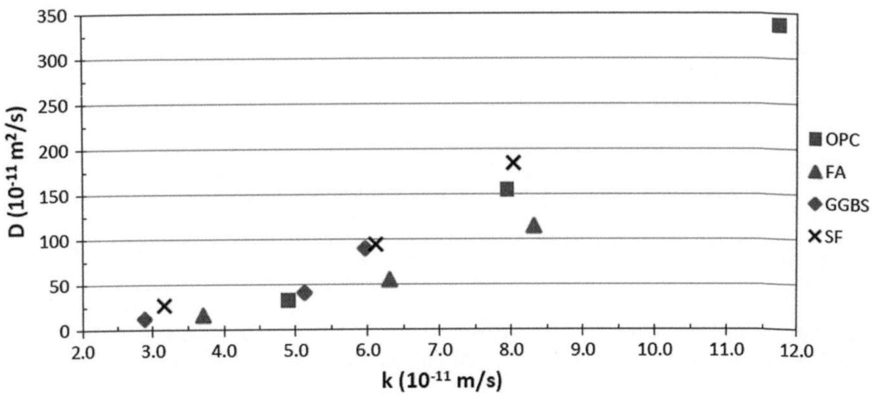

Fig. 8.17 Permeability versus effective dry diffusion coefficient (OPC = plain CEM I concrete, FA = 70 % CEM I and 30 % fly ash, GGBS = 50 % CEM I and 50 % blastfurnace slag, SF = 90 % CEM I and 10 % silica fume) [27]

clear that different concretes reasonably fall into the same band. This is very useful for the carbonation prediction model based on the permeability coefficient.

8.3.2 *Performance Specifications Using OPI Values*

The environments that require OPI values to be specified in the South African context are XC3 (Moderate humidity (60–80 %)) and XC4 (Cyclic wet and dry), with XC4 considered the more critical because steel corrosion can occur faster under these conditions. Since about 2005, durability specifications have been included in certain national infrastructural construction projects in South Africa. A typical specification for limiting OPI values, used by the South African National Roads Agency Limited (SANRAL) for the construction of highway bridges is shown in Tables 8.4 and 8.5.

As a general rule, concrete in the as-built structure can be expected to be of lower quality compared with the same concrete placed and cured under controlled laboratory conditions. To account for the improved performance of laboratory concrete over site concrete, the characteristic values for the durability indexes of the laboratory concrete are generally assumed to be higher. Further details on durability specifications in national infrastructure programmes in South Africa are discussed in the literature [29, 30].

Table 8.4 Typical permeability specifications used in South Africa (extract) (SANRAL, typical highway bridge construction contract)

Environmental class		XC3		XC4	
Cover depth (mm)[a]		40	50	40	50
OPI (log scale)	Recommended	9.4	9.1	9.6	9.3
	Minimum	9.0	9.0	9.2	9.0

[a]Absolute minimum cover, since this is a prediction model value

Table 8.5 Oxygen permeability index acceptance ranges for environmental exposure XC4 (SANRAL, typical highway bridge construction contract)

Acceptance category	OPI (log scale)	
	40 mm cover	50 mm cover
Concrete made, cured and tested in the laboratory	>9.6	>9.3
Full acceptance of in situ concrete	>9.6	>9.3
Conditional acceptance of in situ concrete (with remedial measures as approved by the engineer)	9.2–9.6	9.0–9.3
Rejection	<9.2	<9.0

8.4 Autoclam Permeability System (UK)

8.4.1 Functional Purpose of the Autoclam Permeability System

The Autoclam Permeability System can be used to measure the air and water permeability and the water absorption (sorptivity) of concrete and other porous materials, both in laboratory and on site. Using this equipment, the rate of decay of air pressure is recorded for the air permeability test, whereas the volume of water penetrating into the concrete, at a constant pressure of 0.02 bar and 0.5 bar are recorded for the sorptivity and the water permeability tests respectively. These tests, which can be carried out quickly and effectively on site without prior planning, are essentially non- invasive in nature and a skilled operator is not needed.

8.4.2 Choice of Test Method for Measuring Permeation Characteristics

Whilst it has been recognised that the choice of test method should be primarily to obtain the intrinsic permeation characteristics of concrete, such as sorptivity, permeability and diffusivity, rather than an index of these characteristics, as the base for comparisons, the Autoclam Permeability System and its sister product, Permit Ion Migration Test, were developed to provide an index of these characteristics. This is due to the reason that in most practical situations the test method should be chosen to be appropriate to the predominant mechanism acting on the concrete under consideration [31]. Therefore, for an offshore concrete structure the dominant mechanism could be gas diffusion, water vapour diffusion, water absorption, water permeability and/or ionic diffusion depending on the location of concrete and its exposure environment. For instance, an absorption type of test would be suitable for studying the long-term performance of concrete in the tidal zone, whereas a pressure differential water permeability test would be more appropriate for investigating the behaviour of concrete subjected to deep submersion. Therefore, the choice of Autoclam air permeability test, water permeability test and/or water absorption (sorptivity) test should be based on the exposure condition of the concrete that is being tested.

8.4.3 Principle of Operation of the Autoclam Permeability System

As the moisture content of the test material influences the permeation mechanisms, tests with the Autoclam Permeability System also depend on the moisture content of the specimen [32, 33]. However, if tests are carried out on specimens either preconditioned to remove the moisture or sheltered for long periods [a period of at least 2 weeks in warm weather conditions], the moisture effect can be minimised [34]. Basheer and Nolan [32] have concluded that the Autoclam permeation tests cannot distinguish the quality of concrete if the internal relative humidity of concrete in the cover zone is greater than 80 %. Therefore, it is recommended that the internal relative humidity of concrete in the cover zone up to a depth of 10 mm is measured before carrying out any of the Autoclam permeation tests and testing should proceed only if the internal relative humidity is less than 80 %.

The basic principle of the operation of the Autoclam Permeability System can be explained with reference to Fig. 8.18a. The base ring isolates a test area with a diameter of 50 or 75 mm when it is fixed onto the surface under test. Either the bonding type ring (Fig. 8.18b) or the bolt on type ring (Fig. 8.18c) can be used for this purpose. The use of a 50 mm internal diameter ring is recommended for both the air permeability and water permeability tests. For the sorptivity (water absorption) tests the use of either a 50 mm internal diameter or a 75 mm internal diameter ring is recommended, the latter for testing low absorbent test surfaces, such as surface treated concretes. The bonding type ring is fixed on to the test surface using a suitable epoxy adhesive. Three holes of 6.5 mm diameter are drilled into the concrete to fix the bolt on type ring. As there is a mark left on the test

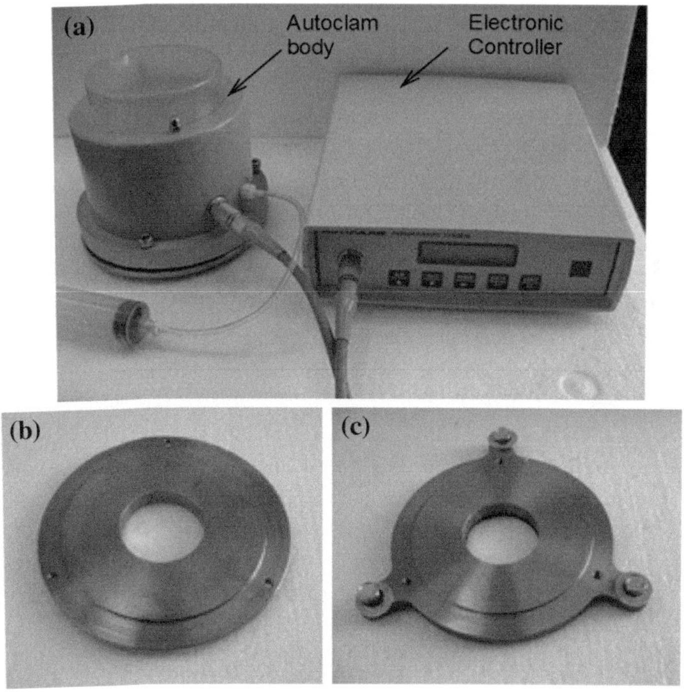

Fig. 8.18 **a** Autoclam permeability system, with bonding type base ring, **b** bonding type ring, and **c** bolt on type ring

surface upon the removal of the bonding type ring, this is not recommended for testing façades and decorative surfaces.

In order to carry out an air permeability test, the pressure inside the apparatus is increased to slightly above 0.5 bar (50 kPa) and the decay in pressure is monitored every minute from 0.5 bar (50 kPa) for 15 min or until the pressure has diminished to zero. A plot of natural logarithm of pressure against time is linear, hence the slope of the linear regression curve between the 5th and 15th minute for tests lasting for 15 min is used as an air permeability index, with units of ln(Pressure)/min. When the pressure becomes zero before the test duration of 15 min, the data from the beginning of the test is used to determine the slope. For concrete manufactured with Portland cement, the Autoclam air permeability index (API) in ln(pressure)/min can be expressed in terms of intrinsic air permeability, k_a (m^2) using the following formula [34]:

$$k_a = (API)^{0.8754} \times 8.395 \times 10^{-16} \tag{8.14}$$

The water absorption (sorptivity) test can be carried out at the same location, but at least 1 h needs to have elapsed after the air permeability test. Water is admitted into the test area through a priming pump with the air escaping through the bleed tube. When the test chamber is completely filled with water the priming pump automatically switches off and the micro pump pressurises the test area to 0.02 bar (2 kPa) above atmospheric. The test then starts. At this pressure water is considered to be absorbed into the capillary pores rather than via pressure induced flow. As water is absorbed by capillary action, the pressure inside would tend to decrease, hence it is maintained constant by the pump and the control system. The volume of water delivered is measured and recorded every minute for a duration of 15 min, the quantity of water absorbed during the test is recorded. A plot of the quantity of water absorbed and the square root of time elapsed is linear (this relationship depends to some extent on the type of concrete tested, however for all practical purposes a square root time plot may be employed). The slope of this graph is reported as the sorptivity index with units of $m^3/min^{0.5}$ and if the portion of the graph between the 5th minute and the 15th minute is utilised, setting up errors are minimised. The rate of inflow of water (water absorption) during the 10th–11th minute is used also to calculate an initial surface absorption at 10 min in mL/m^2/s according to BS 1881: Part 5 [35].

The water permeability test is conducted at a separate test location using the same test procedure as that for the water absorption (sorptivity) test. In this case after priming the system, the pressure inside is increased to 0.5 bar (50 kPa). Again the quantity of water flowing into concrete plotted against the square root of time is linear, and hence as in the case of water absorption test, the slope of the square root time plot between 5 and 15 min is used to report a water permeability index with units $m^3/min^{0.5}$. The major difference between the sorptivity test and the water permeability test is that in the former case capillary absorption causes the penetration of water whereas in the latter one the applied pressure also contributes to the rate of flow. That is, this is not a steady state water permeability test.

The number of each type of tests to be carried out depends on the variability of the material under investigation. However, at least three tests are recommended in order to reduce the effect of random variability of the test material.

8.4.4 *Classification of Concrete Using the Autoclam Permeability System*

In order to develop classification criteria for the Autoclam Permeability System, numerous laboratory investigations were carried out. A typical set is reported in Tables 8.6, 8.7 and 8.8. The mix combinations were decided after carrying out trials to check for their viability, i.e. mixes which either honeycombed or segregated were not included. Tests carried out in each series are also reported in these tables. In all these investigations, the Autoclam Permeability System was used to measure the air permeability, sorptivity and water permeability.

The freeze-thaw test in test series 1 was carried out in accordance with Procedure B of ASTM C666 [36]. This test regime was considered to have simulated XF3 condition in EN 206. A computer controlled environmental cabinet was

Table 8.6 Variables and properties investigated in test series 1

Test variables					Properties investigated
W/C	A/C[a]	Aggregate size (mm)			
0.40	3.16	6	10	20	Air permeability
	4.65	6[b]	10[b]	20[b]	Sorptivity
	6.14	6[b]	10[b]	20[b]	Water permeability
0.55	3.16	6	10	20	Freeze-thaw deterioration
	4.65	6	10	20	
	6.14	6[b]	10[b]	20[b]	Depth of carbonation
0.70	3.16	6	10	20	
	4.65	6	10	20	
	6.14	6	10	20	

[a]A/C: aggregate-cement ratio
[b]Mixes added with a melamine-formaldehyde based superplasticiser

Table 8.7 Variables and properties investigated in test series 2

Test variables				Properties investigated
W/C	A/C	Cover to steel (mm)		
0.45	4.65	25	40	Sorptivity
0.55	4.65	25	40	Chloride penetration
0.65	4.65	25	40	Corrosion initiation time

Table 8.8 Variables and properties investigated in test series 3

Variables								Properties investigated
FA/CA	W/C	A/C						
0.5	0.4	3	4	5				Air permeability
	0.5		4	5	6	7		Sorptivity
	0.6			5	6	7	8	Salt scaling resistance
	0.7				6	7	8	Depth of carbonation

used to carry out the test and as a consequence the test regime was slightly modified from what is given in the ASTM standard. The weight of the sample at the end of each 8-cycle period was noted and the total change in weight of the specimens (i.e. the difference between the highest weight due to the absorption of water and the lowest weight as a result of the deterioration) at the end of 304 cycles, expressed as a percentage of the original saturated weight is reported as the freeze-thaw deterioration.

In test series 3, 100 mm diameter cores were tested for determining the salt scaling resistance of concrete in accordance with RILEM test procedure [37]. The specimens were subjected to a cycle of freezing and thawing at every 12 h and at the end of each 2 cycles, the specimens were taken out, loose particles removed by means of an ultrasonic bath and the specimens weighed. The collected loose particles were dried at 100 °C for 24 h and were weighed. These measurements were continued for a total of 28 cycles. This test was considered to have simulated XF4 condition in EN 206.

In order to study the carbonation resistance of concretes, an accelerated carbonation test was carried out in test series 1 and 3. In test series 1, the samples were placed in the carbonation chamber and allowed to be carbonated in a carbon dioxide rich atmosphere of 20 % concentration and 85 % relative humidity for 2 weeks at room temperature (18 ± 2 °C). However, in the case of test series 3, test specimens were placed in an electronically controlled carbonation chamber in an environment of 5 % carbon dioxide, 20 °C and 65 % RH for 3 weeks. Both conditions were considered to have simulated XC3 regime in EN 206. At the end of the carbonation period, the specimens were split longitudinally and the freshly broken surface was sprayed with the phenolphthalein indicator solution. After 24 h the depth to the pink colouration was measured to the nearest millimetres at 6 different locations. An average of these values was reported as the depth of carbonation, in millimetres.

A cyclic chloride ponding test was carried out in test series 2 to simulate concretes subjected to cyclic wetting and drying regime (XD3 and XS3 conditions in EN 206). The concrete specimens were subjected to a weekly regime consisting of ponding a 15 % sodium chloride solution (approximately 0.3 M) for 3 days at 20 °C, removing the solution and then rinsing the surface with fresh water, and then storing the specimens at 20 °C and 55 ± 2 % RH for 4 days. This regime was repeated up to 44 weeks and at the end of both 10 and 44 weeks chloride samples were drilled out from the concrete at a depth of 25 and 40 mm from the test surface. As these specimens contained steel rods of 10 mm diameter at both 25 and 40 mm (to act as anode) connected electrically to another layer of 10 mm diameter rods kept at the bottom of the specimens (to act as cathode) the time to initiation of corrosion was also noted.

Figures 8.19 and 8.20 show inter-relationships between Autoclam air permeability index and durability parameters (depth of carbonation and salt scaling) measured by using tests described above. The data were transformed by using appropriate transformation functions in order to make them normally distributed before developing the graphs. In Fig. 8.21, the relationship between the Autoclam water permeability index and freeze-thaw deterioration is presented. Figures 8.22, 8.23, 8.24 and 8.25 present relationships between the Autoclam sorptivity index and various durability parameters obtained from the three test series. Although there existed a reasonably satisfactory correlations between the parameters reported in each of these figures, a closer scrutiny of the relationships would suggest that:

i. the Autoclam air permeability index is related very well to both the carbonation depth (XC3) and the salt scaling (XF4) and the relationship of these two durability parameters with the Autoclam sorptivity index was not that good;

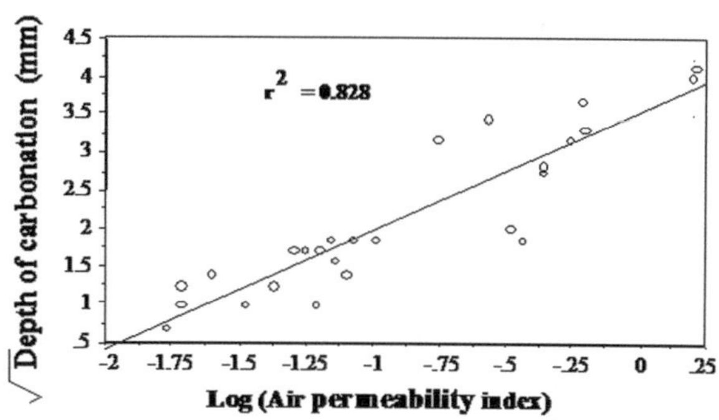

Fig. 8.19 Relationship between Autoclam air permeability index and depth of carbonation in test series 1 (XC3 regime)

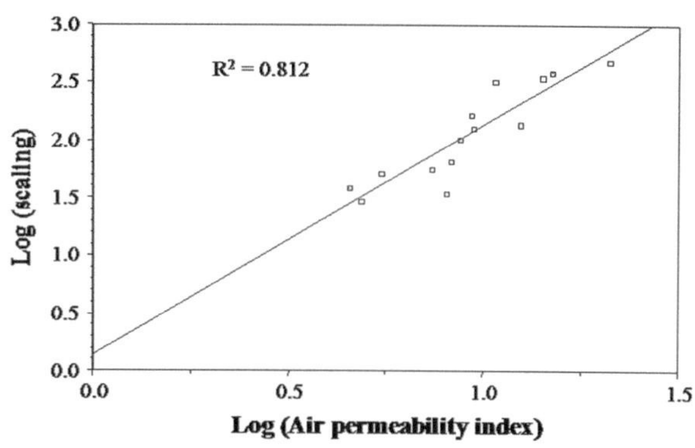

Fig. 8.20 Relationship between Autoclam air permeability index and salt scaling in test series 3 (XD3 and XS3 regimes)

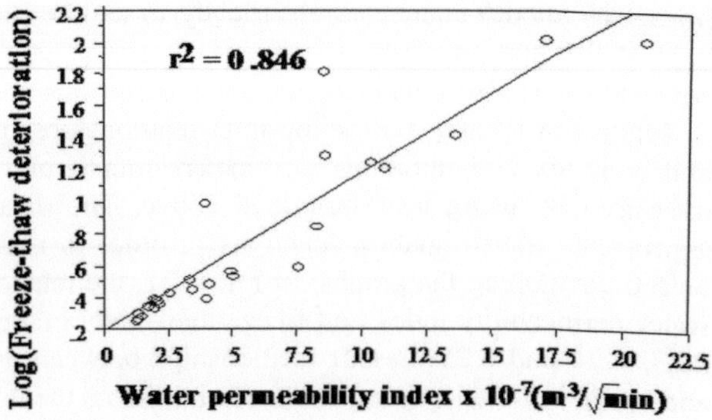

Fig. 8.21 Relationship between Autoclam water permeability index and freeze-thaw deterioration in test series 1

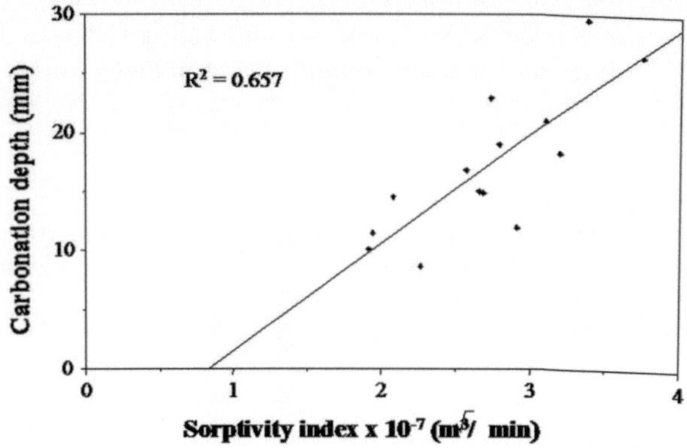

Fig. 8.22 Relationship between Autoclam sorptivity index and depth of carbonation in test series 3

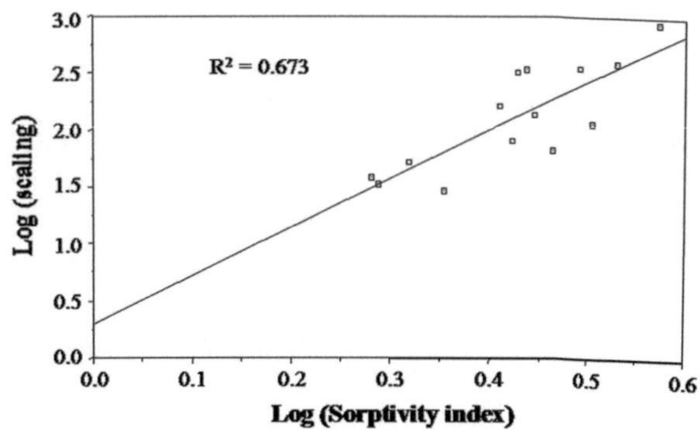

Fig. 8.23 Relationship between Autoclam sorptivity index and salt scaling in test series 3

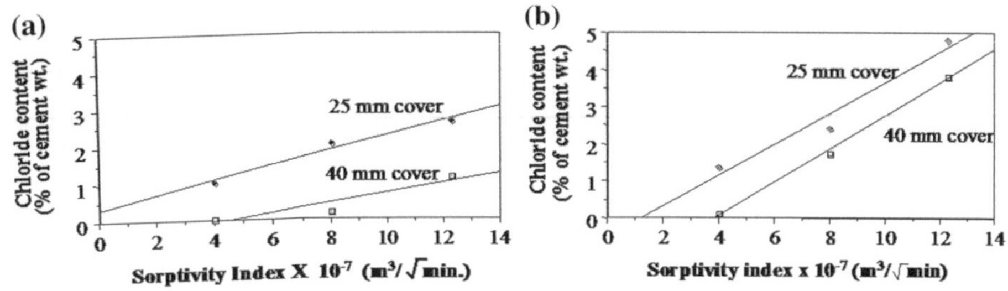

Fig. 8.24 Relationship between Autoclam sorptivity index and chloride penetration in test series 2 **a** after 10 weeks of exposure, and **b** after 44 weeks of exposure

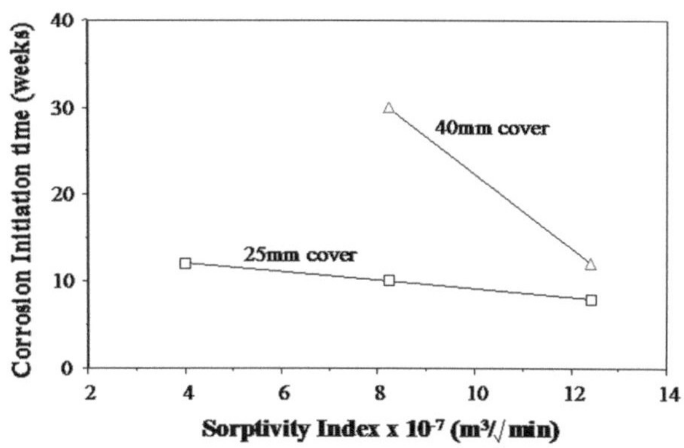

Fig. 8.25 Relationship between Autoclam sorptivity index and corrosion initiation time in test series 2

ii. the Autoclam water permeability index is related better to the freeze-thaw deterioration (XF3) than the Autoclam air permeability index and sorptivity index (the latter two are not presented here);
iii. there is a very good correlation between Autoclam sorptivity index and both chloride penetration and chloride induced corrosion when cyclic ponding regime was used (XD3 and XS3).

On the basis of these relationships and further extensive research, the durability classification criteria were developed for the Autoclam permeation indices, which are reproduced in Table 8.9..

8.4.5 *Typical Field Applications of the Autoclam Permeability System*

The Autoclam Permeability System has been used to classify the potential durability of concrete in notable structures worldwide. Typical applications in China consisted of assessing the durability of nuclear power plants (Fig. 8.26), Bird's Nest

Table 8.9 Durability parameters and acceptance ranges: Autoclam permeability indices [38]

Performance quality	Sorptivity index ($m^3 \times 10^{-7}/min^{0.5}$)	Water permeability index ($m^3 \times 10^{-7}/min^{0.5}$)	Air permeability index (pressure/min)
Very good	≤1.30	≤3.70	≤0.10
Good	>1.30 ≤ 2.60	>3.70 ≤ 9.40	>0.10 ≤ 0.50
Poor	>2.60 ≤ 3.40	>9.40 ≤ 13.80	>0.50 ≤ 0.90
Very poor	>3.40	>13.80	>0.90
Exposure environments	XD1&3 and XS1&3	XD2 and XS2 and XF4	XC1-4

Fig. 8.26 Testing concrete quality with Autoclam in Dayawan nuclear power station by Central Research Institute for Buildings and Construction

National Stadium (Fig. 8.27), and numerous highway and railway bridges (Fig. 8.28). Due to the confidential nature of the data, it has not been possible to report the outcome of these tests.

Fig. 8.27 Testing concrete quality with Autoclam in bird's nest stadium by Central Research Institute for Buildings and Construction

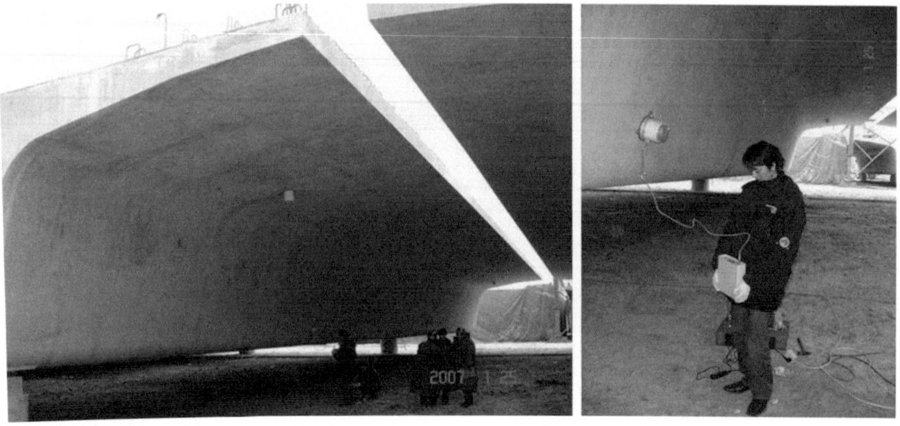

Fig. 8.28 Testing concrete quality with Autoclam in Beijing-Tianjin railway project by Central Research Institute for Buildings and Construction

8.5 Service Life Prediction Using Concrete Resistivity (Spain)

Electrical resistivity, ρ (Ω-m), is the inverse of electrical conductivity. It is the property of the material that reflects its ability to transfer electrical charge. It is a volumetric measurement of the electrical resistance (R_e), which by Ohm's Law is expressed as the ratio of voltage and current applied ($R_e = V/I$).

In the case of concrete, the electrical charge is transferred through the aqueous phase of the pore network by the electrical carriers (ions). The electrical resistivity of water-saturated concrete is therefore an indirect measurement of the concrete pore connectivity. However this relation is not linear, as the tortuosity of the pores makes the relation exponential or potential.

Due to its relation with porosity, the test of electrical resistivity in concrete can be used to assess the potential service life of reinforced concrete structures and to develop limiting values for specifications and quality control. This is largely based on the relationship between resistivity and diffusivity, which was presented by [39–41].

Electrical resistivity is a non-invasive test (NDT) that can be measured by arranging electrode(s) in different ways [42]:

- In specimens by means of the direct method [43] (electrodes placed on two parallel specimen faces) or of the four points method [44].
- In real structures by means of the four point (Wenner) method (see Fig. 8.29), by the one electrode or disc method [45] (see Fig. 8.30) or the two electrode method (provided each probe is calibrated).

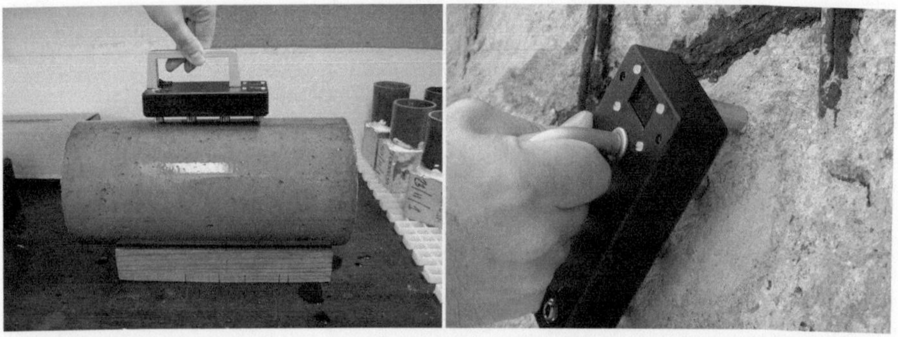

Fig. 8.29 Four points method

Fig. 8.30 One point or disc method

Resistivity measurement results can be used to infer on the following aspects:

- Progress of setting
- Curing degree
- Concrete resistance to chloride ingress (which penetrates through the pore solution) and progress of carbonation (due to the knowledge that carbonation progresses through the empty space i.e. the porosity minus the saturated space of it)
- Reinforcement corrosion

This section focusses on the description of concrete resistivity with respect to its interpretation regarding service life assessment (chloride penetration, carbonation and steel corrosion).

8.5.1 Basis of the Method

The basis of using electrical resistivity for service life prediction of reinforced concrete structures was presented in the references [39, 41, 45]. The electrical resistivity is the ratio between the potential applied by means of two electrodes and the current circulating in the material standardized by a geometric factor K_{geom} which depends on the position of the electrodes (Ohm's law):

$$R = \frac{V}{I} = \rho K_{geom} \tag{8.15}$$

The ability of resistivity to quantify the diffusivity is based on one of the Einstein laws which relates the movement of electrical charges to the conductivity of the medium [39, 40, 46]:

$$D_e = \frac{k}{\rho_{ef}} = k_{Cl}\sigma \tag{8.16}$$

where D_e is the effective diffusion coefficient, k is a factor which depends on the external ionic concentration, ρ_{ef} is the "effective" resistivity (in this case of the concrete saturated with water), and σ is the conductivity (inverse of resisitity).

A value of k_{Cl} of 20×10^{-5} can be used for external chloride concentrations of 0.5–1.0 M [39]. This expression only accounts for the transport of chloride ions, and the effect of chloride binding has to be taken into account separately. This can be done by introducing a reaction or binding factor, r_{Cl}. This reaction factor acts as a "retarder" in the penetration of chlorides. The above equation maintains its mathematical structure but can now be presented as follows (where D_{ap} is an "apparent" diffusion coefficient in saturated conditions, ρ_{ef} is the effective resistivity and ρ_{ap} is the apparent resistivity):

$$D_{ap} = \frac{k_{Cl}}{\rho_{ef} \times r_{Cl}} = \frac{k_{Cl}}{\rho_{ap}} \tag{8.17}$$

Equation (8.17) can also be applied to the case of carbonation provided another constant k_{CO_2} is considered for the atmospheric exposure. Relating r_{CO_2} to the amount of alkaline material able to bind CO_2, it can be written as:

$$D_{CO_2} = \frac{k_{CO_2}}{\rho_{ef} \times r_{CO_2}} \tag{8.18}$$

Other parameters that need to be incorporated in the model are:

- ageing factor q and
- environmental factor k, which will be described in the following sections.

8.5.2 The Reaction Factor

The reaction factors r_{Cl} and r_{CO_2} [47] depend on the type and amount of cement and therefore on the reaction of the penetrating substance with the cement phases. They can be calculated either by direct measurement, or indirectly using the relation between the effective and apparent diffusion coefficients, or by calculation based on the cement composition. Table 8.10 presents examples of r_{Cl} values that were calculated based on test results obtained with the multi-regime chloride test [48].

8.5.3 The Environmental Factor

The environmental factors k_{Cl} and k_{CO_2} depend on the exposure conditions. Table 8.11 presents values that were calculated by inverse analysis of test results obtained on real structures.

8.5.4 Ageing Factor

The apparent resistivity evolves with time due to the progression of hydration, the combination of the cement phases with the chlorides or carbon dioxide, which usually decreases the porosity, and by drying of the concrete (depending on the environment), which is accounted for by introduction of an "ageing" factor q due to the refinement of the concrete pore system results in an increase of resistivity with

Table 8.10 Examples of values of the reaction factor of chlorides, r_{Cl} (unitless), for 3 types of cement

Cement	r_{Cl}	Standard deviation
CEM I	1.9	1.3
CEM I + silica fume	1.5	0.5
CEM IIA (with pozzolan and fly ash, in ≤20 %)	3.0	2.1

Table 8.11 Values of environmental factors, k_{Cl} and k_{CO_2}, following the exposure classification of EN206

Exposure class	k (cm^{3}Ω/year)
X0	200
XC1	1000
XC3	3000
XS1 (d > 500 m distance to the coast line)	5000
XS1 (d < 500 m distance to the coast line)	10000
XS2	17000
XS3	25000

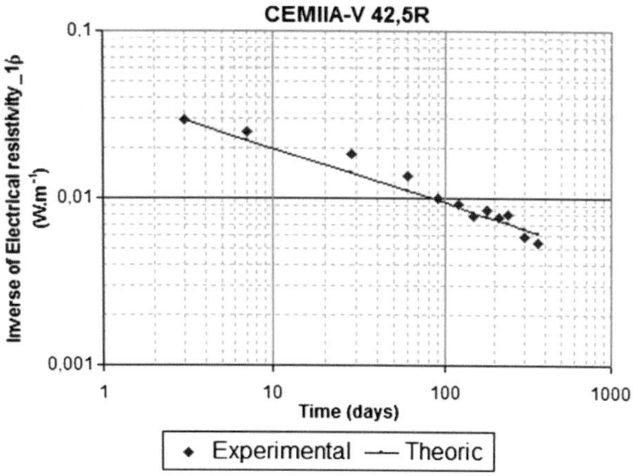

Fig. 8.31 Representation of the inverse of resistivity with respect to time

time. If the inverse of resistivity is plotted as a function of time (Fig. 8.31), the apparent evolution of resistivity can be expressed by a function in which the power exponent q, which is the slope of the straight line, may have different values for OPC and blended cements [49]:

$$\rho_t = \rho_0 \left(\frac{t}{t_0} \right)^q \tag{8.19}$$

where ρ_t is the resistivity at any age t, and ρ_0 is the resistivity at the age of the first measurement t_0.

Values of q for different cement types are given in Table 8.12.

The relationship between q and the ageing factor n of the diffusion coefficient gives the expression [47]:

$$q = 0.8n \tag{8.20}$$

8.5.5 *Propagation Period*

Based on the variation of the concrete resistivity with the degree of water saturation, steel corrosion is proportional to the resistivity value (Fig. 8.32).

Table 8.12 Values of the ageing factor q

Cement	q	Standard deviation
I	0.22	0.01
II/A -P	0.37	0.06
II/A-V	0.57	0.08

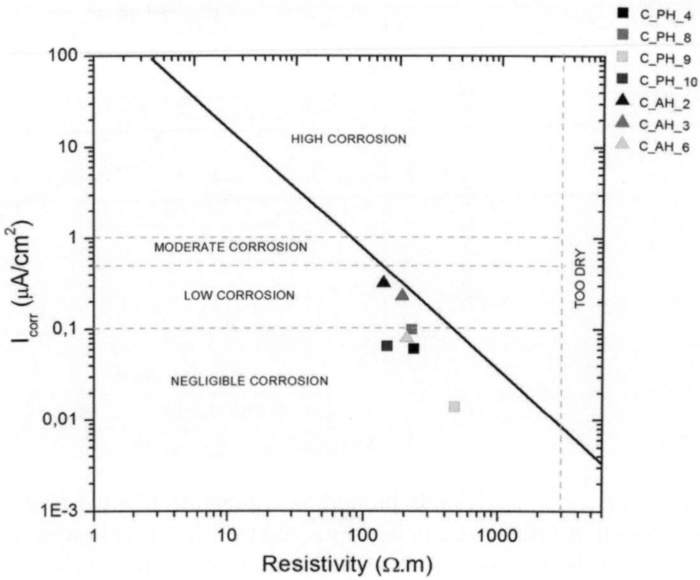

Fig. 8.32 Graph of I_{corr}-ρ_{ef} indicating the relationship between I_{corr} and the porosity plus the degree of concrete saturation. The *symbols* are examples of site measurements

This graph relates to the following equation [50]:

$$I_{corr}\left(\frac{\mu A}{cm^2}\right) = \frac{26}{\rho_{ef} \times (k\Omega - cm)} \tag{8.21}$$

The equation for service life prediction can be then formulated as follows:

$$t_l = \frac{P_{corr} \times \left(\rho_{ef}\left(\frac{t}{t_0}\right)^q \xi\right)}{K_{corr} \times 0.00116} \tag{8.22}$$

where P_{corr} is the steel cross section at time t_p, ρ_{ef} is the resistivity at 28 days in saturated conditions, q is the ageing factor of the resistivity (Table 8.12), ζ is the environmental factor of the corrosion rate (it can be of 10 ± 2 for carbonation and 30 ± 5 for chlorides), and K_{corr} is a constant value of 26 μA/cm^2·kΩ-cm = 26 mV/cm relating resistivity and corrosion rate I_{corr}.

8.5.6 *Calculation of Service Life and Application Example for the Initiation Period*

The resulting expression of the service life model using concrete resistivity for the initiation and the propagation periods is:

$$t_l = \frac{x^2 \times \rho_{ef}\left(\frac{t}{t_0}\right)^q}{k_{Cl,CO_2}} r_{Cl,CO_2} + \frac{P_x\left(\rho_{ef}\left(\frac{t}{t_0}\right)^q \xi\right)}{K_{corr} \times 0.00116} \quad (8.23)$$

For the initiation period the application of the above theory can be shown using an example, assuming a concrete with a cover depth of 50 mm made with cement type II/A to be placed in exposure class XS2 (submerged conditions). Considering a service life of 100 years, the values of the reaction, as well as the environmental and ageing factors are presented in Table 8.9. The calculations indicate that the resistivity needed at 28 days of age, measured in saturated conditions, is 87.6 Ωm.

$$5 = \sqrt{\frac{17,000}{\rho_0\left(\frac{10}{0.0767}\right)^{0.3} \times 1.8} \times 100} \Rightarrow \rho_0(\Omega \cdot \mathrm{cm}) = 8760 \rightarrow \rho_0(\Omega \cdot \mathrm{m}) = 87.6 \quad (8.24)$$

Example for the calculation of the length of the propagation period using input data in Tables 8.13 and 8.14:

$$t_l = \frac{0.01 \times \left(8.760 \times \left(\frac{100}{0.0767}\right)^{0.3} \times 1\right)}{26 \times 0.00116} = 12 \text{ years} \quad (8.25)$$

8.5.7 Compliance Testing

The application of this method to service life prediction should be based on the same statistical principles as for mechanical strength. That is, a characteristic value with a limiting probability of occurrence has to be defined and this value should be

Table 8.13 Input data for calculation of the concrete resistivity (example)

Cement type II/A	r_{Cl} = 1.8
Exposure class (XS2)	K (cm^{3}Ω/year) = 17000 (XS2)
Service life	t (years) = 100
Cover depth	X_{Cl} (cm) = 5
Ageing factor during 10 years	q = 0.3

Table 8.14 Input data for the propagation period

Limiting loss in diameter, P_{corr}	100 μm = 0.01 cm
ρ_{ef} at 28 days (in saturated conditions)	8.760 kΩ-cm
q applied during 10 years	0.3
ξ in saturated conditions	1.0

fulfilled with the same sampling frequency as for mechanical strength. Additionally, just as the specimens for strength have to be wet cured, it is recommended to measure the resistivity on the same specimens just prior to testing strength.

On a voluntary level, the Spanish Committee AENOR CTN-83/SC10 "Durability" proposes that for durability design regarding reinforcement corrosion, the following procedure be followed:

- Monitoring the resistivity at 3, 7, 14 and 28 days in order to obtain the ageing factor q and the nominal resistivity at 28 days.
- Measurement of the porosity and chloride diffusion (ponding test lasting 90 days after 28 days curing) in order to obtain the reaction factor r_{Cl}, r_{CO_2} for the particular cement and mix.

After verifying the stability of the characteristic resistivity values and provided the cement type is not changed, for production control, it is sufficient to fulfill the effective resistivity ρ_{ef} requirement at an age that is less than 28 days.

Additionally, it can be considered an "equivalent durability performance" when apparent resistivity, ρ_{ap} at 28 days of curing (or at any other age) of a concrete is statistically the same (characteristic values to be compared) as another concrete whose durability is likely to be considered equivalent.

$$\rho_{ap}(t) = r_{Cl,CO_2} \times \rho_e \left(\frac{t_0}{t_n}\right)^q \tag{8.26}$$

where ρ_{ap} is the apparent resistivity, r_{Cl,CO_2} is the reaction factor of the concrete, ρ_e is the effective resistivity (specimen at 28 days of wet curing), t_0 is the age at which the test is performed (usually 28 days), and t_n is the life time to declare the equivalence in service life.

8.5.8 Resistivity as a Durability Indicator

Concrete resistivity can be used as a durability indicator if a classification of exposure classes and the cover depth are considered. An example is presented in Table 8.15.

8.5.9 Site Determination in Existing Structures

In existing structures, the resistivity can be measured (see Fig. 8.29) and its remaining service life calculated after establishing the front of the aggressive species relative to the bar position using Eq. (8.23).

For example consider a structure with a cover of 30 mm in which the aggressive carbonation front has not reached the reinforcement, being at 10 mm after 22 years.

Table 8.15 Example of values of resistivity used as durability indicator assuming type II cement (up to 20 % of mineral additions)

Cover (mm)	Apparent resistivity-characteristic value ρ_{ap} in (Ω·m) under saturated conditions at 28 days of curing	
	Carbonation (unsheltered from rain)	Chlorides (submerged)
20	250	2500
30	120	1110
40	63	625
80	15	160

The measured on-site resistivity is of 12 kΩ-cm. For calculating the remaining service life (Table 8.16):

Applying Eq. (8.23) results in:

$$t_l = \frac{3^2 \times 12,000\left(\frac{10}{0.0767}\right)^{0.22}}{3,000} \times 1.9 = 199\,\text{years} \tag{8.27}$$

Then, the remaining life with that resistivity exceeds 100 years.

Another example would be to find the time to corrosion initiation and loss in steel diameter in a corroding structure in exposure class SC2 due to the measured site resistivity is of 5,000 Ω·cm (Table 8.17).

The first step is to calculate when the chloride threshold reached the bar position and then calculate the loss in steel cross-section due to the propagation period.

Table 8.16 Input parameters for example 1 in existing structures

Cement type I	$r_{Cl} = 1.9$
Exposure class (XC3)	K (cm$^3\Omega$/year) = 3000
Time life	t (years) = 22
Cover depth	X_{Cl} (mm) = 30
Ageing factor until 10 years	$q = 0.22$
Remaining service life	??

Table 8.17 Input parameters for example 2 in existing structures

Cement type I	$r_{Cl} = 1.9$
Exposure class (XS2)	K (cm$^3\Omega$/year) = 17000 (XS2)
Cover depth	x_{Cl} = 30 mm
Ageing factor during 10 years	$q = 0.22$
Resistivity	5,000 Ω·cm
Position of chloride threshold at 25 years	40 mm
Time of initiation and corrosion proceed from then	??

$$t_l = \frac{3^2 \times 5,000\left(\frac{10}{0.0767}\right)^{0.22}}{17,000} \times 1.9 = 14.7\,\text{years} \tag{8.28}$$

This is the time taken for the chlorides to reach the bar position. Then the corrosion (loss in steel diameter) produced during 25 − 14.7 = 10.3 years is:

$$P_{corr} = \frac{10.3 \times 26 \times 0.00116}{5 \times 2.92} \times 1.9 = 212\,\mu\text{m} \tag{8.29}$$

8.5.10 *Summary*

Resistivity is a very comprehensive NDT to assess concrete characteristic which facilitates flexible and simple calculations regarding both the initiation and propagation periods and to characterize the overall durability in a parallel manner to the mechanical strength. The expression to be applied is Eq. (8.30):

$$t_l = \left(\frac{x^2 \times \rho_{ef}\left(\frac{t}{t_0}\right)^q}{k_{Cl,CO_2}} r_{Cl,CO_2}\right) + \left(\frac{P_x\left(\rho_{ef}\left(\frac{t}{t_0}\right)^q \xi\right)}{K_{corr} \times 0.00116}\right) \tag{8.30}$$

8.6 Surface Resistivity (USA)

In the USA the surface resistivity test is in the process of being introduced as a quality assurance test and also as an acceptance test for concrete. The test is essentially a four-point (Wenner) test on the side of a core. Many states already use permeability specifications for Portland cement concrete, but typically in the past the permeability has been tested either by ASTM C1202 (Rapid Chloride Permeability) [14] or by ASTM C642 (Boil test) [51]. Extensive research [52–54] has shown that there is a very strong correlation between the rapid chloride permeability test and the surface resistivity test. The surface resistivity test is far simpler and quicker to perform which means significant cost savings for contractors and Department of Transport Laboratories alike. In July 2011 an AASHTO Technical Implementation Group (TIG) approved the test method as a draft standard TP 95-11 "Surface Resistivity Indication of Concrete's Ability to Resist Chloride Ion Penetration" [55].

The same test procedure has also been defined independently by both Florida Department of Transport (FDOT) FM 5-578 [56] and Louisiana Department of Transportation and Development (LADOTD) TR 233-11 [57].

The test procedure is based on the use of standard cylinders cured under controlled conditions. The same cylinders are subsequently used for compressive strength testing in the press, following the surface resistivity test. TP 95-11 defines 5 classes of Chloride Ion Penetrability for two sizes of cylinder (Table 8.18).

This corresponds to 56-Day Rapid Chloride Permeability as shown in Table 8.15 taken from LADOTD research [58] (Table 8.19).

LADOTD are now in the process of introducing this test method as a quality assurance method and as a method for acceptance of concrete mixes. Permeability requirements, to be determined by the surface resistivity test have been defined for 6 different classes of concrete. The intention is to carry out this test as a standard test along with compressive strength, determination of water/cement ratio, air content and slump.

In FDOT, where the test has been in use longer, permeability requirements have already been introduced for specific concrete types in their "Standard Specifications for Road and Bridge Construction 2010" (FDOT). In particular it is used to define a minimum resistivity for concrete classes IV, V and VI:

> When the use of silica fume, ultrafine fly ash, or metakaolin is required as a pozzolan in Class IV, Class V, Class V (Special) or Class VI concrete, ensure that the concrete exceeds a resistivity of 290 Ω·m at 28 days, when tested in accordance with FM 5-578. Submit three 4 × 8 inch cylindrical test specimens to the Engineer for resistivity testing before mix design approval. Take the resistivity test specimens from the concrete of the laboratory trial batch or from the field trial batch of at least 3 yd^3 (2 m^3). Verify the mix proportioning of the design mix and take representative samples of trial batch concrete for the required plastic and hardened property tests. Cure the field trial batch specimens similar to the standard laboratory curing methods. Submit the resistivity test specimens at least 7 days prior to the scheduled 28-day test. The average resistivity of the three cylinders, eight readings per cylinder, is an indicator of the permeability of the concrete mix.

Table 8.18 Correlation between surface resistivity and chloride ion penetrability (Note that *a* is the inter-electrode distance for maximum aggregate up to 38 mm)

Chloride ion penetrability	Surface resistivity test	
	100 mm × 200 mm (4 in. × 8 in.) Cylinder (Ω·m) a = 37.5 mm	150 mm × 300 mm (6 in. × 12 in.) Cylinder (Ω·m) a = 3.75 mm
High	<120	<95
Moderate	120–210	95–165
Low	210–370	165–290
Very low	370–2540	290–1990
Negligible	>2540	>1990

Table 8.19 Correlation between permeability classes, RCPT values and surface resistivity [58]

Permeability class	56-day rapid chloride permeability charge passed (coulombs)	28-day surface resistivity (Ω.m)
High	>4000	<120
Moderate	2000–4000	120–210
Low	1000–2000	210–370
Very low	100–1000	370–2540
Negligible	<100	>2540

8.7 Two Electrode Resistivity Method (The Netherlands)

The principles of concrete resistivity testing have been described in Sect. 4.6.1. Briefly, resistivity testing involves applying a small potential difference, preferably from an alternating current source to avoid electrode polarisation, across a concrete sample, measuring the resulting current and dividing the potential difference by the current, yielding the ohmic resistance [42]. The resistivity is calculated by multiplying the resistance by the cell constant, which for a rectangular specimen is the ratio of the surface area and the length. A rectangular specimen has its axis perpendicular to its base.

$$R = \frac{V}{I} \tag{8.31}$$

where R is the resistance in Ω, V is the potential difference in V, and I is the current in A.

$$\rho = \frac{RA}{L} \tag{8.32}$$

where ρ is the resistivity in Ω·m, A is the specimen surface in m^2, and L is the specimen length in m.

The resistivity of concrete itself contains insufficient information to be used as the sole parameter for service life design. To use it for that purpose, additional information is needed, see Sect. 8.5. However, there is a good correlation between resistivity and chloride diffusion of concrete. Correlation between chloride transport in concrete and its inverse resistivity (conductivity) was theoretically underpinned and practically demonstrated for a wide range of binders in the 1990s [39, 58–60].

The Dutch guideline for service life design [60] specifies the two electrode method (TEM) for concrete resistivity testing as the test method for production control. The method must be applied to standard concrete cubes for compressive strength testing after wet curing in the lab (at 20 °C) at an age of 28 days. Positive experience was gained using this method for quality control of cast in situ concrete for parts of a large tunnel project, see below.

8.7.1 *Test Method*

A particular form of resistivity testing is applied in the Two Electrode Method (TEM). It involves placing a specimen, either a cube or a cylinder, between two metal plates provided with pieces of wetted cloth for electrolytic contact (Fig. 8.33).

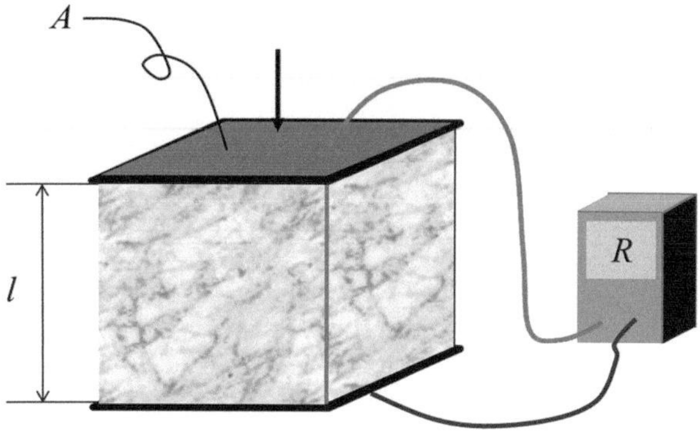

Fig. 8.33 Setup for two electrode method resistivity testing (TEM)

The specimen is a concrete cube of side 150 mm, water cured at 20 °C and tested at 28-day age. Electrodes are steel plates with wetted cloth and pressed to the concrete by a top weight of 5 kg. The measuring frequency is 120 Hz. The resistivity is calculated using Eq. (8.32).

8.7.2 *Limiting Values*

As suggested by [61] the Two Electrode Method (TEM) for concrete resistivity testing can be used for routine production quality control. Data analysis has shown that the correlation between inverse resistivity and chloride migration coefficient holds over a wide range of values, as illustrated in Fig. 8.34. However, this correlation may not apply to each particular concrete mix with sufficient accuracy for quality control. In other words, limiting values have to be determined empirically.

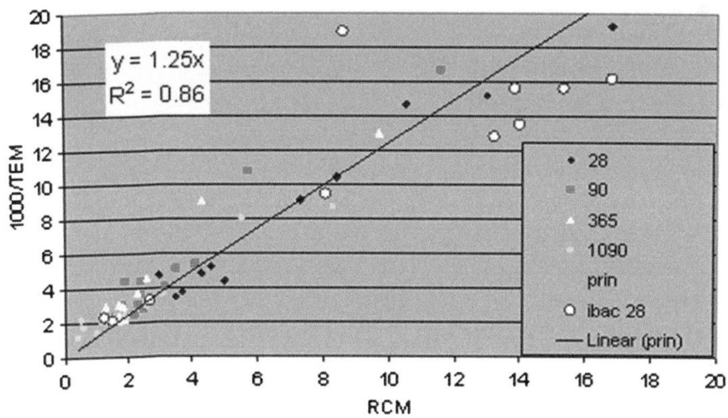

Fig. 8.34 Correlation between inverse concrete resistivity (measured by TEM, in 1000/Ω·m) and rapid chloride migration coefficient (NT Build 492, in 10^{-12} m^2/s) [52, 60, 63]

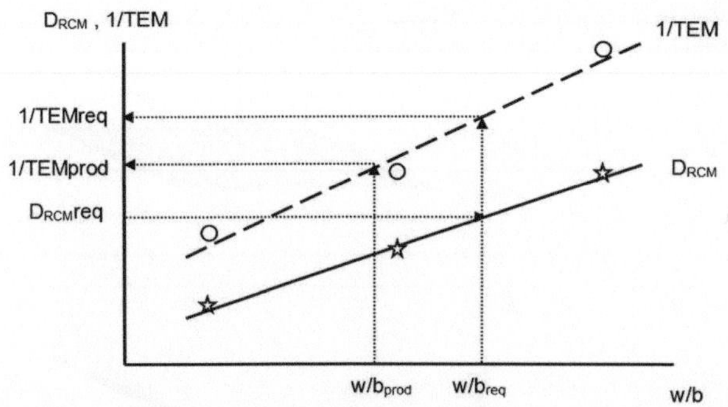

Fig. 8.35 Determination of maximum value for 1/TEM from RCM and TEM testing; subscripts *req* and *prod* refer to required (limiting) value and target production value, respectively, from [60]

Consequently, the Dutch guideline states that the correlation has to be determined for a particular mix or a family of mixes during the prequalification stage [60, 62]. For this purpose, mixes have to be made with the same cement type and content and aggregate mix, for three w/b ratios, that is, the target (prequalified) mix, one with a slightly lower and one with a slightly higher w/b. The difference in w/b between mixes must be 0.03; except for mixes with CEM III/B, where the difference must be 0.05. RCM and TEM are tested at 28 days for these three mixes and their correlation is determined. The required maximum 1/TEM (minimum resistivity) is determined from Fig. 8.35.

8.7.3 Example of Production Control by Resistivity Testing

Production control based on TEM (resistivity) testing was applied to the Green Heart Tunnel, built between 2000 and 2004, which will be briefly described here as an example. Details are provided in [64]. It should be noted that this case was designed and built well before the Guideline [60] was issued. In fact, the positive experience gained with production quality control by resistivity testing was the basis for adopting this method in the Guideline.

The Green Heart Tunnel (GHT) is a bored tunnel in the High Speed Train Link (HSL) between Amsterdam and Brussels. It has a length of about 8.6 km, with the main tunnel constructed by boring a single tube of 14 m inner diameter, with a lining of precast concrete segments. The ramps, sidewalls, rail beds, safety kerbs and the centre wall as well as three ventilator shafts and two technical buildings were made of reinforced concrete cast in situ. Figure 8.36 provides schematic cross sections. The quality control described here relates to the in situ concrete only.

Service life design was based on chloride penetration up to a critical (threshold) value with about 3 % probability of failure at an age of 100 years. Using the DuraCrete degradation model, the requirements were: a maximum chloride

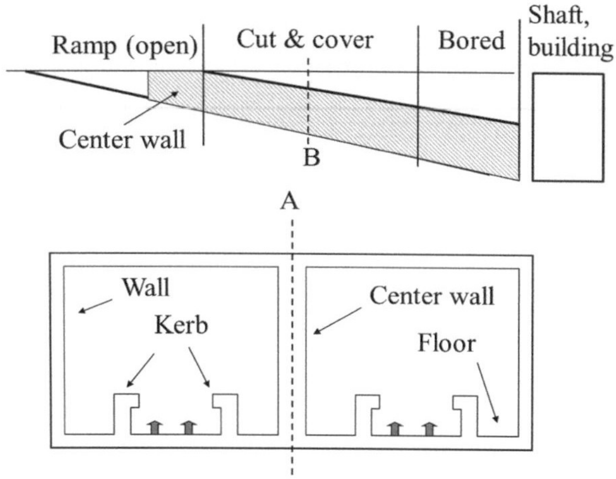

Fig. 8.36 Green heart tunnel, schematic longitudinal cross section of ramp, cut and cover part, bored tunnel and shaft/building (*top*) and schematic transverse cross section with center wall, kerbs, floor and wall inside the cut and cover part (*bottom*)

diffusivity of 5×10^{-12} m^2/s, to be tested using the RCM method at 28 days; a minimum cover depth of 45 mm (35 mm for kerbs, 50 mm if non-inspectable). In situ concrete should be made with blast furnace slag cement with high slag content, CEM III/B.

Several trial mixes produced RCM values between 3 and 4×10^{-12} m^2/s at 28 days. The mixes that were used were based on 360 to 400 kg CEM III/B LH HS (blast furnace slag cement with c. 75 % slag) per cubic meter, a w/c ratio of 0.44 and maximum aggregate size of 32 mm.

To make sure that RCM-values below the maximum specified value (5×10^{-12} m^2/s) were indeed maintained in the production phase, it was proposed to test control cubes made at the production site for 28 days strength verification, hydrated under water at 20 °C for resistivity.

Concrete production was aimed at a maximum 28 days RCM-value of 3.5×10^{-12} m^2/s. Based on the general correlation and considering the statistical variation of production, this relates to a minimum target value for TEM of 260 Ω·m.

A volume of 30,000–40,000 m^3 cast in situ concrete was produced over a total of about 1,000 production days. Each day production was sampled at least once for strength and resistivity testing at the mixing plant. Results for eight selected batches are reported in Table 8.20.

As can be seen in Table 8.20, standard deviations for TEM results are rather small relative to mean values; Coefficients of Variation's (CoV's) ranged from 0.02 to 0.06. Multi-laboratory testing has also shown low CoV's for resistivity testing [65]. These low CoV's suggest that the variability within each day production was small. The batch mean resistivity was higher than the requirement of 260 Ω·m in all cases, with a lowest value of 275 Ω·m. Consequently, according to the quality control testing, all investigated batches complied with the TEM requirement.

Table 8.20 TEM results at 28 days age for concrete measured during production

Batch code/test area	Concrete code	TEM results (Ω·m)										μ_{TEM} (Ω·m)	σ_{TEM} (Ω·m)
1	B35 IS 5D	376	355									366	15
2	B35 IS 5D	375	381	414	414	410	422					403	20
3	B35 IS 5D	279	285									282	4
4	B35 IS 5D	348	373	340	340							350	16
5	B35 IS 5E	478	443									461	25
6	B35 IS 5D	266	283									275	12
7	B35IS Ramp	327	339	366	352	395	364	366	351			358	20
8	B35 IS 5D	351	323	317	324	332	357	313	313	312	308	325	17

After the project was finished, validation of the TEM quality control of the in situ concrete with regard to durability was sought for. This was done by taking 96 cores from eight test areas and determining RCM values. From these values, 28-day values were calculated using Eq. (8.38) (see Sect. 8.9) and n-values between 0.25 and 0.30; the best fit was obtained with $n = 0.27$. Analysis showed that the concrete in all test areas complied with the requirement for RCM at 28 days. It was concluded that the production control scheme using TEM testing had worked well.

It should be noted that this procedure only applies to the quality of concrete produced in the mixing plant; it does not apply to concrete as placed and cured. Further work in that area is needed.

8.8 Chloride Conductivity Index (South Africa)

In South Africa, the potential durability of reinforced concrete structures in marine environments is assessed using the chloride conductivity test. The test apparatus is described in Chap. 4. Chloride conductivity decreases with the addition of fly ash, slag, and silica fume in concrete, extended moist curing and increasing grade of concrete. While the test is sensitive to construction and material effects that are known to influence durability, results are specifically related to chloride ingress into concrete.

8.8.1 *Correlation Between Chloride Conductivity and Chloride Diffusion*

Correlations between 28-day chloride conductivity results and diffusion coefficients after several years of marine exposure have been shown to be good over a wide range of concretes [26]. The philosophy behind examining the correlation between diffusivity and conductivity is that conductivity (σ) of saturated materials is linearly related to steady state diffusivity (D_S), while apparent diffusivity (D_a) is a function of steady state diffusivity and the chloride binding capacity (α) [66]. The chloride conductivity test, which is a steady state accelerated test, gives a value of conductivity for concrete as given by Eq. (8.33):

$$\sigma = \frac{It}{VA} \tag{8.33}$$

where σ is the chloride conductivity in mS/cm, I is the measured current in mA, V is the voltage in V, t is the specimen thickness in cm, and A is the cross-sectional area in cm^2.

The chloride conductivity is fundamentally related to steady state diffusivity (D_s) as shown by Eq. (8.34):

$$Q = \frac{D_s}{D_0} = \frac{\sigma}{\sigma_0} \tag{8.34}$$

where Q is the diffusivity ratio, σ is the conductivity of concrete (calculated from Eq. (8.33)), σ_O is the conductivity of the pore solution, D_s is the steady state diffusivity of chloride ions through concrete in m^2/s, and D_O is the diffusivity of chloride ions in the equivalent pore solution in m^2/s.

The conductivity of the pore solution (σ_O) results from both the saturating salt solution and also from mobile ions such as K^+, Na^+ and OH^-, which are present in concrete pores. To measure the conductivity of the concrete pore solution would involve pore expression measurements which are difficult and impractical for routine rapid testing, and hence the value of σ_0 in the South African Chloride Conductivity test is assumed to be that of the 5 M NaCl saturating solution. Equation (8.35) gives the diffusivity of chloride ions in the pore solution (D_O) and is only applicable under ideal conditions and a relatively dilute solution, i.e. ion–ion interaction is limited (linear potential difference and constant temperature) [66]:

$$D_0 = \frac{RT}{zF}\frac{LJ_c}{Ec} \tag{8.35}$$

where D_o is the ion diffusivity in m^2/s, R is the gas constant in J/(mol·K), T is the absolute temperature in K, z is the electric valency of ion (for chloride z = 1), F is the faraday constant in (C/mol), J_c is the ion flux in steady state in mol/(m^2.s), c is the concentration in mol/m^3, m and E/L is the gradient of the electric field per length in V/m.

In summary, the effective diffusion coefficient is given as:

$$D_s = \frac{D_0 \sigma}{\sigma_0} \tag{8.36}$$

The apparent diffusion coefficient (D_a) in Fick's second law of diffusion is a function of the steady state diffusion coefficient and the chloride binding capacity (ε) as shown by Eq. (8.37) [67]:

$$D_a = \frac{D_s}{\varepsilon + (1 - \varepsilon)\frac{\partial S_b}{\partial_c}} \tag{8.37}$$

where ε is the volumetric porosity, c is the molar concentration, and S_b is the bound chloride in kgCl$^-$/m^3 of solid. The value of $\partial S_b / \partial_c$ depends on the quantity of binder, the binder type as well as the chloride concentration.

Having established a correlation between the apparent diffusion coefficient and the chloride conductivity makes it possible for the designer to use the Fickian service life models directly and input the appropriate conditions (cover depth, environmental classification, desired life, and material) to give material specifications in terms of the diffusion coefficient value that should be achieved using the chloride conductivity test. The correlation between the apparent diffusion coefficient and the chloride conductivity can also be used to establish chloride conductivity index limits used in durability design.

Mackechnie and Alexander [68] carried out experimental correlations between the conductivity values and long-term performance, with the intention of showing how the chloride test can be used to control covercrete quality through specifying limits to chloride conductivity values at a suitable age. For this purpose, they established a correlation between 2-year diffusion coefficient (D_{2years}) and the 28-day conductivity result from the chloride conductivity test using 2 different techniques:

- Correlation tests between the 28-day conductivity index values and chloride ingress in structures in the Western Cape Province of South Africa
- Laboratory-based experimental correlations between 28-day conductivity index values and chloride diffusion coefficients. The specimens in the study covered a range of binder types, water/binder ratios and curing regimes

From the correlations, [69] established that the chloride conductivity could be used as a criterion to assess construction quality, and thus developed a nomogram (Fig. 8.37), in which the apparent diffusion coefficient at 2-years (D_a) is determined from the 28-day chloride conductivity value. The modified chloride conductivity value referred to in the nomogram allows for long-term effects such as chloride

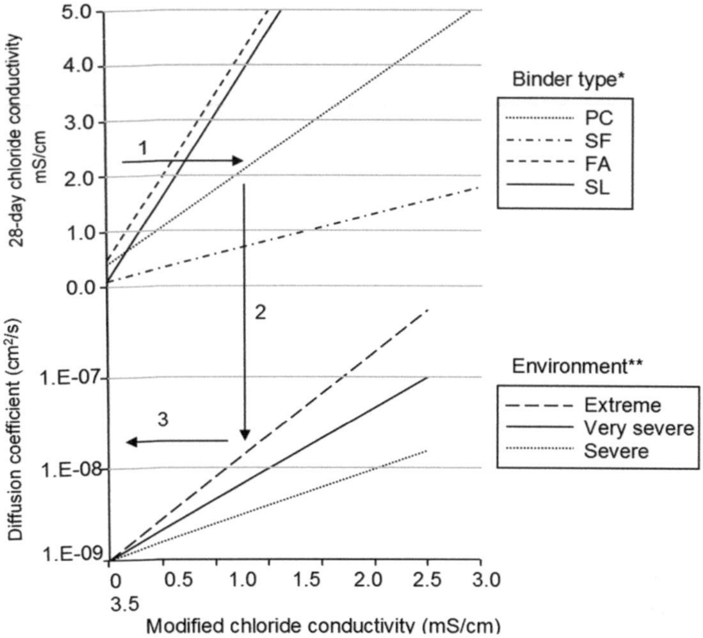

Fig. 8.37 Determining the apparent diffusion coefficient from the chloride conductivity value * (PC = 100 % Portland cement, SF = 90 % PC + 10 % silica fume, FA = 70 % PC + 30 % fly ash, SL = 50 % PC + 50 % slag, extreme), **(Extreme = marine tidal and splash zone, structure exposed to wave action and/or abrasion, very severe = marine tidal and splash zone, structure exposed to little wave action, severe = marine spray zone) [69]

binding and continued cementing reactions. Note that the South African experience with silica fume is that it binds chlorides significantly less than fly ash or slag, and also less than Portland cement. The relative slopes for the cement extenders in the top half of the diagram (Fig. 8.37) roughly reflect their chloride binding capacity, hence the position for the silica fume line below the Portland cement line.

8.8.2 Performance Specifications Using Chloride Conductivity Values

Two possible approaches for specifying chloride conductivity values are a deemed-to-satisfy approach and a rigorous approach. The former is considered adequate for the majority of reinforced concrete construction and represents the simpler method in which limiting chloride conductivity values are obtained from a design table, based on South African binder types and exposure classes, for a given cover depth of 50 mm. Table 8.21 presents chloride conductivity limits for common structures (in this case, for 50 year service life). Note that the use of plain CEM I for marine environment is not allowed in the South African Durability Index approach, which is why the table does not contain values for concretes made with CEM I only.

The rigorous approach will be necessary for durability–critical structures, or when the design parameters assumed in the first approach are not applicable to the structure in question (e.g. when cover depths other than 50 mm are used). Using this approach, the specifying authority would use the relevant service life models (developed in the concrete durability research programme in South Africa). The designer can use the models directly and input the appropriate conditions (cover depth, environmental classification, desired life, and material). The advantage of this approach is its flexibility as it allows the designer to use values appropriate for the given situation rather than a limited number of pre-selected conditions.

As an example of practical implementation of the rigorous approach, consider the case of specifying a marine structure for a 50-year design life, subject to the environmental conditions given in Table 8.21. Combining the relevant durability index of chloride conductivity with the appropriate service life model yields the data given in Table 8.21. It should be noted that the limiting chloride conductivity values are presented here for purposes of illustration only. The relative values are more important than the absolute values as these will vary in response to regional and environmental variations (Table 8.22).

The table shows the trade-off between material quality (i.e. chloride conductivity) and thickness of concrete cover, with lower quality (represented by a higher conductivity value) allowable when cover is greater. The dependence of the conductivity on binder type is also illustrated, with higher values permissible for blended binders at any given cover, based on their superior chloride ingress resistance. These higher values translate into less stringent w/b ratios. Therefore, a conservative approach is recommended at present, with mixes for which the concrete grade may be less than 30 MPa, and/or the w/b may be greater than 0.55, not being recommended. However, in these cases, the particular cover and binder can be used, but the conductivity value will be over-specified, i.e. the concrete will have higher durability than required for the specified service life.

Table 8.21 Maximum chloride conductivity values (mS/cm) (illustrative only) for different classes and binder types: deemed to satisfy approach (50 years of service life, cover = 50 mm)

	Binder combination		
EN206 class	70:30	50:50	90:10
	CEMI:FA	CEMI:GGBS	CEMI:CSF
XS1	3.0	3.5	1.2
XS2a	2.45	2.6	0.85
XS2b, XS3a	1.35	1.6	0.45
XS3b	1.1	1.25	0.35

Table 8.22 Limiting DI values (illustrative only) based on rational prediction model: maximum chloride conductivity (mS/cm) (50 year life)

Exposure class	Cover (mm)	Max. chloride conductivity (mS/cm) for various binder types		
		100% CEM I	30% FA	50% GGBS
XS3b	40	0.45	0.75	1.05
	60	0.95	1.35	1.95
	80	1.30	1.80	2.60
XS0b	40	1.00	1.85	2.50
	60	1.85	2.95	3.90
	80	2.50	3.75	4.80
(light shading)	Concrete grade > 60 MPa[a]			
(dark shading)	Not recommended: grades < 30 MPa, and/or w/b > 0.55			
(no shading)	Acceptable mixes. Grades 30 to 60 MPa			

[a]It may be impractical to select 100 % CEMI mixes unless a strength of 60 MPa or more is required for structural reasons

8.9 Rapid Chloride Migration Test (Suggested for Application in The Netherlands)

The principles of non-steady state migration testing have been described in Chap. 4. Briefly, non-steady state migration testing involves applying a potential difference across a concrete sample to accelerate chloride transport in order to determine a transport coefficient. This transport coefficient is subsequently used in a transport model to determine the time until a particular chloride content (the critical chloride content for corrosion initiation) is reached at a particular depth (the depth of the reinforcement). Models used in this way are usually based on a solution to Fick's second law of diffusion.

This approach has been followed in The Netherlands, resulting in a performance and probability based guideline for service life design of civil engineering structures exposed to environmental classes XS (marine) and XD (de-icing salt) [60, 62, 63]. The probability oriented methodology was conceived in the 1980s and developed in the 1990s in European research project DuraCrete [61]. Later developments are reported in [70, 71].

It should be noted that in view of limited experience the requirements of the prevailing Dutch concrete standards should apply as "ceiling values" (based on NEN 8005, the national version of EN 206). This implies particular maximum water-to-cement ratios and minimum cement contents, depending on environmental class. Under these conditions chloride-induced rebar corrosion is likely to be the dominant mechanism determining the service life, whereas carbonation-induced corrosion can be ruled out.

The guideline specifies the Rapid Chloride Migration (RCM) test (NTBuild 492) as the method for prequalification testing of concrete. The guideline specifies the two electrode method (TEM) for concrete resistivity as the test method for production control, described in Sect. 8.7. The service life model used is a modification of the DuraCrete model. A full set of input variables is provided including statistical parameters allowing full-probabilistic calculations [60].

8.9.1 Service Life Model

The limit state function is given by:

$$C(x,t) = C_s - (C_s - C_i)erf\left(\frac{x}{\sqrt{4kD(t)t}}\right) < C_{crit} \tag{8.38}$$

where $C(x,t)$ is the chloride content at depth x at time t in % by mass of binder, C_s is the surface chloride content in % by mass of binder, C_i is the initial chloride content in concrete in % by mass of binder, K is the correction factor, $D(t)$ is the time-dependent apparent diffusion coefficient in m^2/s, and C_{crit} is the critical chloride content for corrosion initiation in % by mass of binder.

The surface chloride content is assumed to be independent of concrete composition. It only depends on the environment: 3.0 % for marine structures [72] and 1.5 % for structures exposed to de-icing salts [73]. The initial chloride content was taken equal to 0.1 % based on typical measured values for "uncontaminated concrete". The apparent diffusion coefficient $D(t)$ is multiplied by a correction factor k to obtain the chloride diffusivity of concrete in real structures. This correction factor depends on binder type, environment and duration of wet curing. The k-values were taken from [61]. The critical chloride content was taken equal to 0.6 % by mass of cement for all binder types, see amongst others [74–76].

The apparent diffusion coefficient $D(t)$ is time dependent due to hydration of the binder, which causes narrowing of capillary pores (especially in binders with slag or fly ash); and drying, which reduces the amount of liquid in the pores, see Chap. 2 The time-dependent diffusion coefficient is calculated using the rapid chloride migration coefficient and a time dependency following [77] by:

$$D(t) = D_0\left(\frac{t_0}{t}\right)^n \tag{8.39}$$

where D_0 is the D_{RCM}-value at reference time t_0 (usually 28 days), and n is the ageing coefficient ($0 < n < 1$).

Based on DuraCrete and additional work [72], n-values were chosen for the Guideline in two groups of environmental classes: very wet (XD2/XS3) and moderately wet (XD1/XD3/XS1), see Table 8.23.

Table 8.23 Ageing coefficients n for different binders in two groups of environmental classes

	Coefficient n	
Environmental classes	Underground, splash and tidal zone	Above ground, marine atmospheric
	XD2, XS3	XD1, XD3, XS1
Type of binder		
CEM I	0.40	0.60
CEM I, 25–50 % slag, II/B-S; III/A, < 50 % slag	0.45	0.65
CEM III/A or/B, 50–80 % slag	0.50	0.70
CEM I with 21–30 % fly ash	0.70	0.80
CEM V/A with c. 25 % slag and 25 % fly ash	0.60	0.70

8.9.2 *Reliability Considerations and Semi-probabilistic Approach*

For a given environment, concrete cover depth and chloride diffusivity, Eq. (8.38) can be used for calculating the time needed for the critical chloride content to reach the reinforcement. Such a calculation, however, is deterministic and yields a mean value. This means that the probability of corrosion initiation at that point in time and space is 50 %. In practice such a high probability of corrosion is unacceptable, as it would mean that weak spots suffer corrosion much earlier and interventions may already be needed well before the intended end of the service life. An acceptable probability of failure for corrosion initiation of reinforcing steel may be 10 % [78].

To obtain a lower probability of failure than 50 %, either the cover depth can be increased or the maximum D_0 can be decreased. For the guideline it was chosen to add a fixed amount to the (deterministically determined) minimum cover depth *as a safety margin*. This is a semi-probabilistic approach, comparable to using a safety factor for a materials property or a load.

Calculations using probabilistic software have shown for a set of example cases that an increase of the cover depth by 20 mm will reduce the probability of corrosion initiation from 50 % to about 10 %. A safety margin of 30 mm produces a probability of 5 %. Such probabilities are considered appropriate for reinforcing and prestressing steel, respectively.

8.9.3 *Limiting Values*

Following the method described above, combinations of required cover depth (including a safety margin to the cover depth of 20 mm for reinforcing steel or

30 mm for prestressing steel) and maximum D_{RCM}-values were calculated for service lives of 80, 100 or 200 years using Eq. (8.38). Table 8.24 presents limiting values for D_{RCM} (at 28 days) and mean cover depth for 100 years.

Two hypothetical examples may illustrate how Table 8.24 can be used.

Example I concerns a reinforced concrete structure in XD1-3/XS1 environment. Type CEM III/B cement with 70 % slag was chosen. The required service life is 100 years. From Table 8.24 it can be seen that with a cover depth of 45 mm, a maximum $D_{RCM,28}$ is required of 6.0×10^{-12} m^2/s. With this cement and a w/b of 0.45, a D_{RCM}-value of 4.0×10^{-12} m^2/s can be obtained rather easily [63]. Going back to Table 8.24 it can be seen that with a D_{RCM}-value of 4.0×10^{-12} m^2/s the cover depth could be reduced to 40 mm.

Example II concerns the same structure as Example I. The cover depth is 45 mm, but now CEM I is used. For CEM I and a cover depth of 45 mm, Table 8.24 gives a maximum $D_{RCM,28}$ of 8.5×10^{-12} m^2/s. Such a value might be hard to achieve with CEM I. It would require quite a low w/b, probably below 0.4, which may cause workability problems. Increasing the cover to 50 mm will allow an increase of $D_{RCM,28}$ to 12×10^{-12} m^2/s, which can be readily achieved with a w/b of about 0.45.

8.9.4 Application in Rijkswaterstaat Projects

During the first 4 years after its release in 2009 the performance and probability based approach developed in the Netherlands [57, 79] has seen limited application in practice [80]. Further work regarding the validity of the semi-probabilistic approach [81], the realistic level of the performance requirements [82], as well as the validity of the chloride migration test [83] are required. As a consequence, this performance and probability based approach is not accepted in contract documents for Rijkswaterstaat projects [84], however the approach adopted in this guideline is considered to have potential for further development.

8.10 In Situ Ionic Migration Test: PERMIT (UK)

Permit is an in situ ionic (chloride) migration test, which can be performed on the concrete surface. The biggest advantage is that the test can be used for determining the quality of in place concrete without removing cores. Hence the damage caused to the structure is minimal. The test is described in detail in Sect. 4.5.2.6. The chapter also give details about the relationship between $D_{\text{in situ}}$ from Permit and coefficients derived from more commonly used test methods. A procedure for developing performance specification for various chloride exposures is outlined in Fig. 8.38 (Table 8.25).

Indicative $D_{\text{in situ}}$ values for structures in XS3 environment are provided in Table 8.19.

Table 8.24 Maximum $D_{RCM,28}$ for various cover depths as a function of binder type and environmental class for a design service life 100 years [60, 62, 63]

Mean cover (mm)		Maximum value $D_{RCM,28}$ (10^{-12} m^2/s)							
Reinforcing steel	Prestressing steel	CEM I		CEM I + III 25–50 % S		CEM III 50–80 % S		CEM II/B-V, CEM I + 20–30 % V	
		XD1/2/3, XS1	XS2 XS3	XD1/2/3, XS1	XS2 XS3	XD1/2/3, XS1	XS2, XS3	XD1/2/3, XS	XS2, XS3
35	45	*3.0*	*1.5*	*2.0*	*1.0*	**2.0**	*1.0*	*6.5*	*5.5*
40	50	**5.5**	*2.0*	**4.0**	*1.5*	**4.0**	1.5	**12**	**10**
45	55	**8.5**	*3.5*	**6.0**	*2.5*	**6.0**	**2.5**	**18**	**15**
50	60	**12**	**5.0**	**9.0**	*3.5*	**8.5**	**3.5**	**26**	**22**
55	65	**17**	**7.0**	*12*	**5.0**	*12*	**5.0**	*36*	*30*
60	70	**22**	**9.0**	*16*	**6.5**	*15*	**6.5**	*47*	*39*

Note Boldface values are practically achievable by present-day concrete technology with currently used w/b; italic values are not achievable (lower values) or not recommended (higher values associated with low concrete quality)

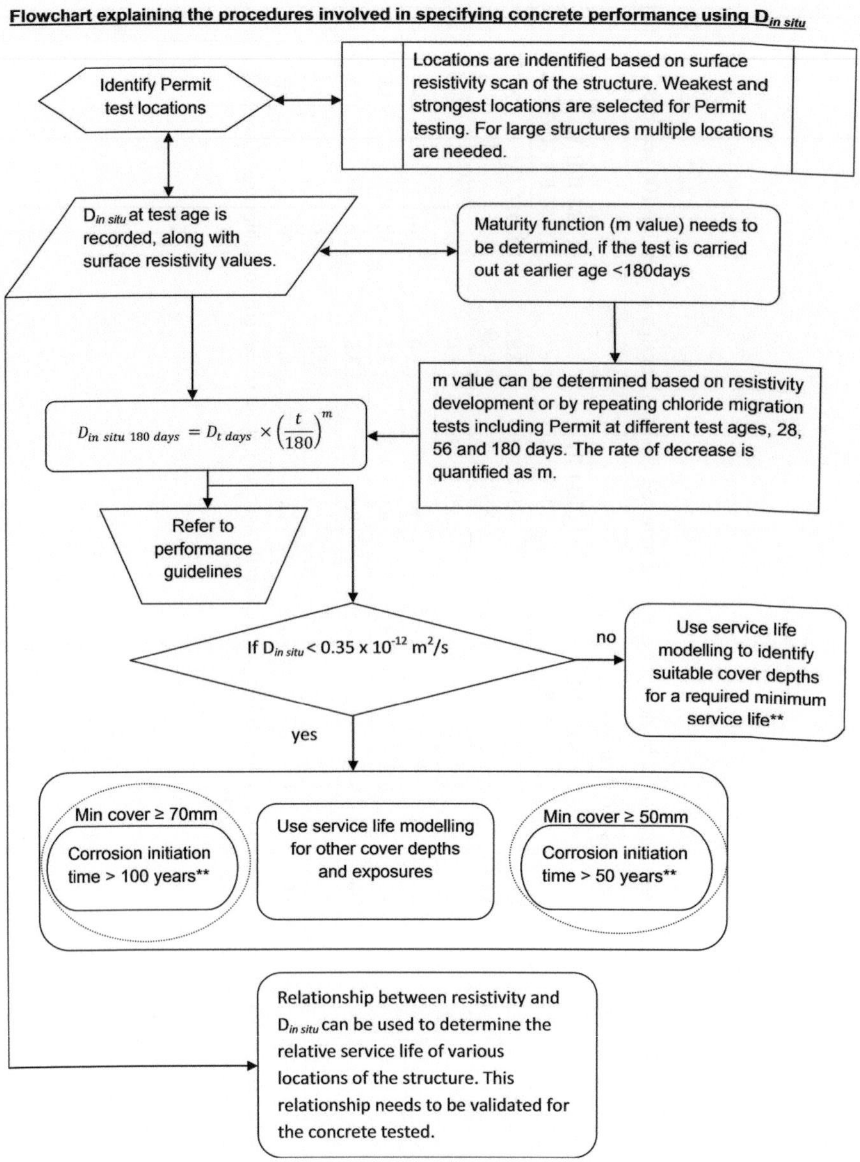

**: Other factors such as critical chloride concentration, tensile capacity of concrete and type/surface of rebar will also influence the corrosion initiation time. Service life includes both corrosion initiation and propagation time and the later may vary based on the use of a structure. Methods to quantify rate of corrosion and other influencing factors need to be employed to predict the corrosion propagation time.

Fig. 8.38 Flowchart explaining the procedures involved in specifying concrete performance using $D_{\text{in situ}}$

Table 8.25 Deemed to satisfy $D_{\text{in situ}}$ values for concrete at an age of 6 months for exposure to chloride environments—XS3—UK

	Common structures	Monumental structures	
Service life	50 years	100 years	100 years
Minimum cover (mm)	50	50	75
$D_{\text{in situ}}$ (10^{-12}m^2/s) (Tested at 6 months)	≤0.30	≤0.15	≤0.35

Note Modelled values based on an assumed critical chloride threshold of 0.1 % by mass of concrete

As $D_{\text{in situ}}$ is measured in steady state condition, the values can be used as input parameter in most models and chloride profiles can be generated for various service life scenarios [85]. Figure 8.39 shows the chloride concentrations after 50 years of exposure to XS2 in North Sea environment [86]. Such information will allow users to (1) select a suitable concrete mix based on the exposure condition and (2) determine the remaining service life of a concrete structure based on testing and modelling. For example, to maintain the chloride concentration at the level of reinforcement (that has a cover depth of 50 mm) to a value below 0.5 % by mass of binder (assumed chloride threshold), a concrete of $D_{\text{in situ}}$ value less than 0.2×10^{-12} m^2/s (at 6 months age) should be selected.

In order to specify concrete based on a 28 days test result, a correction factor as given below can be applied to adjust for test age:

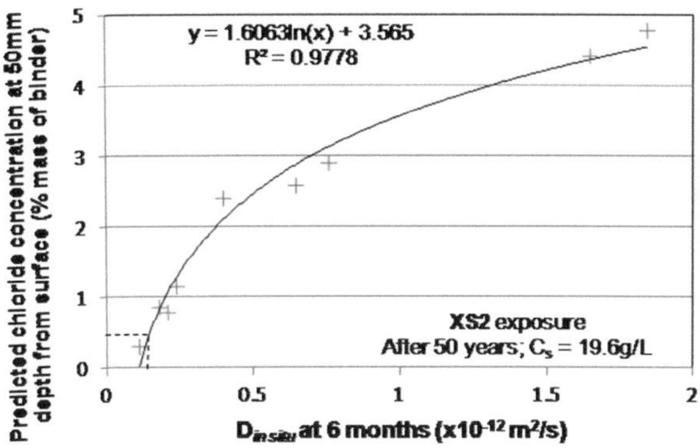

Fig. 8.39 Relationship between $D_{\text{in situ}}$ values at 6 months age and chloride concentration at a depth of 50 mm from the surface for a range of different concrete mixes

$$D_{\text{in situ}}\, at\, 6\,\text{month} = D_{\text{in situ}}\, at\, 28\,\text{day} \times \left(\frac{28}{180}\right)^{m} \tag{8.40}$$

For example, if the maturity function, m = 0.2,

$$D_{\text{in situ}}\, at\, 28\,\text{day} = 1.45 \times D_{\text{in situ}}\, at\, 6\,\text{month} \tag{8.41}$$

Selecting an appropriate maturity function is critical in determining the time dependent reduction in diffusivity. Maturity function will be influenced by factors such as mix proportions, type and quantity of binder, curing condition and exposure environment. Therefore, it is necessary to determine the m value for the type of concrete specified and the exposure environment.

A maturity value (or ageing factor) can be determined by carrying out PERMIT test at several concrete age and the reduction in diffusivity quantified as an *m* value. Alternatively, electrical resistivity based test (Sects. 8.6 and 8.7) can also be used to determine an *m* value. It is important to note that the significance of maturity function may become negligible as concrete mature over 180 days for exposure environments with sufficient moisture. Generally prediction models assume that concrete matures completely at age from 10–25 years.

Table 8.12 gives a range of aging factor (maturity values, *m*) for different concretes based on their exposure environment. Research is ongoing to establish this relationship using PERMIT on site for a range of concrete mixes and test ages [87].

8.10.1 Typical Applications of Permit Migration Test

Typical applications of the PERMIT Migration Test to predict the service life of major construction projects are summarised below. Due to the confidentiality of

Fig. 8.40 Testing Qingdao Bay Bridge with PERMIT by Tsinghua University

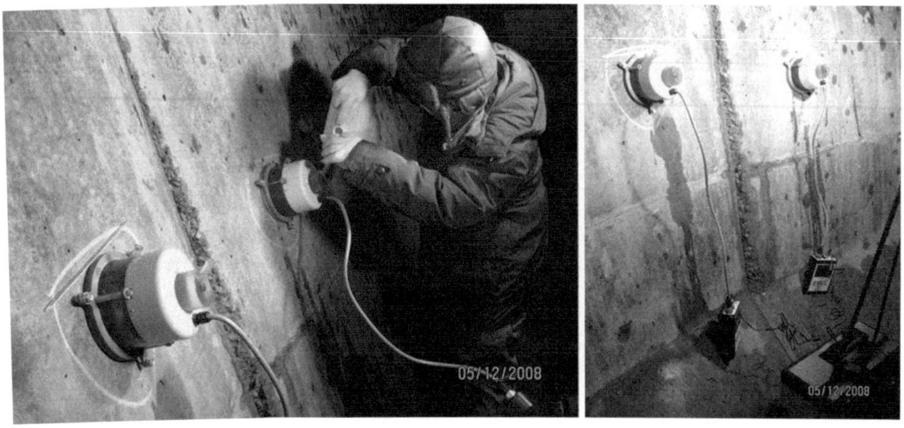

Fig. 8.41 Testing concrete quality with PERMIT in Harbin-Dalian railway project by Tsinghua University

data, details of the concrete and the predicted service life are not available at present (Figs. 8.40 and 8.41).

8.11 Multi-level Prediction of Reinforced Concrete (RC) Durability

8.11.1 Principles of the Multi-level Approach

Multilevel durability design responds to the need to have a coherent integral framework to design durability in such a manner that more advanced models serve for calibrating prescriptive specifications and vice versa. The prescriptive rules should not suggest concrete mixes very different to those deduced from advanced models. The framework discussed in this section is named "multilevel" to express the coherence of the system and to indicate that any level should end in the same set of concrete mix proportions given similar durability. This approach has been incorporated into the latest version of fib Model Code and some countries are developing experimental standards with it.

The number of levels classified may be 3 or 4 and depend on the concepts being considered. As shown in Table 8.26 the levels can be:

- Prescriptive of performance based
- Deterministic or probabilistic
- Having explicit or not (implicit) the time in the mathematical expressions

The several methods to calculate or assess service life have been organized in levels of verification as indicated in Table 8.26 The driving concept in the

Table 8.26 Methods to calculate or assess service life

Durability verification formats			
Category 1	Category 2		Category 3
Deterministic format		Semi-probabilistic format	Probabilistic format
Time implicit		Time explicit	Time explicit
Codes and standards	Durability indicators	Predictive models	Predictive models

organization is based on whether the categories are deterministic, semi-probabilistic or full probabilistic.

Once the required length of service life is defined and the environment identified, the verification that a concrete mix fulfils the target durability can be made by any of the levels or categories specified in the table.

- Category 1 presents the deterministic format that is the traditional method of present codes or standards of specifications of concrete mixes by limiting the maximum w/c ratio, the minimum strength and the maximum initial chloride content and crack width
- Category 2 encounters deterministic or semi-probabilistic formats where the durability is verified by fulfilling threshold values of the so called "durability indicators" which are time implicit or by using predictive models which are time explicit
- Category 3 refers to a full probabilistic format using time explicit models.

8.11.2 Procedure for Verification of Durability

Figure 8.42 presents the procedure for durability verification. The first decisions of the designer correspond to the selection of the length of service life, the level of reliability, and the requirements of the structure. Then, the environment where the structure will be built and the possible degradations processes have to be identified. This leads to the establishment of a first concrete mix and the need or not to use preventive or additional protection methods.

The next step is the definition of the limit state and the selection of the format of verification (deterministic, semi or full probabilistic). The format of verification determines the use of either prescriptive specifications, durability indicators, or predictive models. With these procedures the service life is verified for the concrete mixes selected. If the requirement are fulfilled the process is finished but if the selected concrete mixes do not satisfy the requirements, then new formulations have to be selected and the process starts again or additional preventive methods are incorporated (cathodic protection for instance).

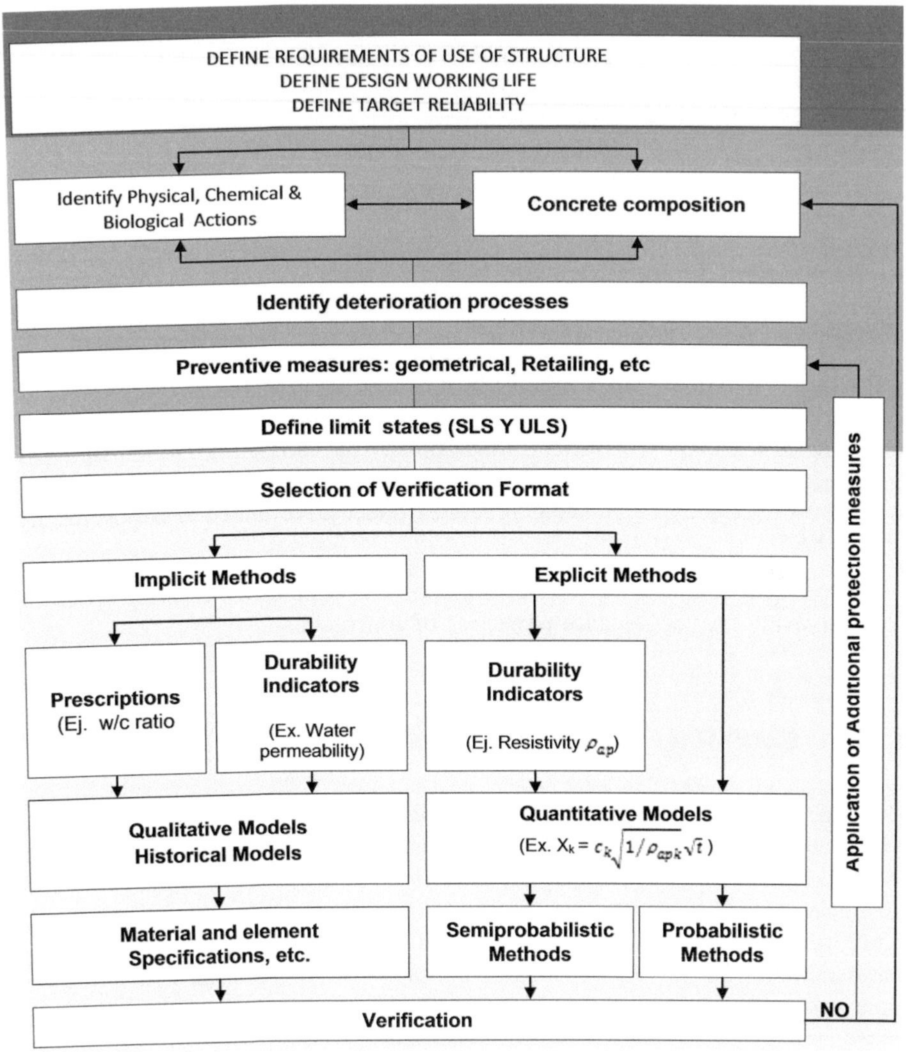

Fig. 8.42 Procedure for verifying durability

In the multilevel methodology there are considered the so-called durability indicators (DIs), which are key material properties with regard to durability [88–90]. A system of classes of "potential" durability with respect to (carbonation- and chloride-induced) reinforcement corrosion has been proposed for each DI. These five classes—very low (VL), low (L), medium (M), high (H) and very high (VH) "potential" durability—can be used for example for mixture comparison or quality control. The evaluation of the "potential" durability of a given mix will consist in comparing the values of the measured DIs to the limits of the associated classes. Further details on this approach can be found in the literature [88, 91–93] (Fig. 8.42).

8.12 Portuguese Technical Specifications

The Portuguese Standard NP EN 206-1:2007 includes in its National Document of Application two Technical Specifications concerning the durability design of concrete structures subjected to carbonation and chloride induced corrosion:

- LNEC E 464:2005—"Concrete. Prescriptive methodology for a design working life of 50 and 100 years under environmental exposure"
- LNEC E 465:2005—"Concrete. Methodology for estimating the concrete performance properties allowing to comply with the design working life of reinforced or pre-stressed concrete structures under environmental exposures XC and XS"

The present section describes the methodologies established in those specifications involving the performance properties of concrete.

8.12.1 Equivalent Performance Concept

The Specification LNEC E 464 allows the use of compositions and cements (or combinations of cement with additions) other than those presented in the prescriptive approach, through the application of the equivalent performance concept (EPC). In order to do so, the specification establishes limits for the ratios between the properties of the candidate concrete and the properties of a reference concrete that complies with the requirements of the prescriptive approach, both using the same aggregates and the corresponding proportions. The properties to be determined and compared are presented in Table 8.27.

For the accelerated carbonation, oxygen permeability and capillary absorption tests, the concrete specimens should be cured for 7 days and then, for the first test, the specimens should be 7 days at 20 ± 2 °C without humidity exchanges, followed by 14 days at 20 ± 2 °C and at 65 ± 5 % RH, and for the last two tests, after wipe the excessive water from the specimens surfaces with a cloth, they should dry for 3 days in a ventilated oven at 50 ± 2 °C, followed by 17 days at 50 ± 2 °C and 1 day at 20 ± 2 °C, without humidity exchange in these last 18 days. The determinations should begin at 28 days and the tests repeated at least every 3 years.

Besides the mixes of the candidate concrete and of the reference concrete (principal mixes), two secondary mixes obtained by varying ± 5 % the binder content of each principal mixture should also be tested.

Table 8.27 Properties, methods and test specimens for EPC

Exposure class	Properties to be determined	Test methods	Number and type of specimens (mm)	Limit candidate/reference ratio
XC1 XC2 XC3 XC4	Accelerated carbonation	LNEC E 391	1 specimen 150 × 150 × 600 mm^3	≤1.3
	Oxygen permeability	LNEC E 392	3 specimens ϕ 150 mm; h = 50 mm	≤2.0
	Compressive strength	NP EN 12390-3	3 specimens of 150 × 150 × 150 mm^3	≥0.9
XS1/XD1 XS2/XD2 XS3/XD3	Chloride diffusion coefficient	LNEC E 463	2 specimens ϕ 100 mm; h = 50 mm	≤2.0
	Capillary absorption	LNEC E 393	3 specimens ϕ 150 mm; h = 50 mm	≤1.3
	Compressive strength	NP EN 12390-3	3 specimens of 150 × 150 × 150 mm^3	≥0.9

The performance of the candidate concrete is then considered equivalent to that of the reference concrete if the overall average values of the properties of the three candidate mixes are equal or better than those of the reference concrete, and if the individual values from each corresponding principal or secondary mix satisfy the limits of Table 8.27.

8.12.2 Methodology for Estimating Design Working Life

The Specification LNEC E 465 incorporates the semi-probabilistic methodology presented in RILEM Report 14 [94], using the service-life models developed in Europe during the 1990s. It complies with the general rules presented in EN 1990:2002, regarding the partial factor approach, establishing a safety factor γ that affects the intended working life of the structures t_g through the following expression:

$$t_d = \gamma t_g = \gamma\left(t_i + t_p\right) \Leftrightarrow t_{ic} = \gamma t_i = \gamma\left(t_g - t_p\right) \tag{8.42}$$

where t_d is the design working life, t_i is the corrosion initiation period, t_p is the corrosion propagation period, according to Tuutti's model of reinforcement concrete deterioration under the environmental actions XC or XS/XD, and t_{ic-} is the design corrosion initiation period.

This methodology consists basically of calculating the propagation period t_p and then determining the minimum value of the relevant concrete property that ensures the initiation period t_{ic} obtained from Eq. (8.42). For the calculation of γ the reliability classes of EN 1990 are considered, and the reliability indexes (β) of 1.2,

1.5 and 2.0 for classes RC1, RC2 and RC2 are established, leading to safety factors of 2.0, 2.3 and 2.8, respectively. It is assumed that the working life is lognormal distributed with a coefficient of variation of 50 %. The Serviceability Limit State concerning the durability is defined as the beginning of cracking of concrete due to reinforcement corrosion.

LNEC E 465 divides class XC4 in two regions: a dry region, located on the south of River Tagus, in the Hot Region of Douro, in the centre/south zone of Castelo Branco region and on the southern coast of Madeira; and a wet region located north of the Tagus River excluding the zones previously mentioned, in the centre/north zone of Madeira and Azores. Table 8.28 shows the relative humidity and time-of-wetness of concrete assumed for each environmental exposure class. The time-of-wetness is defined as the yearly average number of days with rainfall equal to or higher than 1 mm divided by 365.

For the estimation of the initiation period due to carbonation, two models are presented. One is based on Fick's first law of diffusion, assuming a stationary CO_2 flow with a constant concentration of 0.7×10^{-3} kg/m^3 in the atmosphere:

$$R_{C65} = \frac{1.4 \times 10^{-3} \cdot t_{ic}}{R^2} k_0 k_1 k_2 \left(\frac{t_0}{t_{ic}}\right)^{2n} \tag{8.43}$$

where R_{C65} is the carbonation resistance of concrete exposed to 5 % of CO_2 and 65 % of relative humidity in kg.year/m^5, R is the concrete cover depth in mm, k_1 and n are factors that consider the influence of the relative humidity and dry/soaking over the time (values presented in Table 8.29), k_0 is a factor of value 3 when the test conditions are those of LNEC Specification LNEC E 391—*Concrete. Determination of resistance to carbonation*, t_0 is the reference period (1 year), and k_2 is a factor that takes into account the concrete curing conditions, by assuming a

Table 8.28 Relative humidity and the soaking time of the concrete for each exposure class

Exposure class	Relative humidity	Time-of-wetness
XC1 (dry/always wet)	Dry environment: 60 % Wet environment: 100 %	0.05 1[a]
XC2 (wet, rarely dry)	90 %	0.8
XC3 (moderate humidity)	70 %	0.1
XC4 (cyclic wet and dry)	Dry region: 80 % Wet region: 80 %	0.18 0.24
XS1 (air with sea salts)	80 %	0.6
XS2 (permanent immersion)	100 %	1[a]
XS3 (tidal and splash zone)	100 %	1

[a]Absence of oxygen for the corrosion process

Table 8.29 Values of parameters k_1 and n to calculate R_{C65}

	XC1	XC2	XC3	XC4
k_1	1.0	0.20	0.77	0.41
n	0	0.18	0.02	0.09

value of 1 for standard curing conditions and a factor of 0.25 when the formwork is of controlled permeability and the curing period is three days or more (i.e. demoulding age).

The other model is based on the model of air permeability described in [95]:

$$k_{60} = \frac{R^{2.5} c^{1.25}}{(ak_{2.5})^{2.5} t_{ic}^{2.5p} m} \tag{8.44}$$

where k_{60} is the coefficient of air permeability measured by the CEMBUREAU method described in LNEC Specification LNEC E 392—*Concrete. Determination of permeability to oxygen* on specimens with 28 days of age in equilibrium with 60 % RH in 10^{-6} m, m is a factor that relates the coefficient of air permeability of the cover concrete with k_{60}, a is 150, c is the calcium oxide content of the hydrated cement matrix of concrete (depends on the type of cement used and on the exposure class) in kg/m^3, p is an exponent that depends on the relative humidity of concrete and therefore on the exposure class, k_2 is a factor that takes into account the concrete curing conditions, assuming the value 1 for standard curing conditions and 0.5 when the formwork is of controlled permeability and the curing period is 3 days.

The values of the parameters m, p and c are presented in Table 8.30.

For estimating of initiation period due to chlorides penetration the specification provides the following model based on Fick's second law of diffusion:

Table 8.30 Values of the parameters m, p and c to calculate k_{60}

RH (%)	m	p	c (kg/m^3)		
			CEM I[a]	CEM II/III	CEM IV
60	1.00	0.51	460	350	230
65	0.737	0.5	460	350	230
70	0.534	0.48	460	350	230
75	0.382	0.45	470	358	235
80	0.256	0.42	485	365	240
85	0.184	0.37	510	388	253
90	0.117	0.32	535	410	265
95	0.057	0.25	570	430	285
100	0	0.19	615	470	310

[a]Also applicable to CEM II/A-L

$$D_0 = \frac{R^2}{4k_{D,RH}k_{D,T}k_{D,c}t_{ic}\left(erf^{-1}\left(\frac{C_S - C_R}{C_S}\right)\right)^2}\left(\frac{t_{ic}}{t_0}\right)^n \tag{8.45}$$

where D_O is the potential diffusion coefficient determined in the laboratory in accordance with LNEC Specification LNEC E 463—*Concrete. Determination of diffusion coefficient of chlorides from non-steady-state migration test* (similar to NTBuild 492) with the concrete at the reference age of 28 days in m^2/s, R is the concrete cover depth in mm, C_s is the chloride concentration, in % of the binder mass, on the concrete surface, assumed as constant (depends on the exposure class, water/binder ratio, temperature of concrete, distance of the coast line, and the depth of concrete in the water), C_R is the threshold chloride concentration, in % of the binder mass, that causes the depassivation of the reinforcement (Table 8.31), t_O is the reference age (1 year), $k_{d,RH}$, $k_{D,T}$, $k_{D,c}$ are factors that take into account the relative humidity of the environment and the temperature and curing conditions of concrete, respectively (Tables 8.31, 8.32, 8.33, and 8.34), and n is the ageing factor that takes into account the decrease in the chloride ingress over the years (Table 8.35).

The model to estimate the propagation period is based on Faraday's law and on an experimental expression developed by [96] that estimates the reinforcement radius reduction that leads to concrete cracking. The model is given by the following expression:

Table 8.31 Chloride concentration C_R (% of cement mass)

Water/cement	XS1; XS2	XS3
w/c ≤ 0.30	0.6	0.5
0,30 < w/c ≤ 0.40	0.5	0.4
w/c > 0.40	0.4	0.3

Table 8.32 Values of parameter $k_{d,RH}$

Exposure class	$k_{D,RH}$
XS1	0.4
XS2	1.0
XS3	1.0

Table 8.33 Values of parameter $k_{D,T}$

Concrete temperature (°C)	$k_{D,T}$
30	1.5
25	1.2
20	1.0
15	0.8
10	0.75
0	0.4

Table 8.34 Values of parameter $k_{D,c}$

Period of curing, days	$k_{D,c}$
Standard	2.4
In permanent contact with water	0.75
Formwork of controlled permeability and 3 days of wet cure	1.0

Table 8.35 Values of parameter n

Exposure classes	n	
	CEM I/II[a]	CEM III/IV
XS1	0.55	0.65
XS2	0.45	0.55
XS3	0.55	0.65

[a]Except CEM II-W, II-T, II/B-L and II/B-LL

$$t_p = k\frac{\phi_0}{1.15\alpha I_{corr}} \quad (8.46)$$

where, k is the reduction in reinforced cross-section given in percentage and is given by $0.1(74.5 + 7.3R/\phi_{O-1}7.4f_{cd})/(\phi_O/2)$, R is the concrete cover depth in mm, ϕ_O is the initial diameter of the reinforcement in mm, f_{cd} is the tensile splitting strength (2.0–2.5 MPa for carbonation–induced corrosion and 3.0–4.0 MPa for chloride-induced corrosion), I_{corr} is the corrosion intensity in mA/cm^2, and α is 2 or 10 for carbonation or chloride-induced corrosion, respectively.

The estimates of t_p recommended by the Specification LNEC E 465 can be found in Table 8.36.

The concrete is considered to be conform with the specification if the mean values of its performance properties fulfil the requirements of the corresponding models, i.e., the concrete shall have values equal to or less than the mean value of

Table 8.36 Minimum periods of corrosion propagation

Exposure class	t_p estimate (years)	
	t_g = 50 years	t_g = 100 years
XC1	>100	>100
XC2	10	20
XC3	45	90
XC4	15 (dry region) 5 (wet region)	20 (dry region) 10 (wet region)
XS1	0	0
XS2	40	80
XS3	0	0

K_{60} or of D_0 and equal to or higher than the one of R_{C65}. These mean values should be established in initial tests, on at least 9 concrete specimens, obtained in 3 different batches. Furthermore, the maximum deviations obtained in each batch should not be greater than 50 %.

Tests carried out by LNEC (National Laboratory for Civil Engineering) on standard specimens [97] revealed that, in general, concretes complying with LNEC E 464 have better resistance to carbonation than that required by LNEC E 465. For chloride-induced corrosion, the chloride diffusion coefficients of the above concretes seem to be higher than those estimated by LNEC E 465, when using concretes with CEM I or CEM II/A-L and especially for the exposure class XS3.

Recent studies performed on concrete exposed to urban and marine environments for 5 years have shown a good agreement between the predicted and measured carbonation depths [98], suggesting that the models overestimate carbonation depths for concrete with CEM I 42.5 R and concrete under marine environments, and the opposite for concrete with CEM IV and concrete in urban environments. However, tests carried out on concrete over longer times of exposure are needed to properly evaluate the models presented in Specification LNEC E 465.

References

1. Torrent R, Ebensperger L. Studie über Methoden zur Messung und Beurteilung der Kennwerte des Überdeckungsbetons auf der Baustelle, Report No. 506, Bern: Office Fédéral des Routes, 1993 [in German].
2. Torrent R, Frenzer G. Methoden zur Messung und Beurteilung der Kennwerte des Ueberdeckungsbetons auf der Baustelle -Teil II, Report No. 516, Bern: Office Fédéral des Routes, 1995 [in German].
3. Brühwiler E, et al. Applicabilité de la mesure de perméabilité selon Torrent pour le contrôle de qualité du béton d'enrobage, Report No. 587, Bern: Office Fédéral des Routes, 2005 [in French].
4. Jacobs F. Luftpermeabilität als Kenngrösse für die Qualität des Überdeckungsbetons von Betonbauwerken, Report No. 604, Bern: Office Fédéral des Routes, 2006 [in German].
5. Jacobs F, Denariè E, Leemann A, Teruzzi T. Empfehlungen zur Qualitätskontrolle von Beton mit Luftpermeabilitätsmessungen Report No. 641, Bern: Office Fédéral des Routes, 2009 [in German].
6. Torrent R. A two-chamber vacuum cell for measuring the coefficient of permeability to air of the concrete cover on site. RILEM Mater Struct. 1992;25(7):358–65.
7. SIA, Construction en béton – Spécifications complémentaires, Annexe E: Perméabilité à l'air dans les Structures, Norme Suisse SIA 262/1, 2003 [in French].
8. SIA 262:2013, Betonbau, 2013 [in German and French].
9. SIA 262/1:2013, Betonbau - Ergänzende Festlegungen, 2013 [in German and French].
10. Torrent R. Non-destructive air-permeability measurement: from gas-flow modelling to improved testing. Amsterdam, 2012.
11. Jacobs F, Mühlan B. Neubau mit zu geringer Bewehrungsüberdeckung. die Baustellen, Issue Dezember 2012.
12. EN 12390-8:2009 Testing hardened concrete—Depth of penetration of water under pressure, 2009.
13. DIN 1048:1978 Prüfverfahren für Beton, 1978 [in German].

14. ASTM-C1202, Standard test method for electrical indication of concrete's ability to resist chloride ion penetration. Ann Book ASTM Stand. 2010; 4(2).
15. Torrent R, Denarié E, Jacobs F, Leemann A, Teruzzi T. Specification and site control of the permeability of the cover concrete: the Swiss approach. Mater Corros. 2012; 63(12):1127–1133.
16. Imamoto K, Tanaka A, Kanematsu M. Non-destructive assessment of concrete durability of the National Museum of Western Art in Japan. Amsterdam, 2012.
17. Imamoto K, Tanaka A. Evaluation of carbonation progress of existing concrete structure based on air permeability of concrete cover—a case study in Japan. South Africa: Cape Town; 2012.
18. Torrent R, Armaghani J, Taibi Y. Port of Miami Tunnel: carbonation and service life assessment through site permeability tests. Concr Int. 2013; 35(5).
19. Torrent R. Service life prediction: theorecrete. Kyoto: Labcrete and Realcrete Approaches; 2013.
20. Olek J, Lu A, Feng X, Magee B. Performance-related specifications for concrete bridge superstructures. West Lafayette: Joint Transportation Research Program; 2002.
21. CEB-FIP Model Code 1990: 1991: Design Code, 1990.
22. Mackechnie J, Alexander M. Durability predictions using early age durability index testing. Australian Corrosion Association: Brisbane; 2002. 11 pp.
23. EN 1992-1-1:2011 Eurocode 2: design of concrete structures—Part 1-1: General rules and rules for buildings, 1992.
24. EN 206: Concrete: specification, performance, production and conformity, 2013.
25. Torent R, Fernandez Luco LF. Service life assessment of concrete structures based on site testing. São Paulo, Brazil: DBMC 2014 Conference; 1–5 Sept 2014.
26. EN 13670:2009 Execution of concrete structures, 2009.
27. Salvoldi B, Beushausen H, Alexander M. The correlation between oxygen permeability and the carbonation of concrete. Constr Build Mater. 2015;85(2015):30–7.
28. Nilsson L-O, Luping T. Chapter 3—Relations between different transport parameters. In: Performance criteria for concrete durability, RILEM; 1995. pp 15–32.
29. Beushausen H, Alexander M. Application of durability indicators for quality control of concrete members—a practical example. France: Toulouse; 2009. p. 548–56.
30. Beushausen H, Knecht J, Alexander M, Schubert K. Performance-based durability control for precast concrete elements, Washington, 2010.
31. The concrete society technical report no. 31, Permeability testing of site concrete—a review of methods and experience, The Concrete Society, 2007.
32. Basheer PAM, Nolan E, Long AE. Near surface moisture gradients and in-situ permeation tests. Constr Build Mater. 2001;15(2–3):105–14.
33. Basheer PAM, Nolan EA, McCarter WJ, Long AE. Effectiveness of in-situ moisture preconditioning methods for concrete. J Mater Civ Eng. ASCE, 2000; 12(2):131–138.
34. Autoclam permeability system, operating manual, Amphora NDT Limited, 2007.
35. British Standards Institution. Methods of testing concrete, Part 5. London, 40 pp., BS1881:Part 5: 1970.
36. ASTM C666-03. Standard test method for resistance of concrete to rapid freezing and thawing. ASTM Stand Int. 2008; 4(2).
37. Setzer MJ, Fegerlund G, Janssen DJ. CDF test—test method for the freeze-thaw resistance of concrete-tests with sodium chloride solution (CDF). Mater Struct. 1996;29:523–8.
38. Basheer PAM. A brief review of methods for measuring the permeation properties of concrete in situ, 1993. pp 74–83.
39. Andrade C, Alonso C, Goñi S. Possibilities for electrical resistivity to universally characterise mass transport processes in concrete. London: E&FN Spon; 1993. pp 1639–1652.
40. Andrade C. Calculation of initiation and propagation periods of service-life of reinforcements by using the electrical resistivity. Evanston, RILEM Symposium, 22–24 Mar 2004.
41. Andrade C, d`Andréa R. Concrete mix design based on the electrical resistivity, Ancona, Italy, 2010.

42. Polder R, Andrade C, Elsener B, Vennesland O, Gulikers J, Weidert R, Raupach M. RILEM TC-154, Draft RILEM technical recommendation test methods for on site measurement of resistivity of concrete. Mater Struct. 2000;33(10):603–6.
43. PrUNE 83988—1, Part 1: Direct method (reference method).
44. PrUNE 83988—2, Method of four points or Wenner method.
45. Feliu S, Andrade C, Gonzalez J, Alonso C. A new method for in situ measurement of electrical resistivity of reinforced concrete. Mater Struct. 1996;29:362–5.
46. Garboczi EJ. Permeability, diffusivity and microstructural parameters: a critical review. Cem Concr Res. 1990;20:591–601.
47. Andrade C, d' Andrea R. The reaction factor in the resistivity mode, Delft, 2012.
48. Castellote M, Andrade C, Alonso C. Measurement of the steady and non steady state chloride diffusion coefficients in a migration test by means of monitoring the conductivity in the anolyte chamber. Comparison with natural diffusion tests. Cem Concr Res. 2001;31:1411–20.
49. Andrade C, Castellote M, d'Andrea R. Measurement of ageing effect of chloride diffusion coefficients in cementitious matrices. J Nucl Mater. 2011; 412:209–216.
50. Alonso C, Andrade C, González J. Relation between concrete resistivity and corrosion rate of the reinforcements in carbonated mortar made with several cement types. Cem Concr Res. 1988;118(5):687–98.
51. ASTM-C642 Standard test method for density, absorption, and voids in hardened concrete. Ann Book ASTM Stand. 2010; 4(2).
52. Chini A, Muszynski L, Hicks J. Determination of acceptance permeability characteristics for performance-related specifications for portland cement concrete, 2003.
53. Kessler R, Powers RG, Vivas E, Paredes MA, Virmani YP. Surface resistivity as an indicator of concrete chloride penetration resistance, St. Louis, 2008.
54. Vivas E, Boyd A, Hamilton IH. Permeability of concrete—comparison of conductivity and diffusion methods, 2007.
55. AASHTO, Provisional AASHTO test method TP 95-11: standard test method for surface resistivity of concrete's ability to resist chloride ion penetration, AASHTO Technology Implementation Group, 2011.
56. FDOT. Florida method of test for concrete resistivity as an electrical indicator of its permeability, 2004.
57. LADOTD, DOTD TR 233: test method for surface resistivity indication of concrete's ability to resist chloride ion penetration, Baton Rouge, LADOTD, 2011.
58. Rupnow T, Icenogle E. Evaluation of surface resistivity measurements as an alternative to the rapid chloride permeability test for quality assurance and acceptance, 2011.
59. Polder R. The influence of blast furnace slag, fly ash silica fume on corrosion of reinforced concrete in marine environment. HERON. 1996;41(4):287–300.
60. Polder R. Chloride diffusion and resistivity testing of five concrete mixes for marine environment, St-Remy-les-Chevreuses, RILEM 2, 1997.
61. DuraCrete. Duracrete—probabilistic performance based durability design of concrete structures, CUR, Gouda, Gouda: The European Union—Brite EuRam III, CUR, 2000.
62. Polder R, Wegen G, Breugel K. Guideline for service life design of structural concrete—a performance based approach with regard to chloride induced corrosion, Leipzig, 2011. pp 25–34.
63. Wegen G, Polder R, Breugel K. Guideline for service life design of structural concrete—a performance based approach with regard to chloride induced Corrosion, HERON, Issue in preparation, 2013.
64. Rooij M, Polder R, Oosten H. Validation of durability of cast in situ concrete of the Groene Hart railway tunnel, HERON, 2007; 52(4):225–238.
65. CHLORTEST. Workshop resistance of concrete to chloride ingress—from laboratory test to in-field performance, 2005.
66. Hooton R, Thomas M, Stanish K. Testing the chloride penetration resistance of concrete: a literature review, prediction of chloride penetration in concrete. McLean: Federal Highway Administration; 2001.

67. Tang L. Concentration dependence of diffusion and migration of chloride ions: Part 1. Theoretical considerations. Cem Concr Res. 1999; 29(a):1463–1468.
68. Mackechnie J, Alexander M. Marine exposure of concrete under selected South African conditions, 1996.
69. Mackechnie J. Predictions of reinforced concrete durability in the marine environment, 2001.
70. Gehlen C. Probabilistische Lebensdauerbemessung von Stahlbetonbauwerken, Berlin, 2000.
71. fib, Model code for service life design, 2006.
72. Polder R, Rooij M. Durability of marine concrete structures—field investigations and modelling. HERON. 2005; 50(3):133–154.
73. Gaal G. Prediction of deterioration of concrete bridges. Delft: Delft University Press; 2004.
74. Breit W. Critical chloride content—investigations of steel in alkaline chloride solutions. Mater Corros. 1998;49(8):539–50.
75. Breit W. Critical corrosion inducing chloride content—state of the art and new investigation results. Düsseldorf: Verlag Bau Technik; 2001.
76. Polder RB. Critical chloride content for reinforced concrete and its relationship to concrete resistivity. Mater Corros. 2009;60(8):623–30.
77. Maage M, Helland S, Poulsen E, Vennesland O, Carlsen JE. Service life prediction of existing concrete structures exposed to marine environment. ACI Mater J. 1996;93(6):602–8.
78. Fluge F. Marine chlorides—a probabilistic approach to derive provisions for EN 206–1, Third DuraNet workshop on Service life design of concrete structures, from theory to standardisation, Tromsø, 2001.
79. CUR, Duurzaamheid van gewapend beton met betrekking tot chloride-geïnitieerde wapeningscorrosie, Gouda, 2009 [in Dutch].
80. Wegen G, Boutz M. Case studies CUR-Leidraad 1, Cement. 2012; 1:22–26, [in Dutch].
81. Guikers J. Development of guidelines for durability design of reinforced concrete structures. In: Bruegel K, Ye G, Yuan Y, editors. 2nd International symposium on service life design for infrastructure, Delft, 2010. pp 359–368.
82. Gulikers J. Practical implications of performance specifications for durability design of reinforced concrete structures. In: Dehn F, Beushausen H, editors. fib Workshop performance-based specifications for concrete, Leipzig, 2011.
83. Corporaal H, Eldik D. Dé DRCM-waarde bestaat niet. Cement. 2012;1:40–4 [in Dutch].
84. Rijkswaterstaat, Standpunt RWS DI ten aanzien van CUR Leidraad 1, 2011 [in Dutch].
85. Nanukuttan S, Basheer PAM, Basheer L, Holmes N, Srinivasan S, Tang L, McCarter WJ, Chrisp M, Starrs G. Quality control and performance assessment methods for concrete structures. Tampa, ACI Committees 201 (Durability of concrete) and 236 (Materials science of concrete), 2001.
86. Nanukuttan S, Basheer PAM, Holmes N, Tang L, McCarter WJ. Use of performance specification and predictive model for concretes exposed to a marine environment, Edinburgh, 2010.
87. McCarter WJ, Basheer PAM. Developing a performance-based testing methodology for specifying concrete durability, Engineering and physical sciences research council funded project (EP/G02152X/1) 2009–12., 2012.
88. Andrade C. Multilevel (four) methodology for durability design. Proceedings of the international RILEM workshop on performance-based evaluation and indicators for concrete durability, Madrid-Spain, 2006. pp 101–108.
89. UNE, P. 8.-1., Durability of concrete.
90. Baroghel-Bouny V. Concrete design for a given structure service life—durability management with regard to reinforcement corrosion and alkali-silica reaction. Paris: AFGC Scientific and Technical Documents (AFGC); 2007.
91. Baroghel-Bouny V. Durability indicators: relevant tools for performance-based evaluation and multi-level prediction of RC durability, Bagneux, RILEM, 2007. pp 3-30.4.
92. Baroghel-Bouny V. Durability indicators: relevant tools for an improved assessment of RC durability, Tours, 2007. pp 67–84.

93. Baroghel-Bouny V, Nguyen T, Dangla P. Assessment and prediction of RC structure service life by means of durability indicators and physical/chemical models. Cem Concr Compos. 2009;31(8):522–34.
94. RILEM Report, 1st edn. Durability design of concrete structures. In: Sarja A, Vesicari E, editors. CRC Press, 1996.
95. Parrot J. Design for avoiding damage due to carbonation-induced corrosion. ACI Special Publication, 1994.
96. Santiago J, Basagoiti O, Macías J, Arenas J. La corrosion de las armaduras y la vida residual de las estruturas de hormigón, LNEC, 1998 [in Spanish].
97. Gonçalves A, Ribeiro B, Ferreira E. The new LNEC specifications on reinforced concrete durability, Guimarães, 2007.
98. Ribeiro S, Ribeiro A, Gonçalves A. Resistance of concrete to carbonation. Predicted and measured values in natural exposure, Haifa, ConcreteLife, 2009.

Chapter 9
Basis for the Statistical Evaluation of Measured Cover Depths in Reinforced Concrete Structures

A.V. Monteiro, A. Gonçalves, J. Gulikers and F. Jacobs

9.1 Definitions

Actual minimum cover depth
Minimum cover depth achieved in a lot, defined as the characteristic value (percentile) below which a given percentage (5 or 10 %) of all possible values of the cover depth population will fall. This value is used in Procedure A to assess the conformity of the minimum cover depth with the specifications.

Cover depth population
Collection of all theoretically possible measured concrete cover depths of a specific reinforcement layer (usually an outermost layer) of concrete elements. This reinforcement layer should have the same cover depth requirements and detailing of bars, spacers and ties throughout the element and/or within all concrete elements considered.

Defective unit (only relevant for Procedure C)
Unit where the proportion of measured cover depths lower than the specified minimum value (c_{min}) exceeds a predefined percentage.

Limiting quality (only relevant for Procedure C)
Percentage of defective units within a lot which for purposes of sampling inspection is limited to a low probability of acceptance.

A.V. Monteiro (✉)
Laboratório Nacional de Engenharia Civil, Lisbon, Portugal

A. Gonçalves
Laboratório Nacional de Engenharia Civil, Lisbon, Portugal

J. Gulikers
Rijkswaterstaat Bouwdienst, Utrecht, The Netherlands

F. Jacobs
TFB, Wildegg, Switzerland

H. Beushausen and L. Fernandez Luco (eds.), *Performance-Based Specifications and Control of Concrete Durability*,
RILEM State-of-the-Art Reports 18, DOI 10.1007/978-94-017-7309-6_9

Lot
Amount of concrete elements or surface zones (units, in Procedure C) where the cover depth of a specific reinforcement layer, assumed to be from the same population, is subject to inspection. Whenever possible, all the concrete elements, surface zones or units should have the same dimensions and the same method of execution.

Lot size (only relevant for Procedure C)
Total number of units that constitute the lot.

Measurement point
Location at the surface of a reinforced concrete element where a single cover depth measurement is taken over a reinforcement bar.

Percentile
Value below which a given proportion of a collection of values (such as a data sample or a whole population) falls. For example, the 5th percentile of a population corresponds to the value below which 5 % of all theoretically possible values of the population will fall.

Sample
Collection of all cover depth measurements that have been performed (Procedures A and B) or collection of all units that have been tested (Procedure C), which will be used as the basis for evaluation of the conformity of the lot with respect to the specifications.

Spatial autocorrelation
Correlation among observations through space. Its occurrence violates the assumption of independence of observations used in classical statistics.

Tolerance interval
Interval that contains a given proportion of a population, with a certain confidence level. It can be a two-sided interval (aimed to contain a central proportion of the population) or a one-sided interval (aimed to contain an upper or a lower proportion of the population).

Tolerance limit
Upper or lower limit of a tolerance interval.

Unit (only relevant for Procedure C)
A subdivision of a lot. Its geometry and the detailing of bars, ties and spacers should be repeated (approximately) throughout the whole lot. For example, a unit can be a structural element, a clearly defined surface zone in a structure, a portion of a long structural element, etc.

9.2 Introduction

In practice, it is widely recognized that failure to comply of the concrete cover depth with the specifications is one of the main causes of premature deterioration of reinforced concrete structures. However, the great majority of technical standards and codes that deal with durability design and control of execution do not make any provisions for the assessment of concrete cover depth achieved in structures.

Most codes establish that a nominal cover depth c_{nom} shall be specified by the designer on the structural design drawings, assuming a certain tolerance Δc_{dev} appropriate to the construction method, usually between 10 and 15 mm for general structures without special execution requirements, in order to ensure that the required minimum cover depth c_{min} will be achived in the structure, i.e.,

$$c_{nom} = c_{min} + \Delta c_{dev} \tag{9.1}$$

The approach of introducing an "absolute" minimum limit for the concrete cover depth can make it difficult to establish adequate inspection plans, since in most situations it is unpractical to check if that limit is satisfied in the entire structure. For that reason, the European Standard EN 13670: 2009 [1] allows a statistical approach to be adopted in which a predefined proportion of values lower than c_{min} is permitted, however this standard does not provide any information on how to perform this approach.

This chapter provides statistical bases and guidelines for assessing and evaluating the minimum cover depth in concrete structures on the basis of 'inspection by variables' (Procedure A) and 'inspection by attributes' (Procedures B and C). These guidelines are intended to help practitioners to perform adequate estimations of the minimum cover depth achieved in structures, here designated as actual minimum cover depth, and to support decisions on the acceptance of isolated lots regarding this parameter.

It is assumed that c_{min} is a characteristic value (percentile) below which a given percentage of the cover depth population (most often 5 or 10 %, depending on the specifications or the agreement between the owner and the contractor) is permitted to fall.

An effort is made to reduce the sources of uncertainty, to avoid systematic errors and to ensure that there are no major deviations from the theoretical assumptions.

A particular section is dedicated to briefly describe the procedures of a German code of practice [2] for assessing the minimum cover depth in structures, since it was found to be, currently, one of the most comprehensive standardized procedures based on a statistical approach.

At the end, two examples of application of the procedures described in this chapter are presented.

9.3 Procedure A: Inspection by Variables

This section describes a procedure for estimating the actual minimum cover depth of a specific reinforcement layer in a structure or a set of structural elements, and for checking its conformity with the specifications based on an inspection by variables. For its application, it is assumed that the cover depth population is normally or lognormally distributed.

This procedure is not recommended for assessing the cover depth in single surfaces of limited size, surfaces with particular reinforcement detailing (e.g. zones of geometrical discontinuities) or in concrete elements with rigid reinforcement bar cages (e.g. common columns and beams of buildings), due to the fact that the cover depth population may not be reasonably approximated by a well-known statistical distribution. In these cases, an inspection by attributes such as those described in Sects. 9.4 (Procedure B), 9.5 (Procedure C) and 9.6.1 (German qualitative approach) may be more appropriate.

9.3.1 *Steps*

This procedure consists of:

1. selecting the cover depth population (or lot) subject to inspection;
2. defining a sampling method with an appropriate sample size;
3. measuring the cover depth using a properly calibrated measuring instrument;
4. selecting a suitable statistical distribution to describe the cover depth population and checking for potential outliers;
5. estimating the actual minimum cover depth by means of tolerance limit calculation;
6. checking the conformity of the lot with respect to the specifications by comparing the actual minimum cover estimate with the required value

The following sections provide guidelines for performing the above steps.

9.3.2 *Selection of the Cover Depth Population*

The cover depth population (or lot) under inspection should concern a specific reinforcement layer (usually the outermost layer) having the same cover depth requirements throughout. This layer should belong to elements with similar dimensions and have, whenever possible, the same bar and spacers detailing (e.g. stirrups of columns, lateral stirrup layer of beams, top reinforcement layer of slabs, etc.), as well as the same construction method and personnel. All the available information about these aspects should be collected (e.g. structural design drawings, height of spacers, methods of execution, technical data, etc.) and carefully analysed during the inspection planning.

For assessing the cover depth before completion (e.g. for quality control purposes), it may be more appropriate to divide the structure (or structural elements) into multiple lots that are subject to inspection individually and independently from each other. That division should also be considered in existing structures whenever there is suspicion of changes in the method (or quality) of execution.

9.3.3 Sampling Method

The measurement points should be distributed as well as possible over all the concrete surface of the lot. Their location may be randomly chosen over the lot (random sampling). However, adopting a systematic sampling method where the measurement points are regularly spaced over the lot may help dealing with spatial autocorrelation that may occur among measurements.

Due to the continuity and rigidity of the reinforcement bars, it may be expected that the cover depth measurements taken at nearby locations in the structure will show positive spatial autocorrelation, i.e. the measured cover depths showing a positive or negative deviation from the mean value will tend to be followed by neighbouring measurements with also a positive or negative deviation, respectively. This so-called positive spatial autocorrelation violates the assumption of statistical independence among observations, frequently used in classical statistics. If spatial autocorrelation is not taken into account in the statistical analysis, it may lead to biased estimates of the standard deviation of the population and to a gross overestimation of the number of degrees of freedom (usually N or $N-1$, depending on the estimator), resulting in an actual confidence level lower than that adopted to estimate the actual minimum cover depth.

The occurrence of spatial autocorrelation among measurements depends on the sampling method adopted and, even for a systematic sampling method, the pattern of the spatial autocorrelation may greatly vary from element to element, as well as within each element. For that reason, when inspecting a lot, it is difficult to establish a function that satisfactorily describes the autocorrelation among measurements in order to take it into account in the overall cover depth assessment.

The study of methods that deal with spatial autocorrelation, in particular, with bidirectional spatial autocorrelation, is a field where there is still much to develop and, currently there are no studies (at least known to the authors) concerning the evaluation of its influence on cover depth assessment. For that reason, the spatial autocorrelation should be avoided or minimized by adopting, for example, a sampling method that ensures a minimum distance between successive measurement points. In principle, this distance should depend on several factors such as the distance between spacers, the diameter, spacing and detailing of bars, dimensions of the concrete elements, construction method, etc. This large number of factors makes

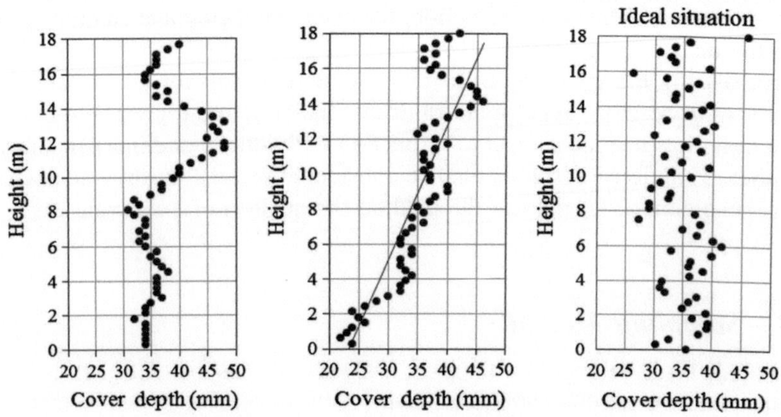

Fig. 9.1 Spatially autocorrelation among the cover depths measured in two columns (*left* and *middle plots*) and an ideal situation of a column with spatially independent cover depths (*right plot*)

it difficult to establish a general rule for the minimum distance between measurements appropriate for most situations.

In practice, the autocorrelation can prevail even over long distances (several meters) and in some cases it may not be possible to avoid it without compromising the sample size (by increasing the distance between measurements). Figure 9.1 shows the cover depth measurements (vertically spaced 30 cm) successively taken in two columns, clearly exhibiting a large positive spatial autocorrelation. In the middle plot it is even possible to notice the overall inclination of the reinforcement bar cages inside the formwork, evidencing that the autocorrelation among measurements can prevail for distances of several meters. Figure 9.1 (right plot) also shows an ideal situation (obtained by simulation) where the cover depth measurements are spatially independent.

Further information about how to detect and deal with spatial autocorrelation among measurements can be found in the Appendix A9.3.

Regarding the sample size, N, its choice usually depends on the importance of the structure, the size of the lot and the inspection costs (more relevant in cases of structures with difficult access). Increasing the sample size not only improves the precision of the actual minimum cover depth estimate and, as a consequence, the degree of certainty of the decision on the acceptance or rejection of the lot, but also facilitates the detection of unexpected multiple cover depth populations that may exist within the lot. The German code, for example, establishes a minimum of 20 measurement points when using an inspection by variables (see Sect. 9.6.2).

9.3.4 Measuring the Cover Depth

The cover depth should be measured according to the instructions given by the measuring instrument manufacturer. Nevertheless, it is highly recommended that the measurements are carried out by an experienced operator in order to avoid measurement errors associated, for example, with the interference of closely spaced bars, presence of metals other than the main reinforcement, etc. [3, 4].

Common measuring instruments may show absolute bias of about 1–3 mm [5]. Therefore, an appropriate calibration of the measuring instrument over its whole working range should be carried out and the cover depth measurements corrected to take into account the possible bias. Methods for calibrating electromagnetic instruments based on cover depth measurements taken in concrete specimens with bars of different diameters located at several well-known depths, can be found in BS 1881 [3].

Two of the main limitations of most cover depth measuring instruments are the cover depth working range (or reliable testing range) and the need for setting the bar diameter. Measuring the cover depth outside the working range or setting an incorrect bar diameter in the instrument may lead to significant systematic measurement errors. For this reason the bar diameter set (when required) in the measuring instrument and its working range should always be reported. Also, all the available information about the reinforcement bar detailing should be carefully analysed prior to the inspection in order to locate the zones with different design cover depth values and bar diameters.

Methodologies for measuring cover depths below the working range of the measuring instrument or when the bar diameter is unknown can be found in BS 1881 [3] and RILEM Report 40 [4].

9.3.5 Statistical Distribution Functions and Detection of Outliers

The normal distribution is often used to describe cover depth populations, mostly because this type of distribution is very well known and owing to the fact that it has proven to reasonably fit a large number of data samples obtained in cast in situ structures and precast elements [6, 7]. However, for certain situations, especially when describing populations with small cover depth values (e.g. with a mean cover depth lower than about 20 mm [6]), the lognormal distribution may be more appropriate and fit the sample data better because it does not allow the occurrence of negative cover depth values. In practice, assuming a lognormal distribution usually leads to actual minimum cover depth estimates slightly higher [7] than those based on the normal distribution.

In any case, the goodness of fit of the chosen distribution should be evaluated, as well as the presence of potential outliers within the sample. For that purpose, a method based on visual analysis of probability plots is described in the Appendix A9.1.

It should be kept in mind, however, that even in case of a normally distributed cover depth population, the following causes may also lead to a major lack of fit of the normal distribution to the sample data:

a. the sample size is too small (less than about 20 measurements);
b. the sample comes from two or more distinct populations (with different means or/and different standard deviations);
c. the cover depth measurements are highly autocorrelated;
d. the measured cover depths are outside or close to the working range limits of the measuring instrument (identifying a case of lack of fit of extreme values);
e. the sample data contain outliers (identifying a case of lack of fit of extreme values).

In cases where the sample data is not approximately normally or lognormally distributed, each of the aforementioned causes should be evaluated carefully. In some situations, the implementation of an inspection by attributes may be appropriate (see Sects. 9.4 and 9.6.1).

In case of detection, the potential outliers can be excluded from the sample data, but the reasons for their occurrence should always be evaluated. These reasons may be, for example:

a. a simple dislodgment of a spacer;
b. a measurement taken over metals other than the reinforcement bars under inspection or over closely spaced bars (high congestion of steel bars can lead to cover depth underestimations since a stronger signal is received by the electromagnetic measuring instrument);
c. an interference of tie wires;
d. a measurement taken outside the working range of the measuring instrument;
e. a measurement taken over a reinforcement layer deeper than that being inspected;
f. a poor detailing of the spacers or/and ties in certain locations of the structure.

An experienced operator should be able to avoid or significantly reduce the occurrence of outliers when related to reasons other than the poor spacers detailing. In this last case, the defective locations should be re-inspected and corrected (if necessary), and, if a systematic occurrence is observed, appropriate actions should be taken to improve the quality of execution on further construction works.

9.3.6 Minimum Cover Depth Estimate

The actual minimum cover depth can be estimated by means of a one-sided tolerance limit (the definition can be found in the Appendix A9.2), as follows:

- for normally distributed cover depth populations,

$$c_{min,estimated} = \bar{c} - k \cdot s \tag{9.2}$$

- for lognormally distributed cover depth populations,

$$c_{min,estimated} = e^{\bar{c}_{ln} - k \cdot s_{ln}} \tag{9.3}$$

where $c_{min,estimate}$ is the actual minimum cover estimate, $\bar{c}$ and s are the mean and the standard deviation, respectively, of the cover depth sample data, $\bar{c}_{ln}$ and s_{ln} are the mean and standard deviation, respectively, of the logarithmic values of the cover depth sample data, and k is a tolerance factor that depends on the percentile on which c_{min} is based, on the sample size N, and on the confidence level $1 - \gamma$ desired for $c_{min,estimate}$.

The percentile is selected based on the proportion of the cover depth population that is permitted to fall below c_{min}. If this proportion is 5 or 10 %, then the 5th or 10th percentile, respectively, should be chosen.

For durability purposes, some technical documents [8, 9] allow 5 % of the cover depth population to fall below c_{min}. However, due to the large values of cover depth standard deviation frequently found on site (8–10 mm [6]), it can be expected that the majority of cast in situ structures (without special execution requirements) designed with a tolerance Δc_{dev} = 10 mm (as recommended by the European execution standard [1]), would fail to comply with the specifications. Therefore, for these structures, a Δc_{dev} of about 13–15 mm seems to be more appropriate. For most precast concrete elements, a Δc_{dev} of 10 mm is likely to be sufficient [7]. For instance, the German code of practice [2] (described in Sect. 9.6) prescribes c_{min} as the 5th percentile of the cover depth population, if Δc_{dev} = 15 mm, and as the 10th percentile, if Δc_{dev} = 10 mm.

The percentile and confidence level should be regarded as conformity criteria and, if not specified prior to the construction works, they should be agreed upon between the owner and the contractor prior to carrying out the measurements.

Values for the tolerance factor k can be found in Table 9.1.

Table 9.1 also includes the k values calculated according to the approach used in the informative Annex D—"Design assisted by testing" of Eurocode 0 (EC0) [11] for the statistical determination of a single property via the characteristic value when the variance is unknown, i.e. on the Bayesian method based on vague prior distributions.

Formulas for calculating the tolerance limits for other statistical distributions can be found in Monteiro and Gonçalves [12].

9.3.7 *Evaluation of Conformity*

For checking the conformity of the actual minimum cover depth with the specifications, the following conformity criterion can be used:

Table 9.1 Tolerance factors for normal and lognormal distributions

Sample size, N	k value for the 5th percentile					k value for the 10th percentile			
	EC0 approach[a]	Confidence level, $1-\gamma$				Confidence level, $1-\gamma$			
		50 %	75 %	90 %	95 %	50 %	75 %	90 %	95 %
10	1.92	1.70	2.10	2.57	2.91	1.32	1.67	2.07	2.35
11	1.89		2.07	2.50	2.81		1.65	2.01	2.28
12	1.87	1.69	2.05	2.45	2.74	1.31	1.62	1.97	2.21
13	1.85		2.03	2.40	2.67		1.61	1.93	2.16
14	1.83	1.68	2.01	2.36	2.61		1.59	1.90	2.11
15	1.82		1.99	2.33	2.57		1.58	1.87	2.07
20	1.76	1.67	1.93	2.21	2.40	1.30	1.53	1.77	1.93
25	1.74		1.89	2.13	2.29		1.50	1.70	1.84
30	1.73	1.66	1.87	2.08	2.22	1.29	1.47	1.66	1.78
40	1.71		1.83	2.01	2.13		1.44	1.60	1.70
50	1.69	1.65	1.81	1.97	2.06		1.43	1.56	1.65
60	1.68		1.79	1.93	2.02		1.41	1.53	1.61
80	1.67		1.77	1.89	1.96		1.39	1.49	1.56
100			1.76	1.86	1.93		1.38	1.47	1.53
120	1.66		1.75	1.84	1.90	1.28	1.37	1.45	1.50
150			1.74	1.82	1.87		1.36	1.43	1.48
180			1.73	1.80	1.85		1.35	1.42	1.46
200			1.72	1.79	1.84		1.35	1.41	1.45
300	1.65		1.71	1.76	1.80		1.34	1.39	1.42
400			1.70	1.75	1.78		1.33	1.37	1.40
∞	1.64					1.28			

[a] $k = \left|T_{N-1}^{-1}(0.05)\right|\sqrt{1+1/N}$, where $T_{N-1}^{-1}(x)$ is the inverse cumulative distribution function of the t-distribution with $N-1$ degrees of freedom [10]

$$c_{min,estimated} \geq c_{min} \tag{9.4}$$

where $c_{min,estimated}$ is obtained by Eq. (9.2) or (9.3), and c_{min} is the required minimum cover depth.

Due to the statistical uncertainty (associated with the limited number of measurement points), the above conformity criterion implies contractor's and owner's risks. The assessment of these risks can be made by means of OC curves as described in Appendix A9.4. It should be noticed, however, that the confidence level $1-\gamma$ of $c_{min,estimate}$ is directly related to these risks, i.e. $1-\gamma$ corresponds to the maximum probability of rejecting a "good" lot (contractor's risk) with an actual minimum cover depth greater than c_{min}, and its complement, γ, corresponds to the maximum probability of accepting a "bad" lot (owner's risk) with an actual minimum cover depth lower than c_{min}. The confidence level should, therefore, be chosen based on the consequences of these two scenarios. In engineering specifications, the confidence level is usually set much above 50 %. However, setting a high confidence level may not always be appropriate due to the consequences of rejecting a "good" lot. In some situations, these consequences may actually be worse than those of accepting a "bad" lot.

In case of non-conformity, the proportion of the cover depth population that falls below c_{min} can still be estimated, using the plots presented in Fig. 9.2. In those plots, k corresponds to the maximum k value that would lead to the conformity of the lot, i.e. $k = \frac{c_{min} - \bar{c}}{s}$, for normally distributed cover depth populations, and $k = \frac{ln(c_{min}) - \bar{c}_{ln}}{s_{ln}}$, for lognormally distributed cover depth populations.

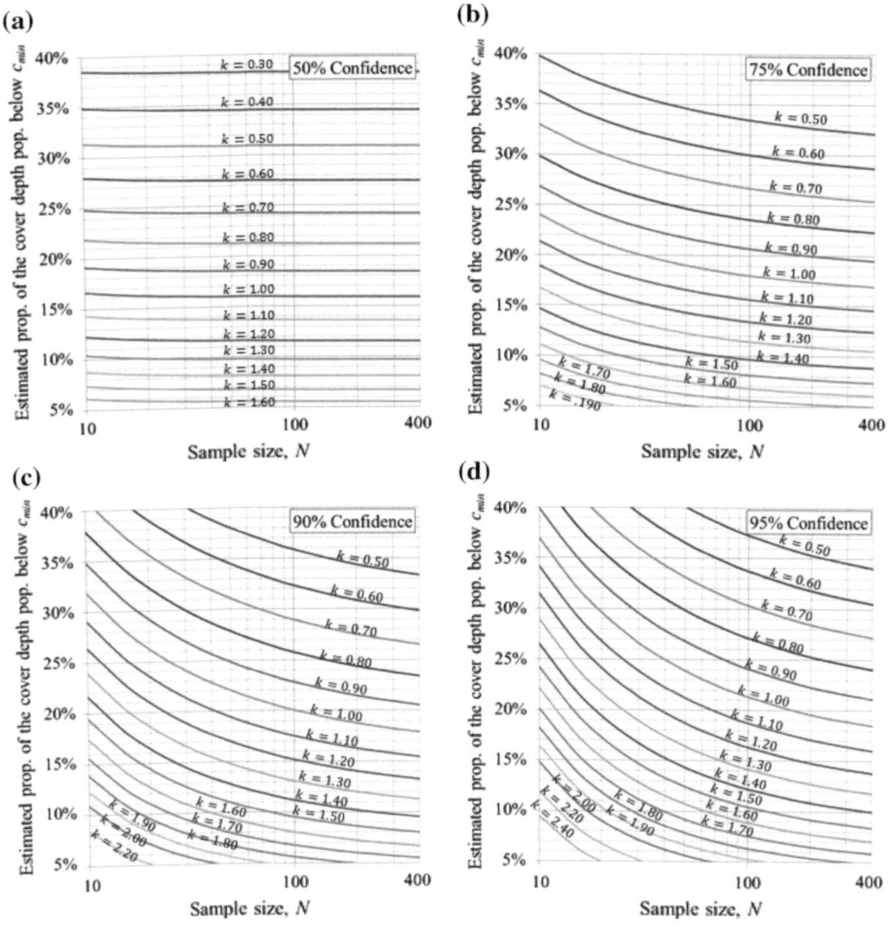

Fig. 9.2 Estimated proportion of the cover depth population below c_{min}. **a** 50 % confidence level. **b** 75 % confidence level. **c** 90 % confidence level. **d** 95 % confidence level

9.4 Procedure B: Inspection by Attributes

This section describes a procedure for checking the conformity of the actual minimum cover depth with the specifications, based on an inspection by attributes. This procedure does not make any assumptions about the statistical distribution of the cover depth population.

9.4.1 *Steps*

This procedure consists of:

1. selecting the cover depth population (or lot) subject to inspection;
2. defining a sampling method with an appropriate sample size;
3. measuring the cover depth using a properly calibrated measuring instrument;
4. checking the conformity of the lot with the specifications by comparing the number of measured cover depths lower than c_{min}, within the sample, with a given acceptance number.

For the above first three steps, the same guidelines provided by Procedure A apply.

9.4.2 *Evaluation of Conformity*

In this procedure the cover depth population is assumed as an infinite collection of "good" and "bad" theoretical cover depth values, where the "good" are those equal or greater than c_{min} and the "bad" are those lower than c_{min}.

For the acceptance of the lot, the number of "bad" measurements within the sample must be equal or less than a given acceptance number Ac which depends on the percentile on which c_{min} is based, on the sample size and on the established owner's risk. These two last factors should be regarded as conformity criteria and, if not specified prior to the construction works, they should be agreed upon between the owner and the contractor prior to carrying out the measurements.

Acceptance numbers for sample sizes up to 250 and for owner's risks between 5 and 50 % are presented in Table 9.2.

The acceptance numbers presented in Table 9.2 were obtained from the cumulative distribution function of the binomial distribution and correspond to the maximum values of Ac that satisfy the following expression:

$$\sum_{i=N-Ac}^{N} \binom{N}{i} P^i (1-P)^{N-i} \leq owner's\ risk \tag{9.5}$$

Table 9.2 Acceptance numbers

Sample size, *N*	Acceptance number, *Ac*							
	5th percentile				10th percentile			
	Maximum owner's risk				Maximum owner's risk			
	50 %	25 %	10 %	5 %	50 %	25 %	10 %	5 %
10	–	–	–	–	0	–	–	–
15	0	–	–	–	0	0	–	–
20	0	–	–	–	1	0	–	–
25	0	–	–	–	1	0	0	–
30	0	0	–	–	2	1	0	0
35	1	0	–	–	2	1	0	0
40	1	0	–	–	3	2	1	0
45	1	0	0	–	3	2	1	0
50	1	0	0	–	4	2	1	1
55	2	1	0	–	4	3	2	1
60	2	1	0	0	5	3	2	1
65	2	1	0	0	5	4	3	2
70	2	1	0	0	6	4	3	2
75	3	1	0	0	6	5	3	2
80	3	2	1	0	7	5	4	3
85	3	2	1	0	7	6	4	3
90	3	2	1	0	8	6	4	4
95	4	2	1	1	8	6	5	4
100	4	2	1	1	9	7	5	4
105	4	3	2	1	9	7	6	5
110	4	3	2	1	10	8	6	5
115	5	3	2	1	10	8	6	5
120	5	3	2	1	11	9	7	6
125	5	4	2	2	11	9	7	6
130	5	4	2	2	12	10	8	7
135	6	4	3	2	12	10	8	7
140	6	4	3	2	13	11	9	7
145	6	4	3	2	13	11	9	8
150	6	5	3	2	14	11	9	8
155	7	5	3	3	14	12	10	9
160	7	5	4	3	15	12	10	9
165	7	5	4	3	15	13	11	9
170	7	6	4	3	16	13	11	10
175	8	6	4	3	16	14	12	10
180	8	6	4	3	17	14	12	11
185	8	6	5	4	17	15	12	11
190	8	6	5	4	18	15	13	11

(continued)

Table 9.2 (continued)

Sample size, N	Acceptance number, Ac							
	5th percentile				10th percentile			
	Maximum owner's risk				Maximum owner's risk			
	50 %	25 %	10 %	5 %	50 %	25 %	10 %	5 %
195	9	7	5	4	18	16	13	12
200	9	7	5	4	19	16	14	12
210	9	7	6	5	20	17	15	13
220	10	8	6	5	21	18	15	14
230	10	8	6	5	22	19	16	15
240	11	9	7	6	23	20	17	16
250	11	9	7	6	24	21	18	16

where $1 - P$ is taken as the proportion of the cover depth population permitted to fall below c_{min} (0.05 for c_{min} as the 5th percentile and 0.10 for c_{min} as the 10th percentile).

9.5 Procedure C: Inspection by Attributes—Large Lots

For large lots or lots with limited access conditions an inspection based on a sampling method where a small number of locations in the lot that are fully tested may be more appropriate. These locations, designated as units, correspond to subdivisions of the lot, such as concrete elements or subdivisions of a long concrete element.

This section describes a procedure for checking the conformity of the actual minimum cover depth with the specifications based on an inspection by attributes where the evaluated items are units instead of individual measurements.

The basic principle is to check whether or not the number of defective units in a sample is greater than an acceptance number, established based on the permitted percentage of defective units within the entire lot. A unit is classified as defective if its proportion of measured cover depths lower than c_{min} is greater than a given percentage (5 or 10 %).

This procedure does not make any assumptions on the statistical distribution of the cover depth population and does not require taking into account the spatial autocorrelation among measurements in the analysis.

9.5.1 Steps

This procedure consists of:

1. selecting the cover depth population (or lot) subject to inspection;
2. defining a sampling method with an appropriate sample size;

3. measuring the cover depth using a properly calibrated measuring instrument;
4. checking the conformity of the lot with the specifications by comparing the number of defective units in the sample with a given acceptance number.

For the first and third step, the same guidelines provided by Procedure A apply.

9.5.2 Sampling Method

First, the lot should be evenly divided into units that have, as far as is practicable, the same dimensions. A sample of these units, well distributed along the whole lot or the production period, should then be randomly selected and the reinforcement layer under inspection should be intensively tested (with test locations uniformly distributed along the whole unit) so that the confidence of the estimated percentage of cover depths lower than c_{min} in each unit cannot be matter for discussion.

9.5.3 Evaluation of Conformity

All sample units with a proportion of measured cover depths lower than c_{min} above the permitted value (5 or 10 %) must be classified as defective.

For the conformity of the lot, the number of defective units must be equal or less than a given acceptance number Ac which depends on the lot size N_{lot} (total number of units that compose the lot), on the sample size N, on the permitted percentage p of defective units in the whole lot and on the established owner's risk.

The acceptance number can be obtained from the cumulative distribution function of the hypergeometric distribution, corresponding to the maximum value of Ac that satisfies the following expression:

$$\sum_{i=N-Ac}^{N} \frac{\binom{N_{lot} - p \cdot N_{lot}}{i} \binom{p \cdot N_{lot}}{N - i}}{\binom{N_{lot}}{N}} \leq owner's\ risk \tag{9.6}$$

It should be noticed that permitting only a small percentage of defective units in the whole lot may be excessively demanding since, in theory, even a lot with a percentage of defective units as high as 50 % may have an overall actual minimum cover depth equal or greater than c_{min}.

Anyway, large percentages of defective units should only be permitted in cases where there are no evidences of changes in the method and quality of execution of the units throughout the whole lot. If these changes are detected, the division of the lot into two or more lots should be considered, as already discussed in Sect. 9.3.2.

Table 9.3 Sample sizes and acceptance numbers according to ISO 2859-2 [13]

Lot size N_{lot}		Limiting quality	
		20 %	32 %
16–25	*N*	9	6
	Ac	0	0
26–50	*N*	10	6
	Ac	0	0
51–90	*N*	10	8
	Ac	0	0
91–150	*N*	13	13
	Ac	0	1
151–280	*N*	20	13
	Ac	1	1
281–500	*N*	20	20
	Ac	1	3
501–1,200	*N*	32	32
	Ac	3	5
1,201–3,200	*N*	50	50
	Ac	5	10
3,201–10,000	*N*	80	80
	Ac	10	18
>10,000	*N*	125	80
	Ac	18	18

To prevent the acceptance of units with cover depths significantly lower than c_{min}, additional criteria may be adopted when classifying a unit as defective. For example, when measured cover depths lower than c_{min} by more than 10 mm are detected over a certain minimum extension of the unit (e.g. a minimum surface area or length). In any situation the owner should always have the right to reject or require the correction of any defective unit.

The international standard ISO 2859-2 [13], which also uses this approach, provides values for the sample size and acceptance number based on the lot size and limiting quality. The limiting quality is defined as the percentage of defective units within a lot which, for purposes of sampling inspection, corresponds to a low probability of acceptance (and thus, also to a low owner's risk). Using the rules given by ISO 2859-2, this probability is in general less than 10 %, but never greater than 13 %. As a reference, in Table 9.3 are provided the sample sizes *N* and *Ac* values by this standard for the limiting qualities of 20 and 32 %.

The conditions for classifying a unit as defective, the permitted percentage of defective units and the owner's risk should be regarded as conformity criteria and, if not specified prior to the construction works, they should be agreed upon between the owner and the contractor prior to carrying out the measurements.

Due to the fact that this procedure does not allow to adequately manage or evaluate the probabilities of acceptance and rejection of a lot as a function of its overall proportion of cover depths lower than c_{min}, it is more demanding in terms of engineering judgment when compared to procedures A and B. However, it may significantly facilitate the inspection plan in certain situations and provide a good comprehension of the overall cover depth quality where systematic defective zones can easily be detected.

9.6 German Code of Practice

The German code of practice [2] establishes two procedures for checking the conformity of the actual minimum concrete cover depth with the specifications: one based on an inspection by attributes (qualitative procedure); and another based on an inspection by variables (quantitative procedure).

The aim of both procedures is to check whether or not the proportion of the cover depth population that falls below c_{min} is greater than a permitted percentage, based on a sample of cover depth measurements. This percentage is established based on the tolerance Δc_{dev} used in the design, as follows:

- 10 % if $\Delta c_{dev} = 10$ mm (required for the exposure class XC1 and for ensuring the reinforcement bond strength), which corresponds to checking the 10th percentile of the cover depth population;
- 5 % if $\Delta c_{dev} = 15$ mm (required for XC2, XC3, and all XS and XD exposure classes), which corresponds to checking the 5th percentile of the cover depth population.

The population is defined as the collection of all theoretically possible measurement points of a measurement surface (or several comparable measurement surfaces) as a part (or parts) of the element surface.

The following surfaces of the concrete elements should be differentiated as measurement surfaces:

- each surface of walls;
- top surfaces of slabs;
- bottom surfaces of slabs;
- each surface of rectangular columns;
- lateral surfaces of the web of beams;
- bottom surface of beams;
- top surface of beams.

Comparable measurement surfaces can be combined together.

The measurement points should be randomly chosen and widely distributed as possible throughout the measurement surface.

The minimum sample size N depends on the procedure used and may also depend on the target percentile of the cover depth population, as described in Sects. 9.6.1 and 9.6.2. If the minimum sample size is not possible to be achieved, then no single cover depth measurement may fall below c_{min}. Whenever the doubt about the conformity of the cover depth of a lot remains after the inspection, the sample size can be increased by means of additional testing in order to increase the degree of certainty about the acceptance or rejection of the population.

The maximum absolute tolerance errors allowed for the measuring instrument are:

- 1 mm, for cover depths up to 40 mm;
- 2 mm, for cover depths between 40 and 60 mm.

9.6.1 Qualitative Procedure

The minimum sample size required to start the process is:

- 10, for checking the 10th percentile of the cover depth population;
- 15, for checking the 5th percentile of the cover depth population.

For the acceptance of the population, the total number of measured cover depths in the sample lower than c_{min} must be equal or less than the acceptance limit given in Fig. 9.3.

If the total number of measured cover depths in the sample lower than c_{min} is greater than the acceptance limit given in Fig. 9.3, the process can continue by adding more measurement points (if possible), provided that no serious defects are detected and the engineering knowledge and experience do not forbid it. The decision on whether the process be continued or the population rejected should be based on a critical engineering judgement on the probability of "justified rejection" presented in Fig. 9.4. In cases where no additional measurement points are available, the process is stopped and the measurement surface has, at the moment, no proof of being in conformity with the specifications. However, if there are at least

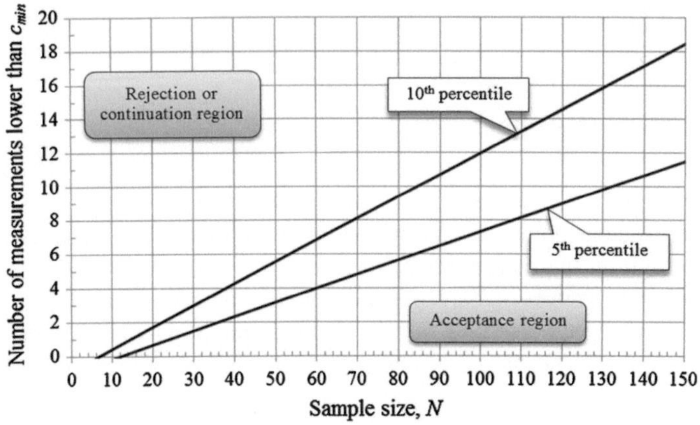

Fig. 9.3 Qualitative confirmation—acceptance limits

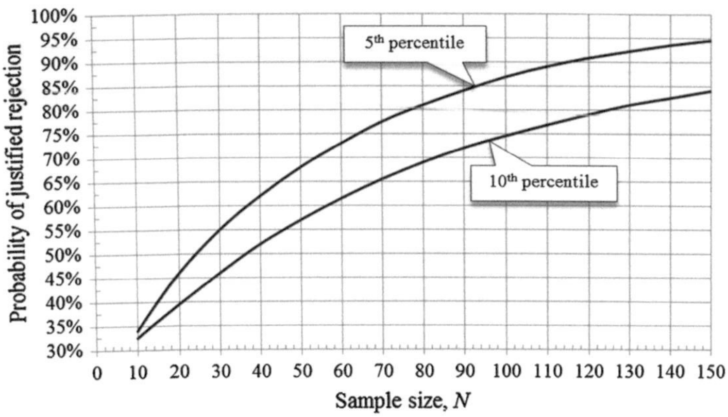

Fig. 9.4 Qualitative confirmation—probability of justified rejection

20 measurement points available, the process can still be continued by using the quantitative procedure described in Sect. 9.6.2.

9.6.2 *Quantitative Procedure*

This procedure uses Neville's distribution as the basis for the statistical evaluation of the cover depth measurements. Like the lognormal distribution, Neville's distribution cannot contain negative values and is positively skewed. Therefore, it can take into account, more realistically than the normal distribution, the different probabilities of occurrence of negative and positive deviations of the cover depth caused by the influence of the unidirectional spacers.

The minimum sample size required to start the process is 20.

The first step consists of calculating the sample median, $\bar{c}_M$, and to identify the smallest measured cover depth in the sample, c_s.

The second step is to calculate the upper limit $2.5\bar{c}_M - 1.5c_s$. All the measured cover depths greater than this upper limit are considered outliers and should be excluded from the sample. In case of the exclusion of any measurement, the median $\bar{c}_M$ must be recalculated based on the sample without the outliers in order to proceed with the analysis.

The conformity of the cover depth population can be evaluated using one of two approaches:

- one based on the estimation of the proportion of the cover depth population that falls below c_{min};
- the other based on the estimation of the target percentile of the cover depth population.

The proportion F_X of the cover depth population that falls below c_{min} is estimated using the cumulative distribution function (c.d.f.) of the Neville's distribution:

$$F_X = \frac{\left(\frac{x}{r}\right)^m}{1 + \left(\frac{x}{r}\right)^m} \tag{9.7}$$

where $r = \frac{\bar{c}+\bar{c}_M}{2}$ is the location parameter, $m = 1.8 \cdot \frac{r}{s}$ is the form parameter, $\bar{c}$, $\bar{c}_M$ and s are the mean, median and standard deviation of the cover depth sample, respectively, and x is taken as c_{min}.

The estimation of the target percentile of the cover depth population $c(x)$ is based on the inverse c.d.f. of the Neville's distribution:

- for the 5th percentile,

$$c(5\,\%) = \frac{r}{19^{\frac{1}{m}}} \tag{9.8}$$

- for the 10th percentile,

$$c(10\,\%) = \frac{r}{9^{\frac{1}{m}}} \tag{9.9}$$

For the acceptance of the population, the value of F_X has to be equal or lower than the permitted proportion of the cover depth population bellow c_{min} (5 or 10 %) or $c(x) \geq c_{min}$.

This procedure complies, in general, with the guidelines presented in Sect. 9.3 (Procedure A). However, two main differences can be pointed out as follows:

- the Neville's distribution instead of a normal or a lognormal distribution;
- the statistical uncertainty, associated with the limited sample size, is not taken into account in the estimations since a cumulative distribution function is used instead of a tolerance limit.

9.7 Actions in the Case of Non-conformity

Whenever doubts on the conformity of the lot remain after an inspection, the possibility to increase the sample size by the addition of further measurement points should be considered in order to increase the degree of certainty of the decision on the acceptance or rejection of the lot.

The actions in case of non-conformity may include:

- checking if the measured cover depths are correct by comparing some of the values with those obtained by direct measurement (e.g. using a caliper on a measuring point after removing the concrete cover);
- reassessing the adequacy of the structure for the actual minimum cover depth estimate, based on the available information on the concrete quality;
- implementing corrective actions;
- renegotiating the execution costs;
- replacing the lot or the defective units by new ones;
- improving the quality of execution of further construction works.

For durability purposes, the corrective actions may consist, for example, of applying coatings for protecting the concrete cover against chloride or CO_2 penetration, keeping in mind, however, that most organic polymer based coatings have very limited lifetime (usually assumed between 10 and 20 years, depending on the sunlight exposure).

In any case, it should be kept in mind that the actual minimum cover depth must also satisfy the limits required for reinforcement bond strength and fire resistance purposes.

9.8 Examples

In this section, two examples are presented using the results from two case studies carried out at the National Laboratory for Civil Engineering, in Portugal.

9.8.1 Cover Depths in a Viaduct Deck Slab (Procedure A)

Lot:
Top reinforcement layer of a deck slab (Fig. 9.5) that comprises 3 spans of a viaduct. Each span is about 32 m long and 11 m wide. The top reinforcement layer is transversal to the viaduct and includes, in general, 100 mm spaced bars with 12 and 16 mm of diameter.

Specifications:
- $c_{min} = 40$ mm;
- proportion of the cover depth population permitted to fall below c_{min}: 5 %;
- maximum owner's risk: 25 %, for both Procedures A and B.

Sampling plan:
- sampling method: grid sampling (equally spaced measurement points);
- sample size: $N = 120$.

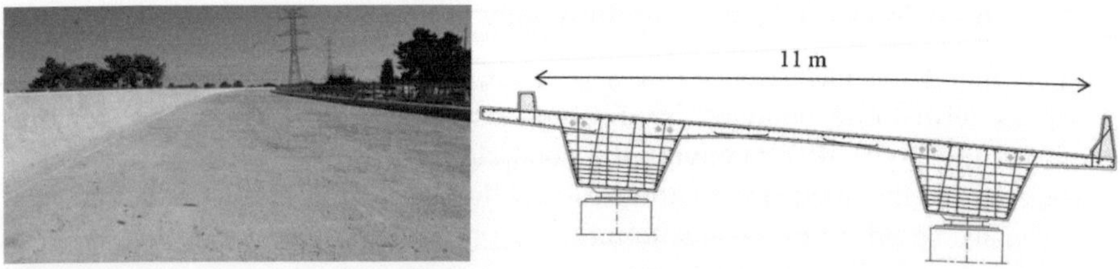

Fig. 9.5 Viaduct deck slab (Example 1)

Span 1								Span 2								Span 3							
64	56	66	76	52	73	69	71	79	76	79	72	76	79	63	53	80	74	72	65	72	69	72	58
61	56	63	69	69	70	74	76	55	72	76	53	57	76	80	28	66	74	62	76	57	63	57	61
57	57	37	40	64	56	60	56	56	48	47	57	54	47	56	56	54	72	64	61	56	45	59	63
49	51	42	51	57	58	57	54	59	63	64	71	48	72	48	39	58	56	49	63	46	79	76	69
58	61	53	49	51	48	55	50	46	49	65	59	63	62	58	47	70	58	58	72	51	62	72	61

Fig. 9.6 Measured cover depths (mm)

Sample data:
Figure 9.6 illustrates the cover depths measured in the slab.

Verification of normality and detection of potential outliers:
Table 9.4 shows the calculated sample (X_i and $X_{ln,i}$) and theoretical quantiles Y_i required for the construction of the probability plot presented in Fig. 9.5 (according to Appendix A9.1). The probability plot for the lognormal distribution is also included in Fig. 9.5 for illustration purposes since, as mentioned in Sect. 9.3.6, it is usually considered only in lots with small cover depths.

According to Fig. 9.7, the normal distribution reasonably fits the sample data except for the larger cover depths (corresponding to the highest quantiles), which is mostly due the low accuracy of the measuring instrument at that range of cover depth values, and for the smallest measurement.

The maximum absolute value of the sample quartiles presented in Table 9.4 is $|Y_1| = 3.100$, which corresponds to the smallest measured cover depth (28 mm). According to Table A9.1, there is no statistical significance to consider this measurement as a potential outlier, since $|Y_1|$ is not greater than the corresponding critical value given by the Grubb's test (3.45).

Thereby, in the further analysis no measurements are excluded from the sample and the cover depth population is assumed to be normally distributed.

Conformity check:
In this analysis, the occurrence of spatial autocorrelation among the measurements is discarded and thus, the effective sample size is taken as $N' = N$.

Since the proportion of the cover depth population permitted to fall below c_{min} is 5 % and the permitted owner's risk is 25 %, the minimum cover depth should be considered as the 5th percentile of the population and the confidence level adopted

Table 9.4 Calculated quantiles

i	Sorted c_i (mm)	$ln(c_i)$	Assuming a normal distrib.	Assuming a lognormal distrib.	$X_i = \Phi^{-1}\left(\frac{i-0.3175}{N+0.365}\right)$
			$Y_i = \left(\frac{c_i-\bar{c}}{s}\right)$	$Y_{ln,i} = \frac{(ln(c_i)-\bar{c}_{ln})}{s_{ln}}$	
1	28	3.332	−3.100	−4.121	−2.532
2	37	3.611	−2.249	−2.608	−2.198
3	39	3.664	−2.060	−2.323	−2.009
4	40	3.689	−1.966	−2.185	−1.872
⋮	⋮	⋮	⋮	⋮	⋮
117	79	4.369	1.720	1.509	2.872
118	79	4.369	1.720	1.509	2.009
119	80	4.382	1.815	1.577	2.198
120	80	4.382	1.815	1.577	2.532
$\bar{c}$ (mm)	60.8				
$\bar{c}_M$ (mm)	59.5				
s (mm)	10.6				
c_{ln}		4.091			
s_{ln}		0.184			

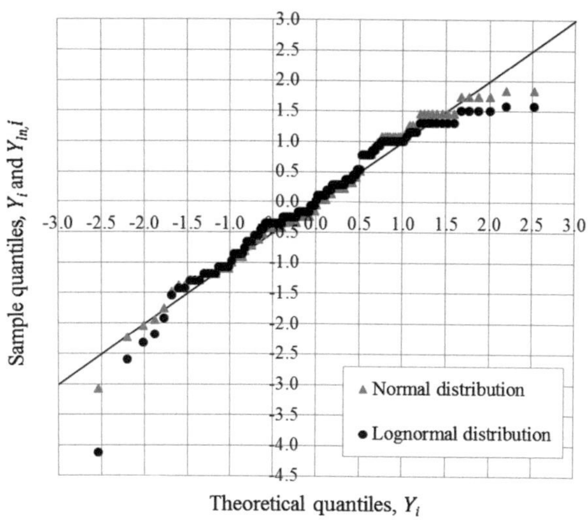

Fig. 9.7 Probability plot

should be 75 %. According to Table A9.1, the tolerance factor k should be taken as 1.75. Using Eq. (9.2), the actual minimum cover estimate is:

$$c_{min,estimate} = \bar{c} - 1.75 \cdot s = 42.3\text{ mm} > c_{min} = 40\text{ mm}\ (Conform) \tag{9.10}$$

which satisfies the conformity criterion given by (9.4).

Procedure B:
According to Table 9.2, and discarding the spatial autocorrelation among the measurements ($N' = N$), the acceptance number is $Ac = 3$

The number of measured cover depths lower than c_{min} is 3 (it can be counted in Fig. 9.6 or in Table 9.4), which is not greater than Ac. Therefore, the lot can be considered to conform to the specifications.

German qualitative approach:
According to Fig. 9.3 the number of measured cover depths lower than c_{min} obtained in the sample is 3, which is clearly in the acceptance region limited by the acceptance limit of 9. Therefore, the measurement surface (lot) can be considered to conform to the specifications.

German quantitative approach:
Considering the upper limit $2.5c_M - 1.5c_s = 2.5 \cdot 59.5 - 1.5 \cdot 28 = 107$ mm, no outliers are detected within the sample (the highest measured cover depth is 80 mm).

Using Eq. (9.7), the following estimate of the proportion of the cover depth population that falls below c_{min} is obtained:

$$F_X = \frac{\left(\frac{c_{min}}{r}\right)^m}{1 + \left(\frac{c_{min}}{r}\right)^m} = \frac{\left(\frac{40}{60.2}\right)^{10.2}}{1 + \left(\frac{40}{60.2}\right)^{10.2}} = 0.015 = 1.5\,\% < 5\,\%\ (Conform) \tag{9.11}$$

where $r = \frac{\bar{c} + \bar{c}_M}{2} = \frac{60.8 + 59.5}{2} = 60.2$, and $m = 1.8 \cdot \frac{r}{s} = 1.8 \cdot \frac{60.2}{10.2} = 10.2$.

Since the estimated proportion is not greater than the permitted percentage (5 %), the measurement surface (lot) can be considered to conform to the specifications.

The same conclusion can be obtained by Eq. (9.8) (alternative approach):

$$c(5\,\%) = \frac{r}{19^{\frac{1}{m}}} = \frac{60.2}{19^{\frac{1}{10.2}}} = 44.7\text{ mm} > c_{min} = 40\text{ mm}\ (Conform) \tag{9.12}$$

9.8.2 Cover Depth of the Columns Stirrups of a Viaduct (Procedure C)

Lot:
Stirrups of 460 columns of a viaduct (Fig. 9.8). The columns have a diameter of 1.5 m and are about 7–14 m high. The diameters of the stirrups are 12 and 16 mm, depending on the column height and the location of the stirrups.

Fig. 9.8 Viaduct columns (Example 2)

Specifications:

- $c_{min} = 30\,mm$;
- proportion of the cover depth population permitted to fall below c_{min}: 5 %;
- limiting quality = 32 %.

Sampling plan:

- unit: one single column;
- defective unit: unit with more than 5 % of measured cover depths lower than c_{min} or with more than two consecutive measured cover depths lower than $c_{min} - 10$ mm;
- lot size: 460;
- sample size: $N = 20$ (from Table 9.3);
- sampling method: the sample units (columns) were randomly chosen and well distributed along the whole lot. The cover depth measurements were vertically spaced by about 30–50 cm along eight vertical alignments equally spaced along the perimeter of each unit (Fig. 9.9).

Fig. 9.9 Sampling method (Example 2)

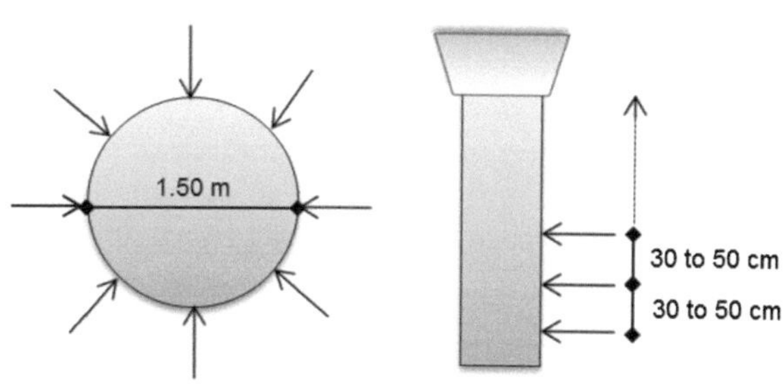

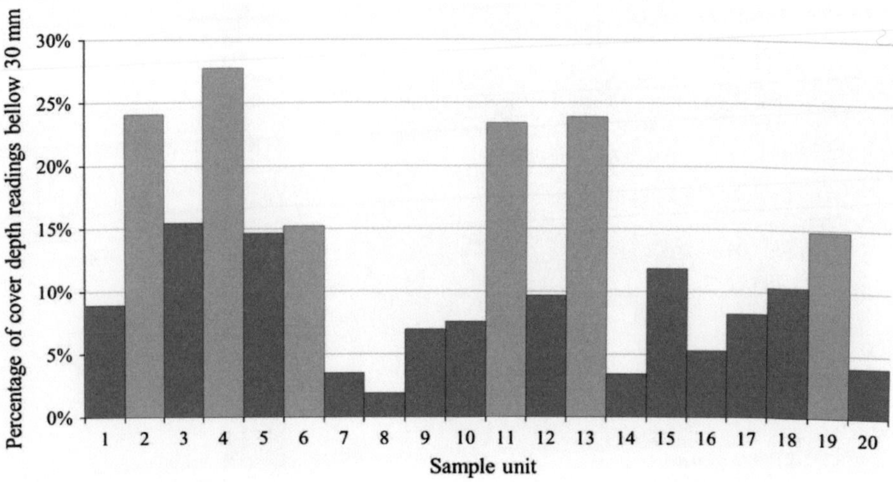

Fig. 9.10 Inspection results (Example 2)

Table 9.5 Number of defective units within the sample for several values of c_{min}

c_{min} (mm)	Number of defective units within the sample
30	16
29	15
28	12
27	11
26	9
25	8
24	6
23	5
22	5
21	5
20	4
19	3

Sample Data:
Figure 9.10 presents the percentages of measured cover depths lower than c_{min} in each tested unit (column). The red bars represent the columns with more than two consecutive measurements with values lower than 20 mm.

Verification of conformity:
Using the reference values presented in Table 9.3, the acceptance number is $Ac = 3$. Based on Fig. 9.10, it is possible to count a total of 16 defective units, which is higher than the acceptance number and, therefore, the lot cannot be considered to conform to the specifications.

In Table 9.5, the quantities of defective units within the sample are presented for several hypothetical values of c_{min}.

It is possible to verify through Table 9.5 that the lot could only be accepted if the required minimum cover depth had not been greater than 19 mm. As an informative note, the 5th percentile of all the cover depth measurements within the sample is 24 mm.

9.9 Final Remarks

The non-conformity of concrete cover depth with the specifications is one of the main causes of premature deterioration of reinforced concrete structures. However, there is a lack of provisions in standards and codes for the inspection of structures regarding this parameter, either during or after the construction works. This chapter attempts to overcome this lack by providing three statistical procedures for the inspection of isolated lots, and describing those standardized in the German code of practice.

Further advances in the presented procedures to improve the reliability of the cover depth evaluation may be needed, such as:

- the inclusion of methods to deal with the spatial autocorrelation among cover depth measurements in Procedures A and B;
- the establishment of an adequate limiting quality for Procedure C, based on results obtained in a large number of case studies;
- the application of Bayesian methods that allow the use of prior information.

It should be noted that the procedures here presented apply to isolated lot inspection. However, developments can be made to extend their applicability to lot-by-lot inspection, which can be useful for assessing the conformity of cover depth during the production of long series of precast elements.

A9.1 Construction of Probability Plots and Detection of Potential Outliers

The construction of normal probability plots may be useful to check whether or not the sample data is approximately normally distributed. Their construction consists of [12]:

1. sorting and numbering the cover depth measurements, c_i, in ascending order;
2. calculating the mean value, $\bar{c}$, and the standard deviation, s, of the N measurements;
3. calculating the sample $\left(Y_i = \frac{(c_i - \bar{c})}{s}\right)$ and theoretical $\left(X_i = \Phi^{-1}\left(\frac{i - 0.3175}{N + 0.365}\right)\right)$ quantiles of each measurement, where $\bar{c}$ and s are the mean value and standard

deviation of the cover depth sample, $\frac{i-0.3175}{N+0.365}$ is the median rank function [14], and $\Phi^{-1}(x)$ is the inverse of the standard normal c.d.f.;
4. plotting a chart with X_i's as abscissas and Y_i's as ordinates.

The assumption of normality is checked by observing the goodness of fit of the results to demonstrate a straight line of unit slope and zero intercept. If the results do not follow approximately this line, the assumption of normality should be rejected.

For checking if the sample data is approximately lognormally distributed, the same procedure can be adopted but using the logarithms of the measured cover depths instead, since the logarithms of a lognormally distributed variable follows a normal distribution. Procedures for evaluating the goodness of fit of other distributions can be found in Monteiro and Gonçalves [12].

One simple way of detecting a single outlier is through Grubb's test [15]. It consists in checking if the largest value $|Y_i|$, associated with the most extreme measurement, is greater than the corresponding critical value presented in Table A9.1. If yes, the corresponding measurement should be considered as a potential outlier.

Other methods for detecting potential outliers (including multiple outliers) can be found in ISO 5725-2 [15], ASTM E178 [16] and ASTM E691 [17].

A9.2 Definition of the Lower One-Sided Tolerance Limit [12]

The lower one-sided tolerance limit corresponds to the limit of the one-sided confidence interval estimated to contain a given upper proportion P of the

Table A9.1 Critical values for Grubb's two-sided test (adopted from ISO 5725-2 [15])

N	Critical value for a significance level of 5 %	N	Critical value for a significance level of 5 %
10	2.29	50	3.13
11	2.35	60	3.20
12	2.41	80	3.31
13	2.46	100	3.38
14	2.51	120	3.45
15	2.55	150	3.52
20	2.71	180	3.57
25	2.82	200	3.61
30	2.91	300	3.72
40	3.04	400	3.80

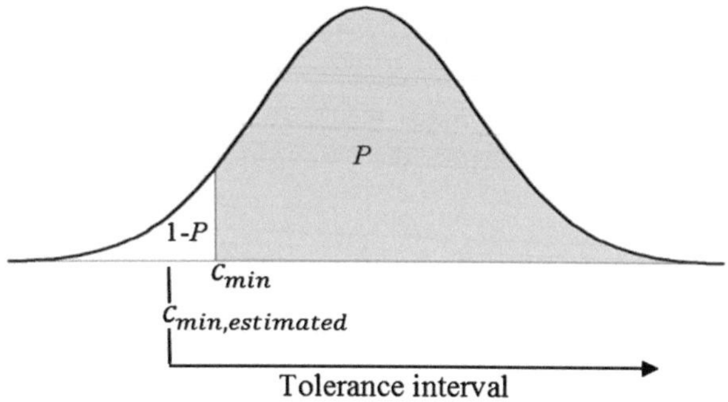

Fig. A9.1 One-sided tolerance interval for the upper proportion P of a population

population with a given confidence level $1-\gamma$. This interval is also often referred as one-sided tolerance interval and is presented in Fig. A9.1.

For example, a confidence level of 90 % can be interpreted as follows: if an infinite number of samples with N measurements is obtained from a population and with each data sample a tolerance interval is calculated, it is expected that 90 % of these intervals will contain the desired proportion, P, of the population and thus, also the respective percentile.

Estimating the 5th or 10th percentile of a population, is the same as estimating the lower one-sided tolerance limit for $P = 0.95$ or 0.90, respectively.

For normally distributed variables, the above tolerance limit can be estimated by the following expression:

$$Tolerance\,Limit = sample\,mean - k \cdot sample\,standard\,deviation \tag{A9.1}$$

where: $k = \frac{T^{-1}_{N-1;\Phi^{-1}(P)\cdot\sqrt{N}}(1-\gamma)}{\sqrt{N}}$ is the tolerance factor; N is the sample size; $T^{-1}_{v;\delta}(x)$ is the inverse cumulative distribution function of the non-central t-distribution with $v = N - 1$ degrees of freedom and a non-centrality parameter of $\delta = \Phi^{-1}(P)\cdot\sqrt{N}$; and $\Phi^{-1}(x)$ is the inverse cumulative distribution function of the standard normal distribution.

For $N \geq 10$, the following approximation can be used (derived from [18]):

$$T^{-1}_{v;\delta}(x) \approx \frac{h\cdot\delta + \sqrt{h^2\cdot\delta^2 + \left(h^2 - \frac{\left(\Phi^{-1}(x)\right)^2}{2\cdot v}\right)\cdot\left(\left(\Phi^{-1}(x)\right)^2 - \delta^2\right)}}{h^2 - \frac{\left(\Phi^{-1}(x)\right)^2}{2\cdot v}} \tag{A9.2}$$

where: $h = \left(1 - \frac{1}{4\cdot v}\right)$.

A9.3 Dealing with Spatial Autocorrelation

As mentioned in Sect. 9.3.3, the occurrence of spatial autocorrelation among measurements violates the assumption of independence of observations and, if not taken into account in the statistical analysis, it may lead to biased estimates of the standard deviation of the cover depth population and significantly compromise the confidence level of the actual minimum cover depth estimate.

A common method for checking the occurrence of spatial autocorrelation in a series of observations equally spaced in space over a single direction is by means of hypothesis testing. It is usually assumed a first order autoregressive model to describe the spatial autocorrelation:

$$c_j = r_1 \cdot c_{j-1} + (1 - r_1) \cdot \mu + \varepsilon_j \tag{A9.3}$$

where ε_j is a series of independent and identically distributed normal random variables with zero mean, μ is the true mean of the cover depth population; c_j is the jth cover depth measurement of the sample ordered according to the spatial location in the structure, and r_1 is the first order autocorrelation coefficient, i.e. the coefficient correlation between the first $c_{j=1,2,\ldots,N-1}$ and the next $c_{j=2,3,\ldots,N}$ measured cover depths, which can be calculated by the following expression:

$$r_1 = \frac{\sum_{j=1}^{N-1}\left(c_j - \bar{c}_{(1)}\right) \cdot \left(c_{j+1} - \bar{c}_{(2)}\right)}{\sqrt{\left[\sum_{j=1}^{N-1}\left(c_j - \bar{c}_{(1)}\right)^2\right] \cdot \left[\sum_{j=2}^{N}\left(c_j - \bar{c}_{(2)}\right)^2\right]}} \tag{A9.4}$$

where $\bar{c}_{(1)}$ and $\bar{c}_{(2)}$ a the arithmetic means of the $c_{j=1,2,\ldots,N-1}$ and $c_{j=2,3,\ldots,N}$ measured cover depths, respectively.

The assumption of independence among measurements can then be checked by testing the hypothesis of positive autocorrelation among measurements (H_1: $r_1 > 0$). For a significance level of 95 %, the following upper limit for r_1 can be used [19]:

$$r_1 < \frac{-1 + 1.645 \cdot \sqrt{N-2}}{N-1} \tag{A9.5}$$

If Eq. (A9.5) is not satisfied, it may be considered that the cover depth measurements within the sample are spatially autocorrelated and that the assumption of independence of observations is violated. In that case, the autocorrelation may be taken into account in the statistical analysis by reducing the number of degrees of freedom adopted for choosing the tolerance factor k (Sect. 9.3.6). This reduction can be calculated using the following expression [20]:

$$N' = N \cdot \left(\frac{1 - r_1}{1 + r_1}\right) \tag{A9.6}$$

where N' is the effective sample size that should be used (rather than N) to estimate the number of degrees of freedom.

It should be noticed, however, that this method is very simplistic and is only unidirectional.

A9.4 Construction of OC Curves [12]

The contractor's and owner's risks associated with the conformity criterion established in Sect. 9.3.7 can be theoretically evaluated by means of operating characteristic (OC) curves. The OC curves relate, for a given conformity criterion, the probability of acceptance of a lot P_a (or rejection, P_r) with its fraction defective θ. In this case, the fraction defective would be the unknown true proportion of the cover depth population (lot) that falls below c_{min}.

Drawing OC curves is usually quite laborious since it implies constructing software routines that are not included in most common computer applications. For the precise drawing OC curves the following expression can be used:

$$P_a = 1 - T_{N-1,-\Phi^{-1}(\theta)\cdot\sqrt{N}}\left(k \cdot \sqrt{N}\right) = 1 - P_r \tag{A9.7}$$

where $T_{N-1,-\Phi^{-1}(\theta)\cdot\sqrt{N}}(x)$ is the c.d.f. of the non-central t-distribution with $N - 1$ degrees of freedom and non-centrality parameter $-\Phi^{-1}(\theta) \cdot \sqrt{N}$.

The following approximation can also be used, provided that

$$\varphi^2 + (1 - \varphi^2) \cdot \left[N \cdot \left(\Phi^{-1}(\theta)\right)^2 - \left(\frac{\Phi^{-1}(\theta)\cdot\sqrt{N}+k\cdot\sqrt{N}\cdot\varphi}{\sqrt{1+N\cdot k^2\cdot(1-\varphi^2)}}\right)^2\right] > 0,$$

with $\varphi = \sqrt{\frac{2}{N-1}} \cdot \frac{\Gamma\left(\frac{N}{2}\right)}{\Gamma\left(\frac{N-1}{2}\right)}$:

$$T_{N-1,-\Phi^{-1}(\theta)\cdot\sqrt{N}}\left(k \cdot \sqrt{N}\right) \approx \Phi\left(\frac{\Phi^{-1}(\theta) \cdot \sqrt{N} + k \cdot \sqrt{N} \cdot \varphi}{\sqrt{1 + N \cdot k^2 \cdot (1 - \varphi^2)}}\right) \tag{A9.8}$$

where $\Gamma(x)$ is the gamma function.

The above expressions are valid for both normally and lognormally distributed cover depth populations. For other distributions, formulas for calculating OC curves can be found in Monteiro and Gonçalves [12].

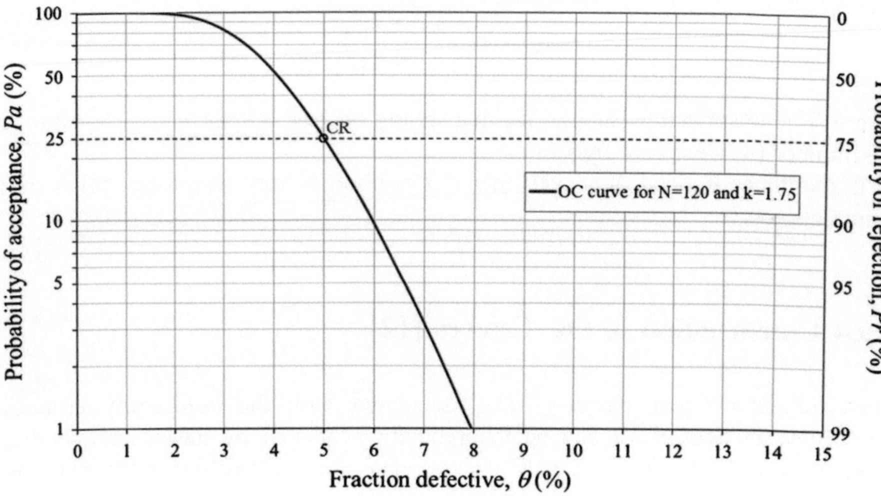

Fig. A9.2 OC curve (Example 1)

Figure A9.2 shows the OC curve associated with the conformity criterion used to check the conformity of the minimum cover depth in the Example 1.

The CR point in Fig. A9.2 is fixed by the confidence level adopted in the estimation of the actual minimum cover depth, and corresponds to a maximum probability of 25 % of accepting a "bad" lot with more than 5 % of cover depths below c_{min} (fraction defective) and, as consequence, to a maximum probability of 75 % of rejecting a "good" lot with less than 5 % of cover depths lower than c_{min}.

References

1. EN 13970, Execution of concrete structures, CEN, 2009.
2. DBV-Merkblatt, Betondeckung und Bewehrung nach Eurocode 2, DBV (in German), 2011.
3. BS 1881, Testing concrete. Part 204. Recommendations on the use of electromagnetic covermeters, London: British Standard Institution, 1988.
4. RILEM Report 40, Non-destructive evaluation of the penetrability and thickness of the concrete cover, state-of-the-art report of RILEM technical committee 189-NEC, R. Torrent and L. Luco, 2007. pp. 133–155.
5. Recommendation of RILEM TC. 189-NEC, Non-destructive evaluation of the concrete cover, comparative test—part ii—comparative test of covermeters. Mater Struct. 2005;38:907–11.
6. fib bulletin 34, Model code for service life design, fib secretariat, Lausanne, 2006.
7. Monteiro A, Gonçalves A. Assessment of concrete cover in structures. Part 2—case studies, fib workshop performance-based specifications for concrete, Leipzig, Germany, 2011. pp. 230–239.
8. Draft ISO/DIS 16204, Durability—service life design of concrete structures, ISO, 2011.
9. Fluge F. Marine chlorides. A probabilistic approach to derive provisions for EN 206-1. In: 3rd DuraNet workshop on service life design of concrete structures, Tromsø, Norway, 2001. pp. 63-83.

10. Caspeele R, Taerwe L. Bayesian assessment of the characteristic concrete compressive strength using combined vague-informative priors. Constr Build Mater. 2012;28:342–50.
11. EN 1990, Eurocode—basis of structural design, CEN, 2002.
12. Monteiro A, Gonçalves A. Assessment of concrete cover in structures. Part 1—statistical tolerance analysis approach. In: fib workshop performance-based specifications for concrete, Leipzig, Germany, 2011. pp. 220–229.
13. ISO 2859-2, Sampling procedures for inspection by attributes – Part 2: Sampling plans indexed by limiting quality level (LQ) for isolated lot inspection, ISO, 1985.
14. Jacquelin J. A reliable algorithm for the exact median rank function. IEEE Trans Electr Insul. 1993;28(2):168–71.
15. ISO 5725-2, Accuracy (trueness and precision) of measurements methods and results—part 2: basic method for the determination of repeatability and reproducibility of a standard measurement method, ISO, 1994.
16. ASTM E178, Standard practice for dealing with outlying observations. American society for testing and materials, West Conshohocken, 2008.
17. ASTM E691, Standard practice for conducting an interlaboratory study to determine the precision of a test method. American society for testing and materials, West Conshohocken, 1999.
19. Abramowitz M, Stegun IA. Handbook of mathematical functions, New York: Dover Publications, 1972. p. 949.
19. Anderson R. Distribution of the serial correlation coefficients. Ann Math Stat. 1941;8(1):1–13.
20. Dale M, Fortin M. Spatial autocorrelation and statistical tests: some solutions. J Agric Biol Environ Stat. 2009;14:188–206.

Chapter 10
Venlo Application Testing (Summary)

L. Fernández Luco, H. Beushausen, F. Jacobs and M. Serdar

10.1 Introduction

RILEM TC 230-PSC organized, along with B|A|S Research & Technology, an Application Test (AT) Program at B|A|S headquarters in Venlo, the Netherlands that consisted of the application of different techniques and testing procedures, mainly non-destructively, to make conformity assessment and/or service life prediction on eight reinforced concrete wall panels.

These panels, representing different types of concrete, curing regimes and outdoor exposure conditions (shielded and unshielded from the rain and wind) were exposed outside the B|A|S facilities, as can be seen in Fig. 10.1, to provide the participants with varied testing situations.

For the interpretation of experimental results, the retaining-wall type panels were assumed to be exposed to extreme carbonation (XC4) and chloride (XD3, XS3) environments, according to EN 206, to determine a reference in exposure condition.

A general measurement programme took place in April 2012, but some of the participants arranged a second set of measurements during July 2012, as they considered the concrete being either too young or too wet during the first programme.

L. Fernández Luco (✉)
University of Buenos Aires, Buenos Aires, Argentina

H. Beushausen
University of Cape Town, Cape Town, South Africa

F. Jacobs
TFB, Zürich, Switzerland

M. Serdar
University of Zagreb, Zagreb, Croatia

H. Beushausen and L. Fernandez Luco (eds.), *Performance-Based Specifications and Control of Concrete Durability*,
RILEM State-of-the-Art Reports 18, DOI 10.1007/978-94-017-7309-6_10

Fig. 10.1 Wall-type panels prepared by B|A|S team

The following sections describe aims and experimental details for the Application Tests, and present an overview of the results. Complementary data on panel casting, concrete composition and general properties of the concrete used in the panels, as well as detailed reports on test results, as supplied by the various participants of the Application Testing programme, are presented in Chap. 11.

10.1.1 Aims and Scope

The aim of the AT was to demonstrate the application of various performance-based approaches of durability assessment on site. The experimental approaches and test methods, based on the measurement of fluid transport properties of concrete, were designed to provide the necessary data to make either a conformity assessment or a service life prediction. All TC members were invited to participate in the AT, using their method of choice for durability assessment.

In addition to using the test methods to make durability assessments, the feasibility of conducting the various tests under on-site conditions (humidity, temperature) was assessed, aiming at identifying the specific limitations of the various approaches applied.

Finally, the AT programme and the associated technical discussions were expected to contribute to the exchange of practical experiences that may help to improve existing approaches towards performance-based durability assessment.

Table 10.1 Concrete wall panels: selected properties, curing conditions and casting dates

Mix	Cement type	w/c	Wall no.	Curing condition	Casting date (2012)
A	CEM I 52,5 N	0.439	1	Air curing	March 28th
			2	7 days plastic cover + air curing	March 28th
B	CEM I 52,5 N	0.537	3	Air curing	March 29th
			4	7 days plastic cover + air curing	March 29th
C	CEM II/B-V	0.396	5	Air curing	March 30th
			6	7 days plastic cover + air curing	March 30th
D	CEM II/B-V	0.586	7	Air curing	April 2nd
			8	7 days plastic cover + air curing	April 2nd

10.1.2 Experimental Program

The B|A|S team prepared eight reinforced concrete walls and exposed them to various external exposure conditions. Four concrete types, named A, B, C and D were used to cast two wall panels each. The difference between the two wall panels for each concrete mix was the curing procedure, with one wall covered with plastic for 7 days and the other fully exposed (Table 10.1). Prior to exposure to these conditions all panels were left for one day in the mould after casting.

The two different curing methods were meant to provide different concrete qualities, with panels cured with plastic for 7 days expected to provide better results. However, the environmental conditions after panel manufacture were cold and humid, with an average RH exceeding 80 % and an average temperature of 7 °C during the period between panel manufacture and testing. In the analysis of test results, the influence of the curing method on durability parameters was therefore found to be negligible. It is suspected that the influence of curing on durability would be much more pronounced in adverse environmental conditions (such as high temperatures and low RH). The test conditions encountered during the AT were therefore not conducive to the evaluation of how curing affects concrete durability.

Once the test panels were exposed to open air, the upper parts of the walls were covered with plastic sheets to prevent the panels from being wetted by rain. The lower half of the panels remained uncovered. This protection was meant to result in dry concrete in the top half of the panels (covered) and moist concrete in the bottom half. As such, the humidity and moisture content of the concretes were supposed to be an additional test parameter in the panel assessment. However, the humidity test results indicated that the assumption of differing moisture conditions between top and bottom halves of the panels was invalid. In practical terms, no significant variation in the humidity content between the upper (covered) and lower (uncovered) parts was detected. This was ascribed to the generally very wet environmental conditions encountered during the time of panel manufacture and testing. In the analysis of test results, the moisture content of the individual panels (i.e. any differences in moisture content between top and bottom of the panels) was therefore not taken into consideration. Test results obtained from top or bottom regions of the respective panels were considered to represent the same environmental condition.

10.1.3 Participants and Test Methods

Researchers from various countries participated in the Application Tests and applied different instruments to acquire the necessary data in line with the aims of the programme. Most of the performance-based test methods and approaches that were used for the AT are discussed in detail in Chaps. 4 and 8. The data obtained with the various methods were required to be able to evaluate conformity criteria (for a certain exposure type or environment) and/or to predict the potential service life of a reinforced concrete structure in the assumed relevant exposure environments. Additionally, some of the applied methods were used to qualitatively rank the panels in relation to their potential durability, since they are not linked to service life models or conformity criteria. For a complete presentation of the outcome of the AT, all methods are included in the discussions in the following sections.

Measurements were originally taken in April 2012 at only about 2 weeks after panel manufacture and consequently some participants considered the panels as too wet and too young to yield conclusive results. Some participants therefore attended a second round of measurements in July 2012. At both testing ages, cores were extracted from the panels and sent to the University of Cape Town, South Africa for evaluation of durability properties according to the Oxygen Permeability Index and Chloride Conductivity Index tests. A list of all attendees and test methods applied is given in Tables 10.2 and 10.3.

10.2 Summary of Test Results

The various participants of the AT supplied detailed test reports, which are presented in Chap. 11. The following sections contain a summary of relevant test results and a general discussion on conformity assessment and service life prediction for the concrete test panels.

Note that a direct comparison of the service life assessments (SLA) made by various participants is not informative as the data analysis is often based on very different assumptions and/or different test conditions (e.g. calibration of test methods against differing country-specific experiences, different assumed environmental exposure conditions for the SLA, different test ages, etc.). Therefore, the test results and analysis of each participant need to be considered in isolation, not in comparison. In cases where the results of various participants are summarized in the

Table 10.2 Attendees and test methods at the first Application Test meeting (April 2012)

Participants	Tests carried out	Reference	Participant affiliation
K. Imamoto	Single chamber method		Tokyo University of Sciences, Japan
O. Shinichiro	Air permeability test		Ehime University, Japan
U. Isao			
Hayashi-Kazuhiko	Surface water absorption test		Yokohama National University, Japan
Komatsu-Satoshi			
Misumi Ai			
P. Paulini	Permeability exponent	Section 4.2.2.5	University of Innsbruck, Austria
Mr. Dix			
R. Torrent	Permea-TORR[a]	Section 4.2.2.4	Materials Advanced Services Ltd, Argentina
M. Serdar	Permea-TORR[a]	Section 4.2.2.4	University of Zagreb, Croatia
H. Beushausen	Oxygen Permeability Index[b]	Section 4.2.2.7	University of Cape Town, South Africa
S. Starck	Chloride Conductivity Index[b]	Section 4.6.2.3	
D. Boubitsas	Rapi-corr[c]		Chalmers University of Technology, Sweden
L. Fernández Luco	Wenner Resistivity[c]	Section 4.6.2.2	University of Buenos Aires, Argentina
S. Nanukuttan	Permit	Section 4.5.2.6	Queen's University of Belfast, Northern Ireland
P. Pouryahya	Autoclam air permeability[d]	Section 4.2.2.3	
	Autoclam water permability[d]	Section 4.3.2.1	

[a] These results are included in the individual reports in Chap. 11 but the service life prediction with results obtained from Permea-TORR (Sect. 11.4) was based on tests carried out by R. Torrent and F. Jacobs in July 2012

[b] Assessment was done on concrete core samples that had been removed from the test panels and pre-conditioned in the laboratory. Due to the time needed for coring and shipping the samples from the Netherlands to South Africa, the test age between the in situ tests (roughly 2 weeks) and core testing in South Africa (roughly 8 weeks) is significantly different

[c] Test results are not linked to specific performance-based criteria for durability

[d] Test results obtained with these methods in April 2012 are not included in the discussions. The tests were repeated in July 2012 and these later results were included in the analysis

same table or figure, this is not meant to indicate a comparison of the various approaches but to merely serve as an overview on the outcomes of the different approaches.

Table 10.3 Attendees and test methods at the second Application Test meeting (July 2012)

Participant	Tests carried out	Participant affiliation
S. Nanukuttan	Autoclam air permeability	Queen's University of Belfast, Northern Ireland
	Autoclam water permeability	
R. Torrent	Permea-TORR	Materials Advanced Services, Argentina
F. Jacobs		T.F.B., Switzerland
H. Beushausen[a]	Oxygen Permeability Index	University of Cape Town, South Africa
	Chloride Conductivity Index	

[a]Participant was not present during the second round of testing in Venlo; the durability assessment was done on concrete core samples that had been removed from the test panels and pre-conditioned in the laboratory

10.2.1 Tests Based on Permeability to Gases

Most of the tests performed during the AT in Venlo concern permeability to gas or air. Some participants performed testing in April 2012 (Table 10.2), while some participants decided to test the panels in July 2012 (Table 10.3), as discussed earlier.

All participants processed their own data according to the specific approach used (see Chap. 11). Some participants limited their assessment to a "durability ranking" (from 1 to 8), assigning 1 to the "best performance predicted" according to the assumed exposure conditions. Others determined ranges for the service life prediction, i.e., more than 25 years, less than 10 years, etc., while some of the participants indicated a specific service life in years, as a deterministic value or as a range of years (optimistic—pessimistic values).

Some of the participants who used gas/air-permeability based tests limited the service life assessment to carbonation (XC exposure class) while others considered the exposure to chloride environments (XS or XD), which further stresses the point that the outcome of the various approaches should not be compared. The data analysis supplied by the individual participants is summarised in Table 10.4. All data sets correspond to a nominal cover of 50 mm. Table 10.5 shows potential durability rankings for the air-permeability based tests performed by some of the participants (for concrete exposed to carbonation).

10.2.2 Tests Based on Capillary Suction of Water

Two tests were based on capillary suction of water: the Autoclam Sorptivity Index and the Surface Water Absorption Test. Note that the data corresponds to measurements taken in July 2012 for the former and in April 2012 for the latter. As discussed in earlier sections of this chapter, the maturity of the concrete, the water

Table 10.4 Service life assessment based on results obtained with permeability-based tests (in years), for 50 mm nominal cover

Method, participant	SCM[a], Imamoto	Autoclam air[b], Basheer	OPI[b], Beushausen	Permea-TORR[b], Torrent and Jacobs	Seal-Test M[a], Okazaki and Shinichiro
Exposure	Carbonation	Carbonation (XC1 to XC4)	Carbonation (XC1 to XC4)	Chlorides	Chlorides
Panel (cement, w/c)					
1 (CEM I, 0.44)	62–71	>25	>100	73–85	95
2 (CEM I, 0.44)[c]	56–91	>50	>100	68–101	220
3 (CEM I, 0.54)	50–57	>25	>100	48–68	38
4 (CEM I, 0.54)[c]	48–61	>25	>100	37–65	50
5 (CEM II, 0.40)	68–77	>50	>100	61–120	165
6 (CEM II, 0.40)[c]	70–85	>50	>100	85–113	180
7 (CEM II, 0.59)	49–66	>25	>100	19–58	28
8 (CEM II, 0.59)[c]	51–75	<10	>100	21–58	25

[a]Based on results obtained in April 2012
[b]Based on results obtained in July 2012
[c]Panels initially cured with plastic sheets for 7 days

content and the temperature at the time of measurements are not comparable and neither are the corresponding results. The outcome of the service life assessment based on the test results is shown in Table 10.6.

10.2.3 *Tests Based on Ion Migration*

The Permit and the Chloride Conductivity Index (CCI) test belong to the group of tests based on ion migration. Table 10.7 summarises the SLA based on the Permit and the CCI tests. Note that the SLA obtained from the test results of the Permit (obtained in April, at an age of about 2 weeks) relates to environmental condition XS3, while the CCI results (obtained in July, at an age of about 4 months) were analysed based on "severe marine exposure" according to the South African exposure classes (which can be considered as equivalent to XS3 plus heavy wave action). Due to the different assumptions and test ages, a comparison of the outcome of the SLA is not valid. In both approaches, the SLA corresponds to a nominal cover of 50 mm.

Table 10.5 Potential durability rankings for carbonation exposure conditions (Permeability exponent)

Method, participant	Permeability exponent[a], Paulini	
Exposure	From in situ values	Corrected for saturation degree
Panel (cement, w/c)		
1 (CEM I, 0.44)	3	6
2 (CEM I, 0.44)[b]	2	2
3 (CEM I, 0.54)	7	7
4 (CEM I, 0.54)[b]	6	5
5 (CEM II, 0.40)	8	8
6 (CEM II, 0.40)[b]	1	1
7 (CEM II, 0.59)	4	3
8 (CEM II, 0.59)[b]	5	4

[a]Panels initially cured with plastic sheets for 7 days
[b]Based on results obtained in April 2012. In first column, the ranking corresponds to the data recorded on site ("from in situ values") while the second column corresponds to the values corrected for the saturation degree of the concrete (which was assessed in parallel with permeability)

Table 10.6 Service life assessment and durability ranking based on results obtained with water sorptivity measurements, for 50 mm nominal cover

Method, participant	SWAT, Hayashi[a]	Autoclam sorptivity index, Basheer[b]
Exposure	Carbonation	Chlorides (XD1 to XD3)
Panel (cement, w/c)	SLA (years)	SLA (years)
1 (CEM I, 0.44)	345	>50
2 (CEM I, 0.44)[c]	368	>50
3 (CEM I, 0.54)	68	>25
4 (CEM I, 0.54)[c]	85	10
5 (CEM II, 0.40)	313	>50
6 (CEM II, 0.40)[c]	409	>50
7 (CEM II, 0.59)	25	<10
8 (CEM II, 0.59)[c]	108	<10

[a]Based on results obtained in April 2012
[b]Based on results obtained in July 2012
[c]Panels initially cured with plastic sheets for 7 days

10.2.4 Electrical Resistivity Measurements

The measurements of electrical resistivity cover several experimental approaches including the Wenner probe and the Rapi-Corr. Tests for resistivity were not aimed at performing a service life assessment. However, the analysis of resistivity measurements leads to practical conclusions and serves to confirm some of the assumptions made in the characterization of the concrete panels.

Table 10.7 Service life assessment based on results obtained with test methods based on ion migration, for 50 mm nominal cover

Method, participant	Permit, Nanukuttan[a]	CCI, Beushausen[b]
Exposure	Chlorides (years)	XS3 (equivalent) (years)
Panel (cement, w/b)		
1 (CEM I, 0.44)	~30	91
2 (CEM I, 0.44)[c]	≥100	87
3 (CEM I, 0.54)	~30	40
4 (CEM I, 0.54)[c]	<20	45
5 (CEM II, 0.40)	~50	>100
6 (CEM II, 0.40)[c]	~50	>100
7 (CEM II, 0.59)	~25	>100
8 (CEM II, 0.59)[c]	~40	>100

[a]Based on results obtained in April 2012
[b]Based on results obtained in July 2012
[c]Panels initially cured with plastic sheets for 7 days

Table 10.8 Average value of electrical resistivity [kΩ.cm], measurements made in April 2012

Method, participant	Rapi-Corr, Boubitsas		Resipod, Serdar	Resipod, Fernandez Luco	
Panel (cement, w/c)	Covered	Uncovered	Covered	Covered	Uncovered
1 (CEM I, 0.44)	38	35	7	8	8
2 (CEM I, 0.44)[a]	33	32	7	8	8
3 (CEM I, 0.54)	23	24	4	4	3
4 (CEM I, 0.54)[a]	33	20	3	4	3
5 (CEM II, 0.40)	35	28	3	5	5
6 (CEM II, 0.40)[a]	25	23	3	5	5
7 (CEM II, 0.59)	17	21	3	4	4
8 (CEM II, 0.59)[a]	15	14	3	3	3

[a]Panels initially cured with plastic sheets for 7 days

Resipod (Wenner, 50 mm) resistivity measurements and Rapi-Corr tests were carried out on covered and uncovered parts of the walls and the comparison of the respective resistivity values allows an assessment of the moisture content in the "assumed" dry areas (covered from rain) and wet areas (fully exposed). Table 10.8 shows the results from resistivity measurements made in April 2012.

From the data summarised in Table 10.8, it is clearly seen that the results obtained with the Rapi-Corr are systematically higher than the ones obtained with the Wenner technique (Resipod).

The results obtained for top and bottom regions of the test panels are equivalent. Considering that the influence of the humidity content on electrical resistivity is very strong, this indicates that the humidity content was relatively constant throughout the panels, despite the attempts to create different conditions.

Table 10.9 Equivalent age of the test panels assessed with resistivity measurements, compared to results obtained with Concremote

Concrete type (cement, w/c)	Equivalent age (Maturity)	Equivalent age (Resistivity)	Difference (%)
A (CEM I, 0.44)	13.1	13.3	1.5
B (CEM I, 0.54)	12.4	12.2	−1.7
C (CEM II/B-V, 0.40)	12.0	10.0	−17.3
D (CEM II/B-V, 0.59)	9.7	10.3	5.9

Electrical resistivity, when measured in close-to-saturation condition, may be linked to the degrees of hydration and maturity as it evolves at similar rates as the microstructure. For this, resistivity measurements on the actual structure are compared to those taken on accompanying cube specimens, accounting for size effects. Concrete maturity assessments on all panels were performed with the Concremote method (based on temperature measurements). Table 10.9 shows the assessment of maturity (expressed as equivalent age) from resistivity measurements (Fernandez Luco, Resipod), compared to the data obtained from Concremote. As indicated in the table, the test results obtained from both methods are generally very similar.

10.3 Further Assessment from Compiled Rankings

Considering that all of the applied performance-based approaches are unique and based on different assumptions, a direct comparison of results is not valid. Nevertheless, the various approaches can all be used to assess the 8 test panels according to their "potential durability ranking". In the following discussion, the most durable panel under the respective exposure class considered receives the ranking value of "1", while the panel with the lowest durability (at the same exposure condition) receives a rank of "8".

Note that the panels were cast at different days and therefore had slightly different ages, which could have had an influence on the outcome of the early testing (in April), which was performed at only about 2 weeks after panel manufacture. For the testing in July, the slight difference in age can be considered negligible.

The overall ranking for each panel is obtained by averaging the durability rankings obtained from the various approaches. Since the concrete maturity and environmental conditions at the two test ages were very different, the ranking is done separately for data obtained in April and July.

10.3.1 Ranking of Test Panels for Carbonation Exposure

For the comparative durability ranking against carbonation exposure, only results obtained in April 2012 are presented as only one single method (OPI) was applied

Table 10.10 Durability assessment for carbonation exposure (April 2012), results and average ranking

Method, participant	SCM, Imamoto	SWAT, Hayashi	Permeability exponent, Paulini	OPI[a], Beushausen	Average ranking
Parameter	API	K	K_{per}	k	
Panel unit	kPa/s	$ml/m^2/s$	$m^2 \times 10^{-17}$	10^{-11} m/s	
1 (CEM I, 0.44)	0.28	0.185	3.89	2.00	3
2 (CEM I, 0.44)[b]	0.29	0.176	8.54	1.92	4
3 (CEM I, 0.54)	0.57	0.548	7.82	9.84	6
4 (CEM I, 0.54)[b]	0.59	0.477	14.4	14.37	8
5 (CEM II, 0.40)	0.22	0.2	2.35	2.17	2
6 (CEM II, 0.40)[b]	0.2	0.162	18.1	1.77	1
7 (CEM II, 0.59)	0.54	0.974	5.88	32.18	7
8 (CEM II, 0.59)[b]	0.41	0.414	5.19	21.76	5

[a]The testing was performed on core samples. Due to the delay in coring and the time taken for transport from the Netherlands to South Africa, the test age was very different compared to the other methods
[b]Panels initially cured with plastic sheets for 7 days

to perform a durability assessment for exposure classes in July 2012. In the first round of tests, four different methods were applied to assess durability of the panels for carbonation exposure, as summarized in Table 10.10 and Fig. 10.4. All methods used for this assessment are based on gas permeability and the ranking obtained from each method is based on a simple comparison of test values (higher values indicating lower durability).

Comparing the results obtained from the different methods it is evident that the rankings are generally similar, except for a number of individual outliers (Fig. 10.4). Omitting results obtained by Paulini, a generally very good consistency in the ranking across the remaining three test methods is observed. Overall, the best performing panels were the ones made with CEM II at a w/c ratio of 0.40, followed by CEM I with a w/c of 0.44. This highlights the dominant influence of water/cement ratio on permeability.

Considering the test values summarized in Fig. 10.4, the influence of binder type on permeability was not significant.

Considering the influence of curing conditions, none of the methods showed significantly improved permeability values resulting from 7-day curing (compare values for the same concrete, i.e. Panels 1&2, 3&4, etc., in Table 10.10). This was expected as the general environmental conditions after casting were moist (average RH 81 %), which minimizes the relative benefits of the applied curing method (plastic cover).

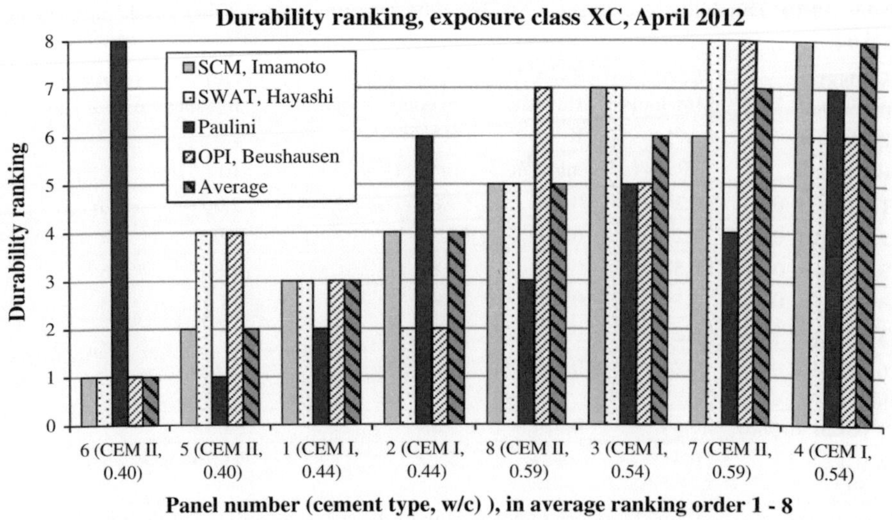

Fig. 10.4 Durability assessment for carbonation exposure (April 2012), individual and average rankings

10.3.2 Ranking of Test Panels for Chloride Exposure

For the comparative durability ranking against chloride exposure, only results obtained in July 2012 are presented as only one single method (Permit) was applied to perform a durability assessment for chloride exposure in April 2012 (the results obtained with the PemeaTORR in April were considered inconclusive due to the young age and high moisture contents of the test panels). In July 2012, four different methods were applied to assess durability of the panels for chloride exposure, as summarized in Table 10.11 and Fig. 10.5. It is worth noting that the test methods are based on different penetrability parameters (air permeability (PermeaTORR, Autoclam air), sorption (Autoclam sorp.), and ion migration (CCI)).

The ranking based on the air permeability and sorptivity test methods is based on direct comparison of test values. In contrast, the analysis of CCI results takes into account the binder type (effects such as pore solution chemistry and chloride binding) and is linked to a service life prediction model). The ranking obtained with this method is therefore based on the predicted service life (in years), not on a simple comparison of test values.

The test methods based on gas permeability and sorptivity generally show a very similar ranking (Fig. 10.5). The influence of water/cement ratio on permeability and sorptivity can clearly be seen, while the influence of cement type appears negligible.

If the CCI (ion migration) values were considered on a pure comparison of test value magnitude (as done with the gas permeability and sorption tests), a relatively consistent ranking across all methods would be observed. However, the

Table 10.11 Durability assessment for chloride exposure (July 2012), test results and average ranking

Method, participant	Autoclam air[a], Basheer	Autoclam sorpt[b], Basheer	PermeaTORR[a], Torrent	CCI[c], Beushausen	Average ranking
Parameter	API	SI	kt	CCI	
Panel unit	ln(mBar)/min	$10^{-7}m^3/min^{1/2}$	10^{-6} m^2	mS/cm	
1 (CEM I, 0.44)	0.202	0.507	0.019	0.27	3
2 (CEM I, 0.44)[d]	0.095	0.355	0.033	0.47	2
3 (CEM I, 0.54)	0.326	1.930	0.610	0.44	5
4 (CEM I, 0.54)[d]	0.363	3.375	0.730	0.30	6
5 (CEM II, 0.40)	0.032	0.758	0.035	0.98	4
6 (CEM II, 0.40)[d]	0.029	0.430	0.017	0.89	1
7 (CEM II, 0.59)	0.418	18.657	4.220	0.91	7
8 (CEM II, 0.59)[d]	0.813	12.007	2.990	0.85	8

[a]Test methods based on air permeability
[b]Test method based on sorption
[c]Test method based on ion migration. Interpretation of results (service life assessment, durability ranking) takes into account the binder chemistry
[d]Panels initially cured with plastic sheets for 7 days

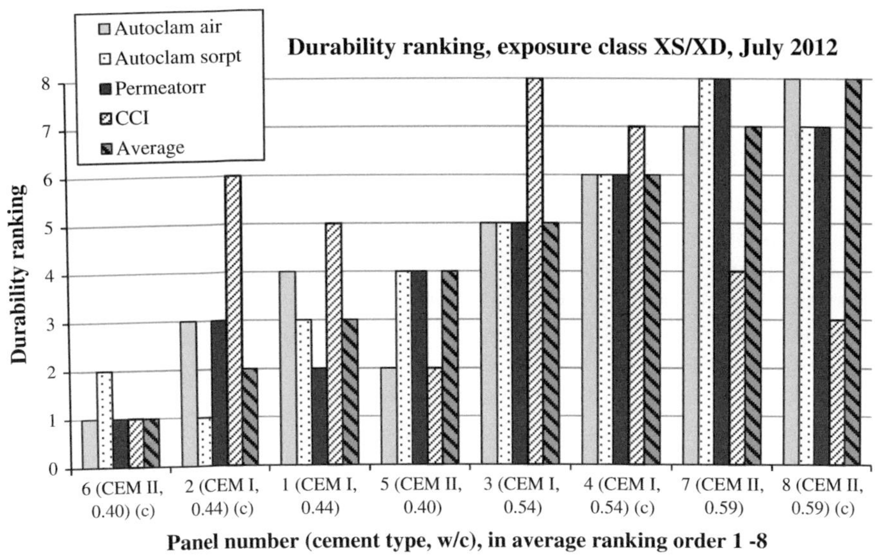

Fig. 10.5 Durability assessment for chloride exposure (July 2012), individual and average rankings

consideration of cement type results in concrete made with CEM I (Panels 1–4) having the lowest rankings in the interpretation of CCI values. Considering that is generally well accepted that concrete made with pure Portland cement has a significantly lower resistance against chloride ingress, compared to concrete made with blended cements, such an assessment outcome for the AT results appears sensible.

10.4 Conclusions

The Venlo Application Tests were successful in relation to the assessment of various test methods for concrete cover quality. In summary, the following specific conclusions are drawn from the AT:

- For the evaluation of concrete cover quality/penetrability on site, various test methods exist, usually based on measuring gas permeability, ion migration or water sorptivity.
- Most of the experimental approaches included in the AT were practical and relatively easy to apply.
- Some of the available performance test methods are linked to service life prediction models while others are based on an indicative evaluation, i.e. "ranking" of results.
- Most of the methods applied show a reasonable discriminating ability to distinguish good from better and bad from worst.
- The durability rankings compiled with different approaches show that the large majority of methods is successful in assessing the influence of w/c ratio on penetrability.
- In situ methods are sensitive to the moisture content in the concrete, which needs to be taken into consideration in the analysis of results.
- It was confirmed that concrete resistivity is a good indicator of moisture content and maturity of concrete.
- Most approaches base the evaluation of test results (e.g. gas permeability or water absorption) on a simple evaluation of the magnitude of penetrability. This presents a limitation as the influence of cement chemistry on actual deterioration processes such as carbonation and chloride ingress is not accounted for.
- There is still a need for further work in developing or refining deterioration and service life prediction models. The various existing models that were used in the AT show a wide range of outcomes with little consistency between different approaches. This also shows that service life models based on performance assessment are (and should be) calibrated with special consideration of local conditions (environmental exposure, binder types, etc.).

Acknowledgments RILEM TC 230-PSC gratefully acknowledges the support provided by Mr. A.J.E.J. van Casteren, CEO B|A|S Research & Technology, Marique Ruijs and all B|A|S Research & Technology staff. Their technical expertise, laboratory facilities, and on-site support greatly contributed to the success of the Application Test. The committee further expresses is gratitude to Joost Gulikers from Rijkswaterstaat Bouwdienst, the Netherlands, who assisted in organizing and financing the test programme.

Chapter 11
Venlo Application Testing (Individual Reports and Additional Data)

H. Beushausen, L. Fernández Luco, F. Jacobs, M. Serdar, M. Basheer, S. Nanukuttan, R. Torrent and K. Imamoto

11.1 Introduction

This chapter contains supplementary experimental data for the concrete panels of the Venlo Application Tests (AT), as well as individual reports on the analysis of test results. The individual reports were prepared by the respective participants, based on the applied test methods and service life prediction models. A requirement for participation in the Venlo AT was that experimental data needed to be linked to service life prediction models or conformity criteria. Consequently, most of the sections in this chapter contain analyses of experimental data based on performance-based approaches for durability. These approaches link to country-specific experiences and research, or may in some cases only relate to the

H. Beushausen (✉)
University of Cape Town, Cape Town, South Africa

L. Fernández Luco
University of Buenos Aires, Buenos Aires, Argentina

F. Jacobs
TFB, Zürich, Switzerland

M. Serdar
University of Zagreb, Zagreb, Croatia

M. Basheer
University of Leeds, Leeds, UK

S. Nanukuttan
Queen's University Belfast, Belfast, UK

R. Torrent
Materials Advanced Services, Buenos Aires, Argentina

K. Imamoto
Tokyo University of Science, Tokyo, Japan

H. Beushausen and L. Fernandez Luco (eds.), *Performance-Based Specifications and Control of Concrete Durability*,
RILEM State-of-the-Art Reports 18, DOI 10.1007/978-94-017-7309-6_11

experience and knowledge of the contributing author. A globally accepted approach for concrete durability performance testing and service life prediction does not exist at this stage. Therefore, the individual approaches presented in the following sections should be considered separately. However, in combination, these approaches present the state of the art of performance evaluation of concrete structures for durability.

The information given in this chapter presents the background data for the summarizing discussions and combined analysis of test results that were provided in Chap. 10. Supplementary information on panel manufacture and general concrete properties is also provided.

11.2 Panel Manufacture and Concrete Properties

For the Venlo Application Tests, four concrete types, named A, B, C and D were prepared and two retaining wall type panels were cast with each concrete. The difference between the two panels made from each concrete was the curing procedure. One of the panels from each concrete type was covered with plastic sheets for 1 week after removal from the formwork, while the other remained exposed to outdoor conditions. The concrete was made with two different cement types and different w/c ratios, in order to provide members with a range of durability properties. A summary of test specimens, details on the concrete mix designs, and selected fresh concrete properties are presented in Tables 11.1, 11.2 and 11.3, respectively.

The panels were steel reinforced with Ø8 mm vertical steel bars with 150 mm spacing and with Ø6 mm horizontal bars with 300 mm spacing. The nominal cover depths on the inner and outer faces of the panels were 50 and 30 mm, respectively. Figure 11.1 shows a photograph of a test panel.

The panels were cast indoors and demoulded after roughly 24 h, subsequent to which they were exposed to an outside environment. A few days before commencement of the AT, the upper halves of the panels were protected from rain and

Table 11.1 Test panel mix characteristics, curing conditions, and casting dates

Mix	Cement type	w/c	Panel no.	Curing	Casting date (2012)
A	CEM I 52.5 N	0.45	1	<1 day in the mould	March 28th
			2	+7 days plastic cover	March 28th
B	CEM I 52.5 N	0.54	3	<1 day in the mould	March 29th
			4	+7 days plastic cover	March 29th
C	CEM II/B-V 42.5 N	0.41	5	<1 day in the mould	March 30th
			6	+7 days plastic cover	March 30th
D	CEM II/B-V 42.5 N	0.59	7	<1 day in the mould	April 2nd
			8	+7 days plastic cover	April 2nd

Table 11.2 Concrete mix design for the test panels (in kg/m^3)

Constituent	Mix A	Mix B	Mix C	Mix D
CEM I 52,5 N	333.4	347.6	–	–
CEM II/B-V 42,5 N	–	–	339.1	338.2
Water	146.4	186.7	134.3	198.2
Superplasticizer (Glenium Sky 640)	4.33	0	4.41	0
Aggregate (coarse + fine)	1920	1829	1928	1782
Total	2404	2363	2406	2318
w/c ratio	0.45	0.54	0.41	0.59

Table 11.3 Selected fresh concrete properties

Property	Mix A	Mix B	Mix C	Mix D
Specific weight (kg/m^3)	2420	2380	2410	2370
Temperature (°C)	22	18	18	19
Slump flow (mm)	480	510	550	560
Air content (%)	1.9	1.3	1.7	1.3

Fig. 11.1 Test panel during experimental investigations

wind with a plastic sheet that allowed for water evaporation. This treatment was meant to provide different moisture conditions in the upper (sheltered from rain) and lower halves of the panels. However, due to the generally high relative humidity of the environment, different moisture conditions across the height of the panels were not achieved.

Table 11.4 Compressive strength development

Mix A		Mix B		Mix C		Mix D	
Age (d)	MPa	Age (d)	MPa	Age (d)	MPa	Age (d)	MPa
2	49.0	4	28.3	4	45.2	2	11.1
5	60.1	5	31.1	5	49.0	3	15.5
7	63.8	7	36.6	6	50.3	8	22.1
14	70.4	14	39.8	14	57.1	14	28.9
28	78.0[a]	28	46.0[a]	28	63.0[a]	28	34.0[a]
AT[b]	70.0**	AT[b]	39.3	AT[b]	55.8	AT[b]	24.9

**Estimated value at the time of the application test
[a]Estimated from strength development during the first 14 days
[b]Strength at the time of the application testing in April was estimated from maturity calculations

The climatic conditions in Venlo between the day of casting of the first panel and application testing can be summarized as follows:

- Average temperature = 7 °C
- Average relative humidity = 81 %
- On average, it rained every 2nd day, with an average precipitation of 2.3 mm per rainy day
- Average wind speed 3.6 m/s

For compressive strength assessment, standard 150 mm cube specimens and cylinders (100 mm/200 mm) were cast and cured under standard conditions. Strength development was tested at various ages during the first 14 days after casing, as shown in Table 11.4. The 28-day strength was estimated based on the strength development during the first 14 days. The compressive strength at the time of the AT in April was based on maturity calculations under consideration of the temperature development inside the panels and air temperature.

11.3 Air Permeability (OPI) and Chloride Penetration (CCI)

This section was provided by Hans Beushausen.

11.3.1 Introduction and Aims of the Testing

The South African Durability Index approach is based on measuring concrete penetrability (oxygen permeability and chloride conductivity) and linking the obtained test values ("Durability Indexes") to service life prediction models. Detailed information on test methods and service life prediction models is presented in Chaps. 4 and 8, respectively. The aim of participating in the Venlo Application

Tests and testing the 8 concrete test panels for Durability Indexes was to provide conformity assessment and make service life predictions for the panels for different environmental exposure conditions (severe carbonation and chloride environments).

The testing of the Venlo test panels for Durability Indexes (Oxygen Permeability Index (OPI) and Chloride Conductivity Index (CCI)) at the University of Cape Town was carried out on core samples removed from the test panels. Two sets of samples were received, following the different dates for in situ testing (April and July). In both instances, the cores were taken approximately 2 weeks after the in situ testing was completed. The procedure for sample preparation was as follows for both test ages:

- Drilling of 2 cores (Ø68 mm) from each test panel (throughout the whole depth of the panels)
- Preparation of 2 test samples per core:
  - From each core end, cut off 5 mm and discard
  - From each resulting core end, cut off a disk with a thickness of 30 mm, which represents the test specimen
  - Dry the specimen in an oven at 50 °C for at least 7 days, to prevent further hydration (and hence ageing) of the concrete
  - Wrap the specimens in plastic (to prevent moisture intake) and courier them to Cape Town
  - In Cape Town, oven-dry the samples for at least another 7 days (at 50 °C) prior to testing

In the analysis of test results, no difference was made between the two different sides of the panel, assuming that the concrete quality was the same on both the 30 and 50 mm cover sides. The samples were tested for OPI and CCI according to the South African Durability Index test procedures (compare Sects. 4.2.2.7 and 4.6.2.3). The samples were first tested for OPI and subsequently the same samples were tested for CCI.

The test results for OPI and CCI were analysed to assess the concrete's potential service life for carbonating and marine environments, respectively, using the South African (SA) prediction models for concrete durability (compare Sects. 8.3 and 8.8).

11.3.2 Oxygen Permeability

11.3.2.1 Summary of Results and General Discussion

Graphical summaries of all test results (mean values) for the coefficient of oxygen permeability (k_{OPI}) and Oxygen Permeability Index (OPI) are presented in Figs. 11.2 and 11.3, respectively. Note that the OPI value represents the negative logarithm of the permeability coefficient. Therefore, a higher OPI represents a less permeable concrete.

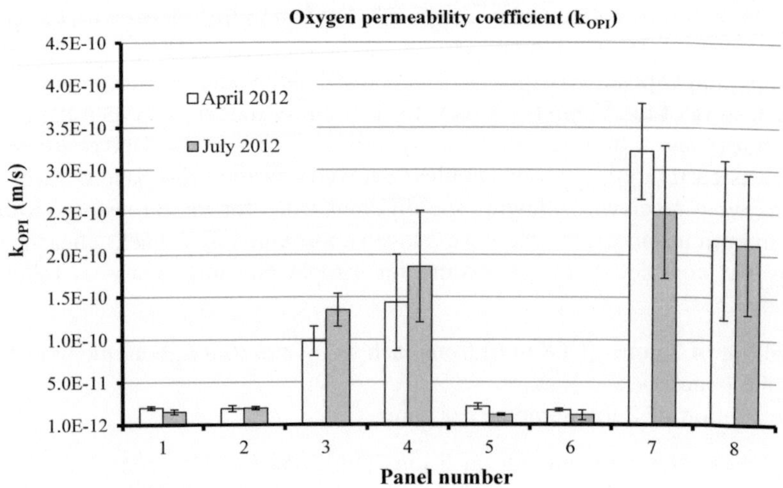

Fig. 11.2 Summary of test results: coefficient of oxygen permeability (k_{OPI}) (error bars indicate plus/minus one standard deviation)

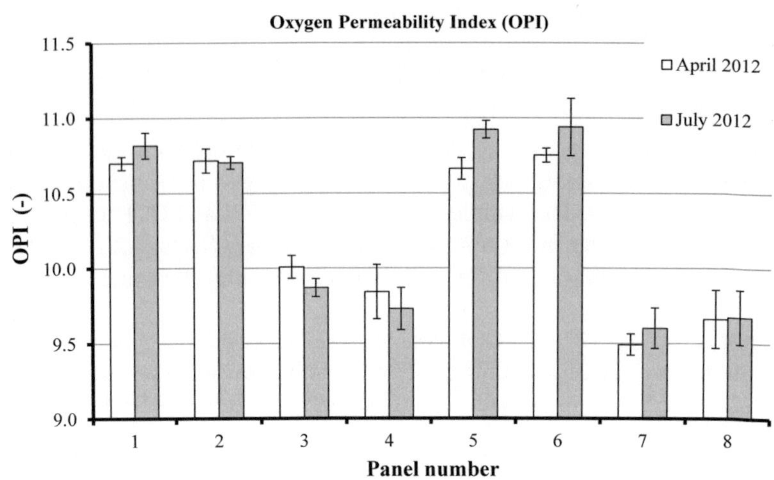

Fig. 11.3 Summary of test results: Oxygen Permeability Index (OPI) (error bars indicate plus/minus one standard deviation)

The following general observations are made from the experimental results:

- The curing conditions (i.e. 1, 3 "no curing" or 2, 4 "7 days wet") were found to have no influence on test values. This results from the relatively wet and cold conditions during early specimen exposure, which left non-protected panels in a favourable environment.

- The superior performance of panels made with lower water/cement ratios (1, 2, 5, 6) was clearly confirmed for all samples.
- The results obtained from the second round of testing generally indicated a better concrete quality (i.e. lower permeability), as expected. However, for the CEM I w/c = 0.59 samples (N. 7, 8) the difference is not significant.
- For concretes with similar w/c ratios, test results for samples made with CEM I and CEM II/B-V were generally of similar order of magnitude.
- For all test panels, the relatively high OPI values of 9.6 and above indicate a dense concrete matrix of low permeability.

11.3.2.2 Conformity Assessment (Carbonation Exposure)

The conformity assessment is based on standard specifications that are currently used in South Africa for large infrastructure projects by the South African National Roads Agency (SANRAL). These specifications are based on the Durability-Index-based service life prediction models used in South Africa and make use of the environmental classes defined in EN-206 (XC classes for carbonation exposure). In the specifications, no difference in limiting OPI values is made for various binder types. The minimum recommended cover in SANRAL specifications for civil engineering structures is 40 mm (absolute value). The minimum OPI value specified by SANRAL for a cover depth of 40 mm is 9.6, for environmental class XC4 and a service life of 50 years. All test panels (Panels 1–8) therefore conform to the durability specifications and would be considered of acceptable quality (for cover depths of 40 mm and more).

11.3.2.3 Service Life Assessment and Carbonation Prediction

The South African service life prediction models link OPI values of in situ structures to mean carbonation depth development for typical South African environmental conditions such as dry inland (60 % RH), coastal (80 % RH), and partly wet (90 % RH). The model accounts for different cement (binder) types and allows the prediction of mean carbonation depth development based on OPI values. Table 11.5 presents the predicted service life duration, based on the time needed for mean carbonation to reach the reinforcing steel (for cover depths of 30 and 50 mm). In addition, Table 11.6 shows the prediction of carbonation depths for two different ages (50 and 100 years).

As shown in Table 11.5, most test panels are predicted to have a service life in excess of 100 years, for both 30 and 50 mm cover depths, independent of exposure conditions. The only exceptions are Panels 7 and 8 at 30 mm cover, as indicated.

Table 11.5 Service life assessment based on OPI values (service life given in years)

Panel	OPI	30 mm cover			50 mm cover		
		60 % RH	80 % RH	90 % RH	60 % RH	80 % RH	90 % RH
1	10.8	>100	>100	>100	>100	>100	>100
2	10.7	>100	>100	>100	>100	>100	>100
3	9.9	>100	>100	>100	>100	>100	>100
4	9.7	>100	>100	>100	>100	>100	>100
5	10.9	>100	>100	>100	>100	>100	>100
6	10.9	>100	>100	>100	>100	>100	>100
7	9.6	**40**	**87**	>100	>100	>100	>100
8	9.7	**53**	>100	>100	>100	>100	>100

Table 11.6 Carbonation depth prediction based on OPI values (for 50 and 100 years), in mm

Panel	OPI	50 years			100 years		
		60 % RH	80 % RH	90 % RH	60 % RH	80 % RH	90 % RH
1	10.8	n.a[a]	n.a[a]	n.a[a]	n.a[a]	n.a[a]	n.a[a]
2	10.7	n.a[a]	n.a[a]	n.a[a]	n.a[a]	n.a[a]	n.a[a]
3	9.9	17	12	6	21	15	8
4	9.7	22	16	8	28	21	10
5	10.9	n.a[a]	n.a[a]	n.a[a]	n.a[a]	n.a[a]	n.a[a]
6	10.9	n.a[a]	n.a[a]	n.a[a]	n.a[a]	n.a[a]	n.a[a]
7	9.6	33	24	12	44	32	16
8	9.7	29	21	11	38	28	14

[a]n.a = for very impermeable concretes (i.e. OPI values of 10.5 and above), the model is unable to predict any carbonation

11.3.3 *Chloride Conductivity Index*

11.3.3.1 Summary of Results and General Discussion

A graphical summary of all test results for the Chloride Conductivity Index (CCI) (mean values and STDV) is presented in Fig. 11.4. The following general observations are made from the experimental results:

- As observed also with the oxygen permeability results, the curing conditions (i.e. 1, 3 "no curing" or 2, 4 "7 days wet") were found to have no influence on test values.
- The superior performance of panels made with lower water/cement ratios (1, 2, 5, 6) was clearly confirmed for all samples.
- The results obtained from the second round of testing generally indicated better concrete quality (i.e. lower conductivity), as expected. This is particularly

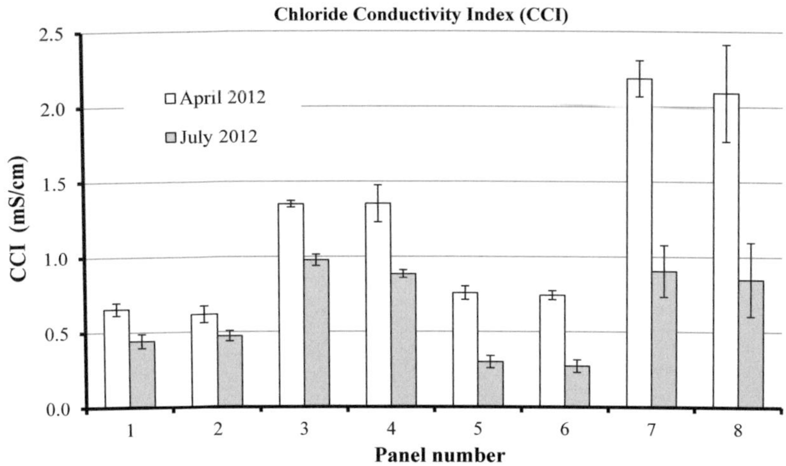

Fig. 11.4 Summary of test results for Chloride Conductivity Index (CCI) (error bars indicate plus/minus one standard deviation)

noticeable for the CEM II/B-V concretes which show a significant decrease in chloride conductivity with time, confirming the notion that FA concretes mature much slower than CEM I concretes. The panels cast in April were subjected to very low temperatures during the first weeks, which explains the originally high CCI values.

- The CEM I concretes generally performed better than expected from South African experience, while the CEM II/B-V concretes performed slightly worse than expected. This could relate to the low temperature during casting and storage and probably differences in cement properties.

11.3.3.2 Conformity Assessment (Chloride Exposure)

The conformity assessment is based on specifications that are currently used in South Africa for large infrastructure projects, as developed by the South African National Roads Agency (SANRAL). These specifications are based on the Durability-Index-based service life prediction models used in South Africa and make use of the environmental classes defined in EN-206 (XS classes for chloride exposure from sea water). In the specifications, different limiting values are given for concretes made with different binder types, which accounts for influences such as pore chemistry and chloride binding mechanisms. The minimum recommended cover in SANRAL specifications for civil engineering structures is 40 mm (absolute value) and a service life of 50 years. The assessment shown in Table 11.7 is therefore based on cover depths of 40 and 50 mm.

Table 11.7 Conformity assessment for chloride conductivity values, based on SANRAL specifications for Civil Engineering structures (a "y" indicates that the panels conform to the specifications, an "n" indicates otherwise)

Panel	CC	40 mm cover			50 mm cover		
		XS1	XS2	XS3	XS1	XS2	XS3
1	0.44	**n** (plain CEM I mixes are not considered (i.e. not allowed) in the specifications)					
2	0.47						
3	0.98						
4	0.89						
5	0.30	y	y	y	y	y	y
6	0.27	y	y	y	y	y	y
7	0.91	y	y	**n**	y	y	y
8	0.85	y	y	**n**	y	y	y

The SANRAL specifications do not allow the use of CEM I concretes in the marine environment and consequently, all test panels made with CEM I would not pass the conformity criteria. Most of the CEM II/B-V concretes pass the criteria (for 40 and 50 mm cover depths), except for Panels 7 and 8 for 40 mm cover in environmental class XS3.

11.3.3.3 Service Life Assessment (Chloride Exposure)

The South African service life prediction models link Chloride Conductivity Index values of in situ structures (typically measured at 28–35 days of age) to chloride ingress for typical South African marine conditions such as "extreme", "very severe", and "severe". These classes relate to the following conditions:

- Extreme: Structure exposed directly to sea water with heavy wave action and/or abrasion
- Very severe: Structure exposed directly to sea water under sheltered conditions with little wave action
- Severe: Structure in a sheltered location within 1 km of the shore

The model accounts for different binder types and is based on the correlation between the CCI and the diffusion coefficient of the concrete, accounting for effects such as chloride binding.

Table 11.8 presents the predicted service life duration, based on the time needed for chloride concentrations to exceed a chloride threshold value of 0.4 % (by mass of binder) at the level of the reinforcement (at covers of 30 and 50 mm).

Table 11.8 Service life assessment based on CCI values (in years)

Panel	CC	30 mm cover			50 mm cover		
		Severe	Very severe	Extreme	Severe	Very severe	Extreme
1	0.44	45	21	21	>100	91	91
2	0.47	43	20	20	>100	87	87
3	0.98	23	9	6	95	40	28
4	0.89	26	10	8	>100	45	34
5	0.30	>100	46	46	>100	>100	>100
6	0.27	>100	51	51	>100	>100	>100
7	0.91	78	7	4	>100	>100	>100
8	0.85	90	9	5	>100	>100	>100

11.3.4 Conclusions

The test panels were measured for Durability Indexes and evaluated with respect to common conformity criteria applied for large infrastructure projects in South Africa. For cover depths of 40 mm or more, all test panels were found to meet the specified limiting permeability values for carbonation exposure classes XC.

Due to the generally poor chloride ingress resistance of concrete made with CEM I, such concrete is not allowed to be used for marine environments in South Africa. Therefore, with respect to chloride exposure classes (XS), test panels made with plain CEM I cement (i.e. Panels 1–4) did not conform to common South African specifications. In addition, Test Panels 7 and 8 (CEM II, w/c = 0.59) were found to not conform to standard limiting values for exposure class XS3, at 40 mm cover.

The data were further used to make service life predictions for all test panels under consideration of standard South African environmental conditions and different cover depths. The service life assessments confirmed the relatively poor carbonation resistance and superior chloride ingress resistance of CEM II concretes, compared to CEM I concretes, and highlighted the generally high importance of cover depths.

The measured Durability Index values were able to clearly distinguish between the durability properties of the various test panels with respect to cement type and water/cement ratio. The two different curing conditions that the panels had been subjected to after casting were found to not have had a significant influence on penetrability properties of the concrete. This was ascribed to the fact that even the non-protected panels were exposed to favourable curing conditions (i.e. high relative humidity).

11.4 Air-Permeability ("Torrent Method") and Cover Depth

This section was provided by Roberto Torrent, Frank Jacobs and Marijana Serdar.

11.4.1 Aims of Testing

The tests were conducted with the aim of assessing the potential service life of the panels assumed to be exposed to extreme carbonation (XC4) and chlorides (XD3) environments.

The service life assessment was done on the basis of measuring on site two fundamental parameters of the cover concrete:

- The coefficient of air-permeability *kT*, measured according to the "Torrent Method"
- The cover depth c measured with an electromagnetic covermeter.

Conformity with the requirements of Swiss Standards SIA 262:2003 [1] and 262/1:2013 Annex E [2], assuming a service life of 50 years, was checked for all panel faces tested.

In addition, the expected service life was computed (for XD3 exposure class), applying the Ref-Exp model described in Sect. 8.2, in particular checking the likelihood that any of the panels would achieve a service life of 100 years under that exposure class.

11.4.2 Testing Program

Two Rounds of measurements were carried out within the framework of RILEM TC 230-PSC's Application Test, as summarized in Table 11.9. Values in italics indicate lack of compliance with prescriptions of [2].

11.4.3 Testing Methods

The following testing methods and instruments were applied on the panels:

- Coefficient of Air-permeability, "Torrent Method" (*kT*)
  MS, FJ and RT2 worked each with their own units of the PermeaTORR instrument. The instruments were conditioned and calibrated before starting the measurements of each day; in some cases they were recalibrated during the day. When the test duration exceeded 6 min, the facility of the PermeaTORR to interrupt the test and record the value at 6 min (kT6) was applied.
- Moisture Content, Electrical Impedance Method (*M%*)
  FJ used an analogical Concrete Encounter (CME) instrument and MS, RT1 and RT2 used the same digital CMEXpert instrument.
- Rebound Hammer (*R*)
  A type N hammer, calibrated a few weeks prior to the measurements, was used.

Table 11.9 Measurements conducted in RILEM TC 230-PSC application test

			Conditions			
			Age	Concrete temperature °C	Concrete moisture %	ΔP^{a}_{cal}mbar
Date	Participant	Instruments	Days			
Requirements Swiss Standard SIA262/1-E [2]			28–90	≥10 °C	≤5.5	≤5.0
	M. Sejdar (MS)	*PermeaTORR*		*1.3–6.1*	4.9–5.6	*8.3, 6.1*
		Resipod				
	R. Torrent (RT1)	*PermeaTORR*		*1.0–6.3*	5.2–5.7	1.0, 1.6, 0.9
		CMEXpert				
		Profoscope				
2nd Round: 9 July 2012	F. Jacobs (FJ)	*PermeaTORR*	*101–108*	–	4.4–5.5	–
		CME				
		Schmidt Hammer				
	R. Torrent (RT2)	*PermeaTORR*		17–21	4.4–5.3	2.6
		CMEXpert				
		Profoscope				
		Resipod				

[a]Maximum pressure increase in the *PermeaTORR*, when applied on the impermeable calibration plate

- Cover Depth, Electromagnetic Method (c)
  The cover depth was measured with a Profoscope instrument. The instrument was set for Ø = 8 mm for vertical bars and Ø = 6 mm for horizontal bars.
- Electrical Resistivity (ρ)
  The Wenner electrical resistivity was measured with a Resipod instrument. The probes' reservoirs were filled with water before starting the readings of each panel, the surface of which was not pre-treated beforehand. The test was applied after all the previous measurements had been completed.

11.4.4 Sampling

The measurements were conducted primarily on the assigned face (30 mm nominal cover) of the 8 panels, although several measurements were also conducted on the opposite face (50 mm nominal cover) of some panels.

At least 6 testing areas were selected on each panel, at least 50 mm from the edges and at least 200 mm from each other, in line with the prescriptions of [2]. The selection of the areas was made primarily at random, although in some cases, e.g. to assess the reproducibility of the instruments, some measurements were deliberately made on the same points used by previous operators.

Figure 11.5 presents a sketch of a typical arrangement of measurements.

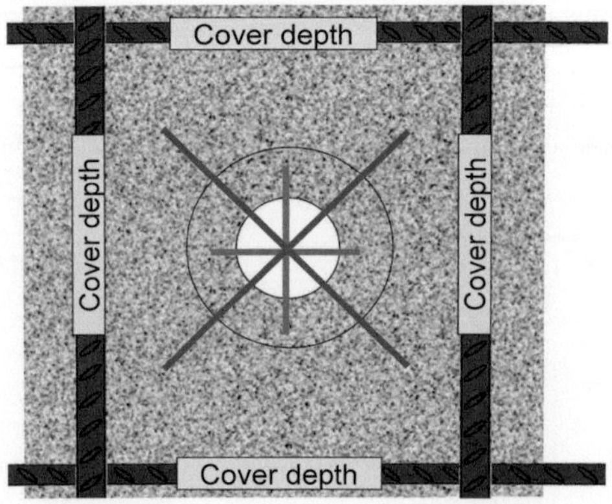

Fig. 11.5 Sketch of the measurements conducted by RT. The *circles* represents PermeaTORR's inner and outer cell location, the *vertical* and *horizontal red lines* the orientation of the CMEXpert and the inclined *blue lines* that of the Resipod. Finally, the approximate places where the cover depth was measured with the Profoscope are also indicated

- Coefficient of Air-permeability (kT)
  One measurement of kT was performed at each test area.
- Moisture Content ($M\%$)
  Two readings of the moisture content were performed at 90° and the average at each location recorded.
- Rebound Hammer (R)
  FJ conducted 25 readings of the Rebound, randomly distributed over the entire surface of the panel.
- Cover Depth (c)
  RT conducted four readings of the cover depth, two for vertical and two for horizontal bars in the vicinity of the area were kT had been measured. For each area the minimum values of vertical and horizontal bars were reported.
- Electrical Resistivity (ρ)
  MS and RT2 conducted 2 readings of the Wenner electrical resistivity, at angles of ±45° to the vertical line in the vicinity of the area were kT had been measured. The average of the two was recorded.

11.4.5 Age and Environmental Conditions

The measurements of the 1st Round (April 2012) were conducted under conditions that were well outside those required by [2], shown in the 2nd row of Table 11.9. In particular, the age of the concrete was too short, its temperature too low and its moisture too high. In the particular case of MS, the calibration pressures ΔP_{cal} were too high, possibly due to the low ambient temperature.

The tests by MS and RT1 were conducted at random locations within the upper half of the panels (bottom part as cast), that had been reserved for "dry" tests and covered by plastic sheets before the tests. The tests were conducted outdoors, under the current ambient conditions.

Since the conditions in April were clearly outside those required by SIA 262/1 Standard, FJ preferred not to participate and reserved himself to conduct the tests under conditions corresponding to the requirements in the standard (in particular when the concretes were mature and dry enough). RT2, due to the same reasons, decided to repeat the tests simultaneously with FJ. This second round of tests took place on 9 July 2012, when the panels' age was slightly above the maximum recommended age. On the other hand, the temperature and moisture were well within the ranges recommended in [2]. It is worth mentioning that, due to bad weather forecast, the organizers decided to move the panels inside a warehouse on July 6, where the tests were finally conducted.

11.4.6 Test Results

11.4.6.1 Results Obtained

Tables 11.10 and 11.11 summarise the results of the measurements obtained by MS and RT1 in the 1st Round of tests, showing the main statistical parameters.

Table 11.10 Test results obtained in the 1st Round by MS

Panel no.- cover (mm)	Number of test areas	Air-permeability		Moisture content		Electrical resistivity	
		kT_{gm}	s_{LOG}	$M\%_m$	$M\%_s$	ρ_m	ρ_s
		10^{-16} m^2	–	% mass	% mass	kΩ. cm	kΩ. cm
1–30	6	0.0023	0.42	5.0	0.18	6.8	1.3
2–30	6	0.0013	0.17	5.0	0.24	6.5	1.3
3–30	6	0.019	0.23	5.3	0.15	3.5	1.2
4–30	6	0.012	0.27	5.2	0.26	2.9	1.0
5–30	6	0.0080	0.14	4.9	0.20	3.3	0.9
6–30	6	0.019	0.23	5.3	0.15	3.4	1.2
7–30	6	0.53	0.32	5.6	0.21	2.5	0.8
8–30	6	0.11	0.17	5.6	0.29	3.4	1.6

kT_{gm} Coefficient of Air-permeability—Geometric Mean
s_{LOG} Coefficient of Air-permeability—Standard Deviation of logarithms$_{10}$
$M\%_m$ Moisture content (% of concrete mass)—Average
$M\%_s$ Moisture content (% of concrete mass)—Standard deviation
ρ_m Electrical Resistivity—Average
ρ_s Electrical Resistivity—Standard deviation

Tables 11.12 and 11.13 summarise the results of the measurements obtained by FJ and RT2, in the 2nd Round of tests, showing the main statistical parameters.

Table 11.11 Test results obtained in the 1st Round by RT1

Panel no.- cover (mm)	Number of test areas	Air-permeability		Moisture content		Minimum cover	
		kT_{gm}	s_{LOG}	$M\%_m$	$M\%_s$	c_8	c_6
		10^{-16} m^2	–	% mass	% mass	mm	mm
1–30	6	0.0027	0.28	5.3	0.08	31.7	35.8
2–30	6	0.0043	0.23	5.4	0.08	33.8	28.7
3–30	6	0.020	0.18	5.5	0.19	32.8	34.8
4–30	6	0.024	0.19	5.7	0.27	30.7	34.2
5–30	6	0.0077	0.12	5.2	0.14	32.8	36.5
6–30	6	0.0037	0.24	5.2	0.09	31.2	35.7
7–30	6	0.58	0.31	5.3	0.25	34.0	38.7
8–30	6	0.17	0.27	5.7	0.40	29.2	33.8
1–50	6	0.0080	0.21	5.3	0.08	52.3	53.8
7–50	6	0.60	0.25	5.3	0.25	48.5	50.8

c_8 Minimum cover depth of Ø = 8 mm vertical bars—Average
c_6 Minimum cover depth of Ø = 6 mm horizontal bars—Average

Table 11.12 Test results obtained in the 2nd Round by FJ

Panel no.- cover (mm)	Number of test areas	Air-permeability		Moisture content[a]		Rebound hammer	
		kT_{gm}	s_{LOG}	$M\%_m$	$M\%_s$	R_m	R_s
		10^{-16} m^2	–	% mass	% mass	–	–
1–30	6	0.019	0.22	5.1	0.1	55	4
2–30	6	0.033	0.37	5.1	0.3	55	2
3–30	7	0.61	0.86	4.9	0.2	48	3
4–30	6	0.73	0.63	5.0	0.3	46	3
5–30	6	0.035	0.19	4.7	0.2	56	2
6–30	6	0.017	0.17	5.0	0.1	56	2
7–30	7	4.22	0.76	4.4	0.2	44	1
8–30	6	2.99	0.51	4.4	0.2	44	1
2–50	6	0.019	0.20	5.5	0.0	58	2
3–50	6	0.061	0.16	5.5	0.1	50	3
4–50	6	0.098	0.29	5.5	0.0	51	4
6–50	6	0.010	0.16	5.5	0.1	55	2
8–50	6	0.21	0.57	4.9	0.3	45	2

Rm Rebound Hammer—Median of 25 individual readings
Rs Rebound Hammer—Standard Deviation of 25 individual readings
[a]The values were reduced by 0.5 from the readings, following a calibration on reference plate

Table 11.13 Test results obtained in the 2nd Round by RT2

Panel no.-cover (mm)	Number of test areas	Air-permeability		Moisture content		Electrical resistivity		Minimum cover	
		kT_{gm}	s_{LOG}	$M\%_m$	$M\%_s$	ρ_m	ρ_s	c_8	c_6
		10^{-16} m²	–	% mass	% mass	kΩ.cm	kΩ.cm	mm	mm
1–30	6	0.020	0.29	5.0	0.2	10.8	1.6	30.2	35.3
2–30	7	0.024	0.39	4.9	0.1	10.6	1.3	33.3	28.3
3–30	7	0.35	0.66	4.8	0.3	4.9	0.7	32.9	34.9
4–30	7	0.45	0.75	5.0	0.5	4.1	0.7	30.3	33.7
5–30	6	0.025	0.26	4.6	0.2	27.5	3.3	32.8	37.0
6–30	6	0.016	0.21	4.9	0.2	25.0	2.8	31.5	35.3
7–30	6	1.76	0.52	4.4	0.2	27.7	11.9	34.3	37.3
8–30	6	0.86	0.58	4.4	0.3	16.0	3.9	29.0	33.0
1–50	6	0.020	0.08	5.3	0.2	13.7	0.5	51.0	52.5
5–50	6	0.021	0.36	4.9	0.2	33.7	3.0	50.3	51.8
7–50	6	0.34	0.51	4.9	0.2	18.3	2.7	49.0	51.3

11.4.6.2 Analysis of the Results

Analysis of the Results of 1st Round (Age: 14–21 Days)

Figure 11.6 presents (in a logarithmic scale) the values of LOG(kT_{gm}) ± *s*LOG obtained in the 1st Round by MS and RT1. The results correspond to Panels 1 to 8, 30 mm cover.

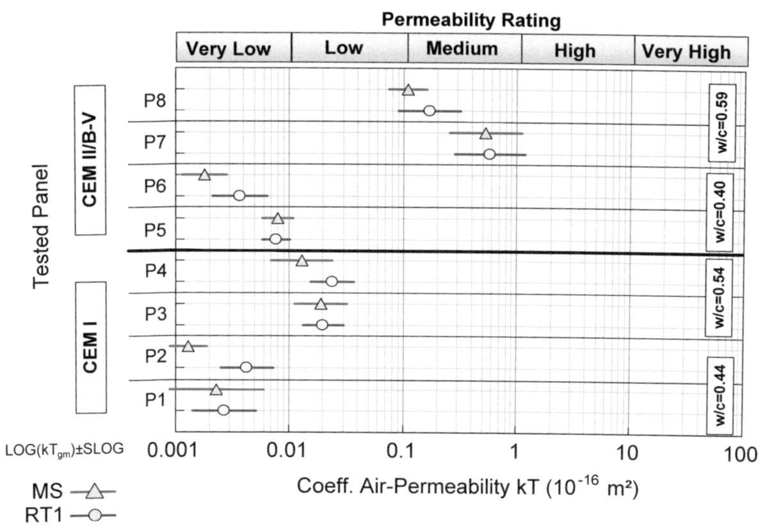

Fig. 11.6 Summary of *kT* results obtained by MS and RT1 in the 1st Round

The results reported by MS and RT1, obtained exactly on the same Test Areas of each panel, are quite coherent. A t-test analysis shows that out of the 8 panels' results, only those of panels P2 and P6 reject the H_0 hypothesis of coming from the same population (5 % significance level), despite the problems of MS to calibrate her instrument. All the results obtained on the panels with w/c = 0.40 and 0.44 fall within the "Very Low" permeability class (the permeability classes, which is rather arbitrary, apply strictly to mature concretes, typically 28-90 days old). Those obtained on the panel made with CEM I and w/c = 0.54 fall within the "Low" permeability class and those obtained on the panel with CEM II/B-V and w/c = 0.59 in the "Medium" permeability class.

Analysis of the Results of 2nd Round (Age: 101–108 Days)

Figure 11.7 presents (in a logarithmic scale) the values of LOG(kT_{gm}) ± sLOG obtained in the 2nd Round by FJ and RT2. The results correspond to Panels 1 to 8, 30 mm cover.

Again, the results reported by FJ and RT, not obtained exactly on the same Test Areas of each Panel, are highly coherent. In none of the panels the null hypothesis H_0 that the results from FJ and RT2 come from the same population can be rejected (5 % Significance level).

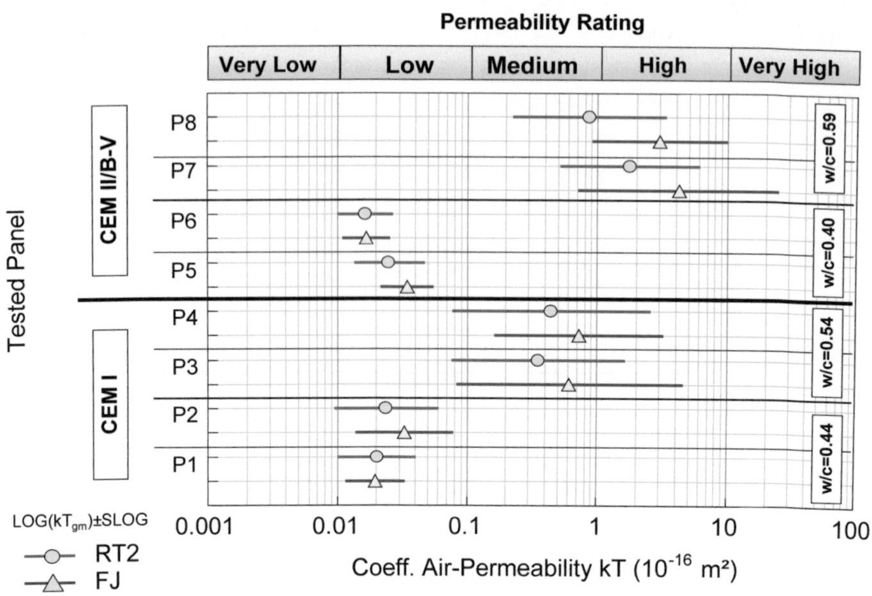

Fig. 11.7 Summary of *kT* results obtained by FJ and RT2 in the 2nd Round

Comparing the results of Fig. 11.7 with those of the 1st Round (Fig. 11.6), it appears as if the values were shifted one permeability class higher by the passage of time.

This is better seen in Fig. 11.8, presenting the kT_{gm} values obtained in the 2nd Round by FJ compared with those obtained by RT2 (abscise). Just for comparison, the results obtained in the 1st Round by MS and RT1 are also plotted. The results correspond to Panels 1 to 8, 30 mm cover.

A very good agreement was found between the results of FJ and RT2, laying very close to the equality line. On the other hand, the results of MS and RT1 lay clearly below the line, which is at first glance paradoxical because the permeability of the still immature concrete should have been higher. This can be attributed to the relatively high moisture content of the young concrete, aggravated by its low temperature during the measurements. However, since the points tend to follow a line rather "parallel" to those obtained during the 2nd Round, the door remains open to conduct *kT* tests of young concrete for early detection of areas of low quality.

The results of FJ tend to be slightly higher than those of RT2; this could be due to a bias between instruments or to the fact that FJ explored the whole height of the panels whilst RT2 restricted himself mostly to the upper half as standing.

Figure 11.9 presents results obtained by FJ and RT2 exactly on the same spots of panels 1 to 4 (30 mm cover), with a delay of about 1–2 h. They show that the reproducibility of the measurements by two operators using two different *PermeaTORR* units is excellent; notice that the values span more than three orders of magnitude of *kT* (from "Very Low" to "Very High" Classes). Therefore the higher *kT* values of FJ, shown in Fig. 11.8, have to be attributed to his testing the bottom of the panels (upper as cast) that are usually "weaker" than the top.

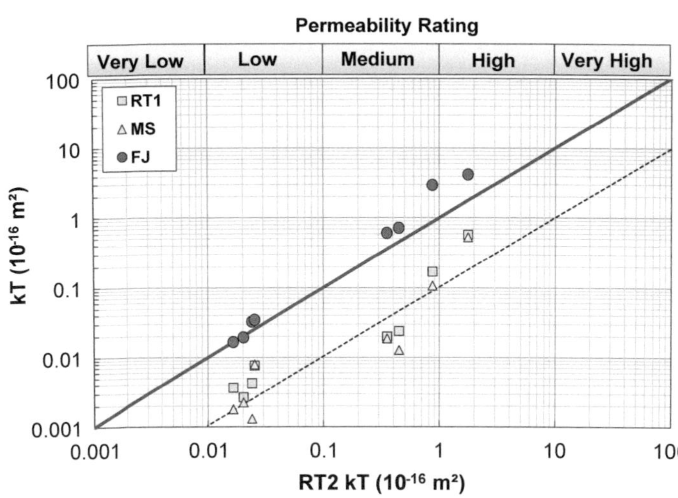

Fig. 11.8 Values of kT_{gm} obtained by FJ, MS and RT1 compared with those obtained by RT2, on the same panels (1–8, 30 mm cover side)

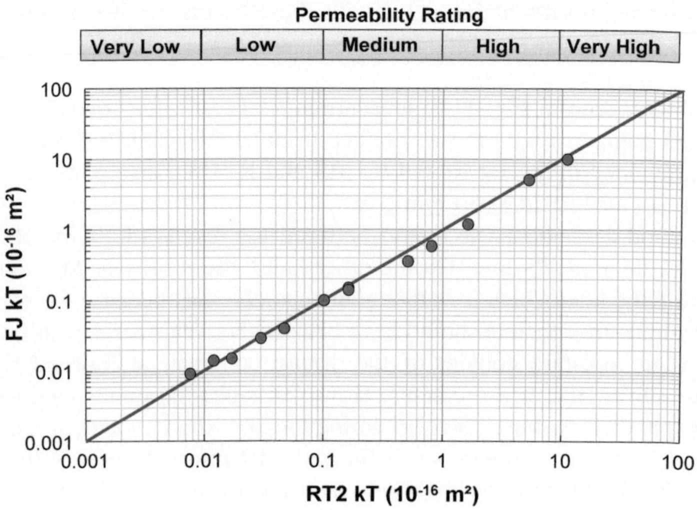

Fig. 11.9 Individual *kT* values obtained by FJ and RT2 on the same spots (Panels 1–4, 30 mm cover side)

Figure 11.10 shows the kT results obtained in the 2nd Round, by the same operator (FJ or RT2), on opposite faces of each of the 8 panels. Almost invariably, the 50 mm face yielded lower *kT* values than the 30 mm face. This can be attributed to the fact that the 50 mm face was openly exposed to the environment, whilst the 30 mm face was protected by a nearby building. The extra curing provided by rain (frequent in Venlo) could account for the lower permeability of the 50 mm face, although the different casting conditions may have also had an influence (e.g. the larger space to consolidate the concrete with 50 mm cover depth).

RT2 measured the cover depth on some areas where he had conducted the same measurements in April (RT1).

The results obtained on both occasions were compared, both for the vertical (Ø8 mm) and the horizontal (Ø6 mm) bars, including the 30 mm cover face of the 8 panels and the 50 mm cover face of panels 1 and 7. The results of the comparison are presented in Table 11.14, where the Mean Quadratic Difference (MQD) and the

Fig. 11.10 Values of *kT* measured on opposite faces of each of the 8 panels (2nd Round)

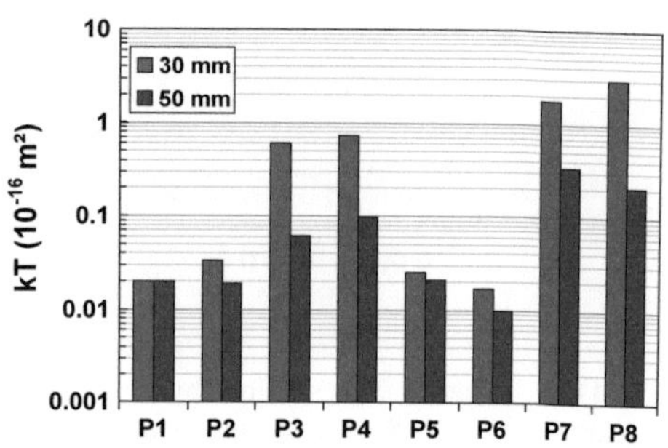

Table 11.14 Repeatability of cover depth in terms of difference between 1st and 2nd Round measurements

	Cover Ø8	Cover Ø6
MQD (mm)	1.8	2.2
Min (mm)	0	0
Max (mm)	5	8
N	42	42

minimum and maximum difference in cover depth of the successive measurements are indicated. The repeatability of the measurements is excellent. At the moment of the preparation of this report, no direct measurements of the cover depth had been conducted, so the accuracy of the estimates cannot be assessed.

11.4.7 Conformity with Swiss Standards

The Application Test called for an assessment of the potential service life of the panels, assuming them being subjected to Exposure Classes XC4 (carbonation-induced corrosion under alternate wet and dry conditions) and XD3 (chloride-induced corrosion due to de-icing salts).

The Swiss Standards specifications include Prescriptive and Performance Requirements (both for tests on Cast Specimens and for Site testing), summarized in Table 11.15 (for an expected service life of 50 years).

Within the frame of TC 230-PSC we will focus on the on-site requirements, i.e. cover depth and air-permeability *kT*. For the conformity evaluation, only the results obtained by FJ and RT2 in the 2nd Round, i.e. on sufficiently mature concrete, are considered.

11.4.7.1 Cover Depth Requirements

Compliance with XC4 Requirements: As shown in Tables 11.11 and 11.13, the faces with nominal cover 50 mm comply always and the faces with nominal cover 30 mm comply in all but 2 panels with the minimum cover depth of 40 mm − 10 mm = 30 mm (Table 11.15).

Table 11.15 Prescriptive and performance requirements established in Swiss Standards [1–3]

Exposure class	Prescriptive [3]		Performance on cast specimens [3]			Performance on site	
	w/c_{max}	C_{min}	$f'_{c\ min}$	K_N	$D_{Cl\ max}$	c_{min} [1]	kT_s [2]
	–	kg/m^3	MPa	mm/a$^{1/2}$	10^{-12} m^2/s	mm	10^{-16} m^2
XC4	0.50	300	30/37	10	–	**40 ± 10**	**2.0**
XD3	0.45	320	30/37	–	10	**55 ± 10**	**0.5**

Compliance with XD3 Requirements: As shown in Tables 11.11 and 11.13, the cover depths on faces with nominal cover 30 mm never comply and on faces with nominal cover 50 mm comply mostly with the cover depths minimum required of 55 mm − 10 mm = 45 mm (Table 11.15).

11.4.7.2 Air Permeability Requirements

The compliance criterion of [2] states that not more than 1 out of 6 test results applied on the same Lot (Panel in our case) may exceed the specified values kT_s, shown in Table 11.15 (there is a second chance if just two of the results exceed kT_s, not considered in this exercise). For more details, please refer to Sect. 8.2.3 of this Report.

Compliance with XC4 Requirements: The compliance criterion of [2] has been fulfilled by all Panels and sides, except Panels 3, 5 and 8 (30 mm side).

Compliance with XD3 Requirements: The compliance criterion of [2] has been fulfilled by Panels 1, 2, 5 and 6 (30 mm side) and Panels 1 to 6 (50 mm side).

Table 11.16 summarizes the conformity analysis according to [1] and [2].Panels 1 and 2 (CEM I; w/c = 0.44) and Panels 5 and 6 (CEM II/B-V; w/c = 0.40) have reached a concrete quality that allows them to comply with the more demanding air-permeability requirement for XD3 Exposure Class. Panel 4 does not comply with XD3 concrete requirements because of a too high permeability on the 30 mm side. Had the cover depth of these panels complied with the minimum of 45 mm, they would have been suitable for use in an XD3 environment with respect to cover thickness and air permeability.

On the other hand, the concrete quality reached by Panels 3 and 4 (CEM I; w/c = 0.54) and Panels 7 and 8 (CEM II/B-V; w/c = 0.59) is not suitable for XD3 Exposure Class in terms of *kT* and questionable also for XC4 Exposure Class.

Table 11.16 Conformity evaluation of the 8 Panels and Sides vis-à-vis Swiss Standards on site performance requirements [1, 2]

Class	Compliance for XC4 exposure				Compliance with XD3 exposure			
Side	c_{nom} = 30 mm		c_{nom} = 50 mm		c_{nom} = 30 mm		c_{nom} = 50 mm	
Requirement	c_{min}	kT_s	c_{min}	kT_s	c_{min}	kT_s	c_{min}	kT_s
Panel 1	Yes	Yes	Yes	Yes	No	Yes	Yes	Yes
Panel 2	No	Yes		Yes		Yes		Yes
Panel 3	Yes	No		Yes		No		Yes
Panel 4	Yes	Yes		Yes		No		Yes
Panel 5	Yes	Yes		Yes		Yes		Yes
Panel 6	Yes	Yes		Yes		Yes		Yes
Panel 7	Yes	No		Yes		No		No
Panel 8	No	No		Yes		No		No

Notice in Table 11.16 the better *kT* performance of the sides with $c_{nom} = 50$ mm, already discussed and shown in Fig. 11.10.

11.4.8 Compliance with 100 Years of Service Life

The following challenge was formulated to the participants of the Application Test:

> Can these panels (simulating a structure) on an individual basis be accepted by the client for a design service life of 100 years, exposure to XD3?
> Required: measurements (preferably non-invasive) at representative locations
> N.B. The evaluation should be performed for the entire panel based on the results obtained for individual points.

The prediction of the service life is based on the Ref-Exp method presented in Sect. 8.2.7. The estimated Service Life is computed at each location as:

$$T_i = 0.0086 \frac{c^2}{\sqrt[3]{kT}} \tag{11.1}$$

where T_i is the time for initiation of corrosion in years (assumed as the service life), *kT is the* coefficient of air-permeability in 10^{-16} m^2 (measured on site), and *c* is the cover depth of the steel in mm (minimum value measured).

The value of T_i was computed with Eq. (11.1) for each of the 6 or 7 areas investigated in each panel, based on the results of *kT* and the minimum value of *c* for the horizontal or vertical neighbouring bars, measured by RT2. The minimum, average and maximum values of T_i, calculated for each panel and side, are presented in Fig. 11.11.

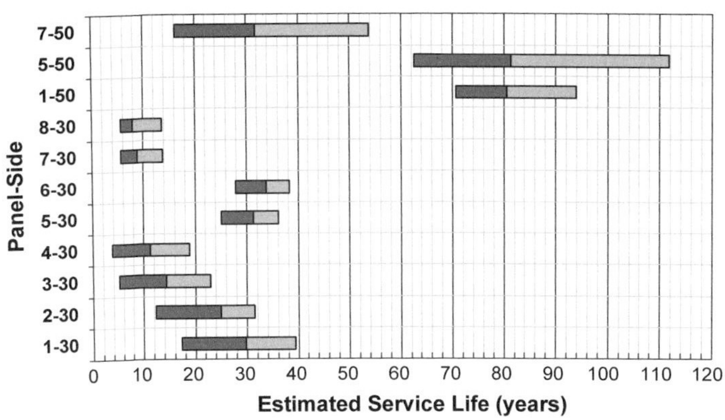

Fig. 11.11 Estimated service life of panel faces investigated by RT2, under XD3 exposure

Table 11.17 Estimated mean service life assuming cover depth = nominal value (30 or 50 mm)

	Estimated mean service life (years)							
Panel no.	1	2	3	4	5	6	7	8
c = 30 mm	29	24	9	9	24	30	5	5
c = 50 mm	79	81	55	47	78	100	31	36

Figure 11.11 shows that the average Service Lives are widely different, ranging from a minimum of 8–9 years for Panels 7 and 8 (CEM II/B-V; w/c = 0.59) −30 mm cover to a maximum of 81 years for Panels 1 (CEM I; w/c = 0.44) and 5 (CEM II/B-V; w/c = 0.40) −50 mm cover. From these results we can conclude that none of the panels faces investigated is expected to last 100 years under XD3 exposure conditions. However, if in Panels 1 and 5 (and possibly Panel 6), the minimum cover were 65 mm, the expected service life would likely exceed 100 years.

In order to compare with the results of other participants, who may have not measured the cover depth, the mean expected service life of each panel face has been computed applying Eq. (11.1), using the kT_{gm} value for the face (Tables 11.12 and 11.13) and the nominal cover depth (c = 30 or 50 mm). The results obtained are presented in Table 11.17, reflecting exclusively the influence of the permeability of the cover, since the cover depth has been assumed as constant.

11.5 Autoclam Permeability Tests

This section was provided by Muhamad Basheer.

11.5.1 Introduction

The Autoclam Permeability tests were carried out on test specimens shown in Fig. 11.12 when they were just over 100 days old. All tests were carried out on the vertical surfaces, as shown in Fig. 11.12. On 31st of July, two air permeability tests were carried out on each test specimen, by following the procedure given in the Autoclam Permeability System manual [4]. In the case of mixes 7 and 8, a third test was carried out. Two Autoclam sorptivity tests were carried out on 1st of August for all mixes, except mixes 7 and 8 (for which an additional test location was considered), again by following the procedure in the manual. The temperature and relative humidity inside a drilled cavity of 10 mm diameter and 10 mm depth at the rear side of the specimens were also noted. These values were used to verify the suitability of test results for assessing the durability of concrete specimens [5].

Fig. 11.12 Testing the vertical surfaces using the Autoclam Permeability System

11.5.2 Test Results

All the data are reported in Tables 11.18 and 11.19 for the Autoclam air permeability indices and the Autoclam sorptivity indices.

As per Basheer and Nolan [5] the Autoclam air permeability indices are not significantly affected by the variations in internal temperature of concrete (a linear decrease of air permeability index by 0.015 Ln(pressure)/min due to an increase in temperature from 10 to 30 °C, or a decrease of API by 0.00075 for each degree Celsius increase in temperature), but significantly influenced by the internal relative humidity (or moisture content). In Table 11.18, it can be seen that the variations in ambient temperature was between 15 and 18 °C and hence the correction to Autoclam Air Permeability Index is between −0.0015 Ln(pressure)/min for 18 °C

Table 11.18 As measured Air Permeability Indices (API)

Wall N°	Mix details	Details of curing	API-1	API-2	API-3	Average API	Temp. ° C	RH %
1	0.44 CEM I	Air cured	0.288	0.119		0.204	18	75
2	0.44 CEM I	Wet cured	0.076	0.118		0.097	18	75
3	0.54 CEM I	Air cured	0.256	0.400		0.328	18	75
4	0.54 CEM I	Wet cured	0.289	0.441		0.365	17	80
5	0.40 CEM II/B-V	Air cured	0.032	0.041		0.036	15	80
6	0.40 CEM II/B-V	Wet cured	0.026	0.039		0.033	15	80
7	0.59 CEM II/B-V	Air cured	0.368	0.494	0.403	0.422	15	80
8	0.59 CEM II/B-V	Wet cured	0.818	0.683	0.951	0.817	15	80

Table 11.19 As measured Sorptivity Indices (SI)

Mix No.	Mix Details	Details of curing	SI-1	SI-2	SI-3	Average SI	Temp °C	RH %
1	0.44 CEM I	Air cured	0.491	0.521		0.506	20	85
2	0.44 CEM I	Wet cured	0.467	0.295		0.381	23	80
3	0.54 CEM I	Air cured	1.519	2.628		2.073	23	80
4	0.54 CEM I	Wet cured	5.336	1.917		3.626	23	80
5	0.40 CEM II/B-V	Air cured	0.880	0.748		0.814	23	80
6	0.40 CEM II/B-V	Wet cured	0.346	0.512		0.429	20	70
7	0.59 CEM II/B-V	Air cured	46.138	7.673	12.382	22.060	27	75
8	0.59 CEM II/B-V	Wet cured	31.439	6.478	4.684	14.200	27	75

to −0.00375 Ln(pressure)/min for 15 °C. The temperature compensated Air Permeability Indices are reported in Table 11.20. It can be seen that there is no significant effect due to these temperature variations compared to the measured values of the air permeability of the concrete specimens.

Table 11.18 shows that the internal RH varied from 75 to 80 % during the air permeability tests. This variation could affect the measured air permeability values, but as per Basheer and Nolan [5] the values of measured Autoclam air permeability index could be used to classify the durability of concrete when the internal RH is

Table 11.20 Temperature compensated air permeability indices

Mix No.	Mix Details	Details of curing	Average Measured API	Temp. °C	Correction	Temp. compensated API
1	0.44 CEMI	Air cured	0.204	18	−0.0015	0.202
2	0.44 CEMI	Wet cured	0.097	18	−0.0015	0.095
3	0.54 CEMI	Air cured	0.328	18	−0.0015	0.326
4	0.54 CEMI	Wet cured	0.365	17	−0.00225	0.363
5	0.40 CEM II/B-V	Air cured	0.036	15	−0.00375	0.032
6	0.40 CEM II/B-V	Wet cured	0.033	15	−0.00375	0.029
7	0.59 CEM II/B-V	Air cured	0.422	15	−0.00375	0.418
8	0.59 CEM II/B-V	Wet cured	0.817	15	−0.00375	0.813

less than 80 %. Therefore, the values in Table 11.18 are not compensated for variations in internal relative humidity of the concrete specimens.

In Table 11.19, the as measured sorptivity indices are reported. The internal temperature and relative humidity values varied substantially during the sorptivity tests and hence it is essential to consider the influence of these factors on sorptivity index before classifying the concretes on the basis of the Autoclam sorptivity values.

According to Basheer and Nolan [5], the correction factor for Autoclam sorptivity indices is obtained by using the following equation:

$$CF_{20} = 1.638 - 0.0388T + 0.00035T^2 \quad (11.2)$$

where CF_{20} is the correction factor to be used to multiply the as measured sorptivity index to obtain an equivalent value at 20 °C and T is the temperature of the concrete in °C. Table 11.21 reports these correction factors and the temperature compensated sorptivity indices.

As per Basheer and Nolan [5] the quality of concrete can be distinguished with Autoclam sorptivity tests if the internal relative humidity is less than 80 %. In Table 11.19, except for mix 1, all other specimens had a value equal or less than 80 %. Therefore, no further correction for internal relative humidity was made before discussing the sorptivity values, bearing in mind that the sorptivity index of mix 1 could be higher than the reported value because this was affected by the slightly higher moisture content at the time of testing.

In order to discuss the results in Tables 11.20 and 11.21, the classification criteria as per the Autoclam manual are reproduced in Table 11.22. Figure 11.13 shows a comparison of the temperature compensated Autoclam air permeability indices between the eight mixes and Fig. 11.14 demonstrates the variation in temperature compensated Autoclam sorptivity indices for the eight mixes. Also

Table 11.21 Temperature compensated Sorptivity Indices (SI)

Mix No.	Mix Details	Details of curing	Average SI	Temp. °C	Correction factor	Temp. compensated SI
1	0.44 CEM I	Air cured	0.506	20	1.002	0.507
2	0.44 CEM I	Wet cured	0.381	23	0.930	0.355
3	0.54 CEM I	Air cured	2.073	23	0.931	1.930
4	0.54 CEM I	Wet cured	3.626	23	0.931	3.375
5	0.40 CEM II/B -V	Air cured	0.814	23	0.931	0.758
6	0.40 CEM II/B-V	Wet cured	0.429	20	1.002	0.430
7	0.59 CEM II/B-V	Air cured	22.060	27	0.846	18.657
8	0.59 CEM II/B-V	Wet cured	14.200	27	0.846	12.007

Table 11.22 Protective quality based on autoclam air permeability index and sorptivity index

Protective quality	Autoclam air permeability index ln (Pressure)/min	Autoclam sorptivity index $m^3 \times 10^{-7}/\sqrt{min}$
Very good	≤0.10	≤1.30
Good	>0.10 ≤ 0.50	>1.30 ≤ 2.60
Poor	>0.50 ≤ 0.90	>2.60 ≤ 3.40
Very poor	>0.90	>3.40

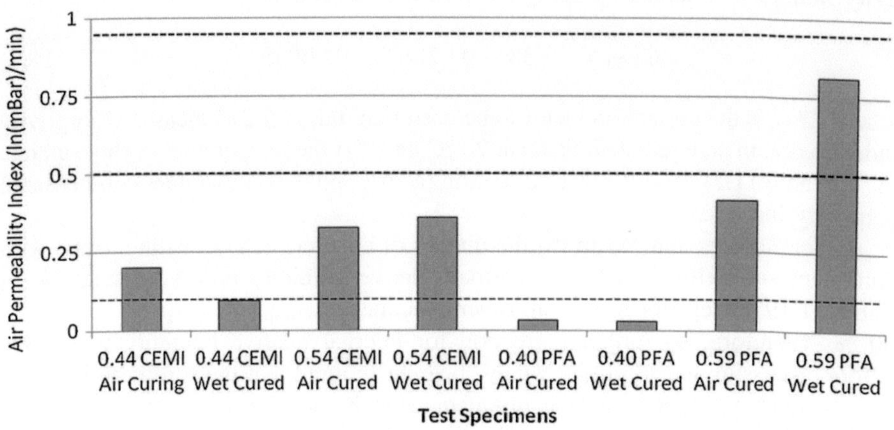

Fig. 11.13 Average Autoclam air permeability indices for all the mixes

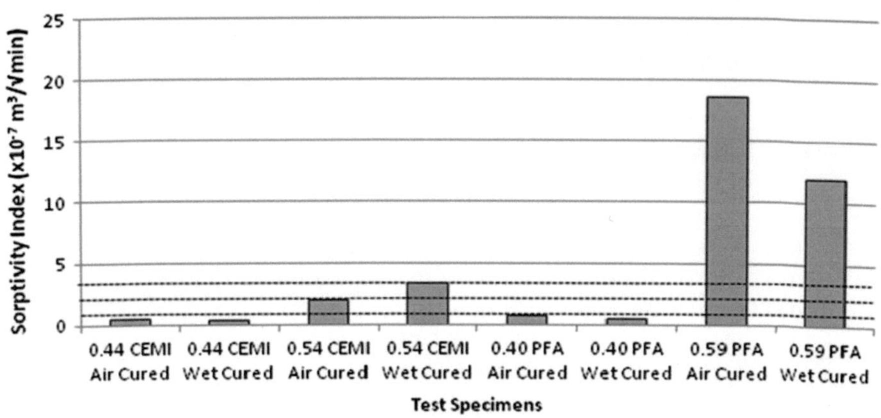

Fig. 11.14 Average sorptivity indices for all the mixes

shown in these figures are the boundaries defined by the values in Table 11.22. A number of observations can be made from these two figures, which are summarised in Table 11.23.

Table 11.23 Classification of concrete mixes based on Autoclam Permeability Indices

Mix No.	Mix Details	Details of curing	Classification based on API	Classification based on SI
1	0.44 CEM I	Air cured	Good	Very good
2	0.44 CEM I	Wet cured	Very good	Very good
3	0.54 CEM I	Air cured	Good	Good
4	0.54 CEM I	Wet cured	Good	Poor
5	0.40 PFA	Air cured	Very good	Very good
6	0.40 PFA	Wet cured	Very good	Very good
7	0.59 PFA	Air cured	Good	Very poor
8	0.59 PFA	Wet cured	Poor	Very poor

Table 11.23 shows that only in four cases both types of tests classified the quality of the concrete mixes the same way, but they were different in other four cases. This was particularly the case for 0.59 w/b PFA concrete mix. As the classification criteria in Table 11.22 are based on resistance to carbonation (XC1 to XC4 exposure classes), freeze-thaw deterioration and salt scaling (XF1 to XF4) in the case of air permeability indices and resistance to chloride induced corrosion (XD1 to XD3 and XS1 to XS3) in the case of sorptivity indices, it is not advisable to compare the two sets of classification. Therefore, each type of tests is discussed further separately.

11.5.3 Temperature Compensated Air Permeability Indices

According to previous tests carried out on 264 mixes involving different types of cements and mineral additions and a wide range of w/c and binder content (see Chap. 8), mixes 5 and 6 and possibly mix 2 could give more than 50 years of service life in XC1 to XC4 exposures and XF1 to XF4 exposures. The results of the classification in Table 11.23 also indicate good durability for mixes 1, 3, 4 and 7 and as such one could expect them to last for at least 25 years. However, mix 8 may deteriorate after a period of possibly 10 years of service life.

According to Table 11.23, mixes 1, 2, 5 and 6 would provide excellent durability in XD1 to XD3 and XS1 to XS3 exposures. This could be considered equivalent of at least 50 years of service life. Mix 3 would give a service life of at least 25 years, mix 4 10 years and mixes 7 and 8 less than 10 years in XD1 to XD3 and XS1 to XS3 exposures. However, it may be noted that a fuller analysis of data of previous mixes analysed (see Chap. 8) relating permeation characteristics to durability would only allow a better interpretation of the measured Autoclam permeation results from the exposure site in Venlo. Therefore, the above interpretation may be viewed with caution.

11.6 Permit Ion Migration Test

This section was provided by Sreejith Nanukuttan.

11.6.1 Introduction

The procedure was applied for the structures tested as part of Application Test programme by RILEM PSC.

The rate of chloride penetration was quantified using surface mount Permit Ion Migration Test, hereafter mentioned as Permit. The output is a migration coefficient determined using steady state migration test principle ($D_{\text{in situ}}$). The test was carried out on all 8 samples at one location (Fig. 11.15). The concrete samples were 14-18 days old when the tests were carried out. The shortest test duration was 1.5 h (samples 3 and 7) and the longest 20 h (samples 1 and 2). Migration coefficients for the 8 samples are presented in Table 11.24. It should be noted that mix design details are indicative and the actual details may vary. Sample numbers were the only reference.

11.6.2 Test Results

The test results obtained from the testing carried out are provided in Table 11.24.

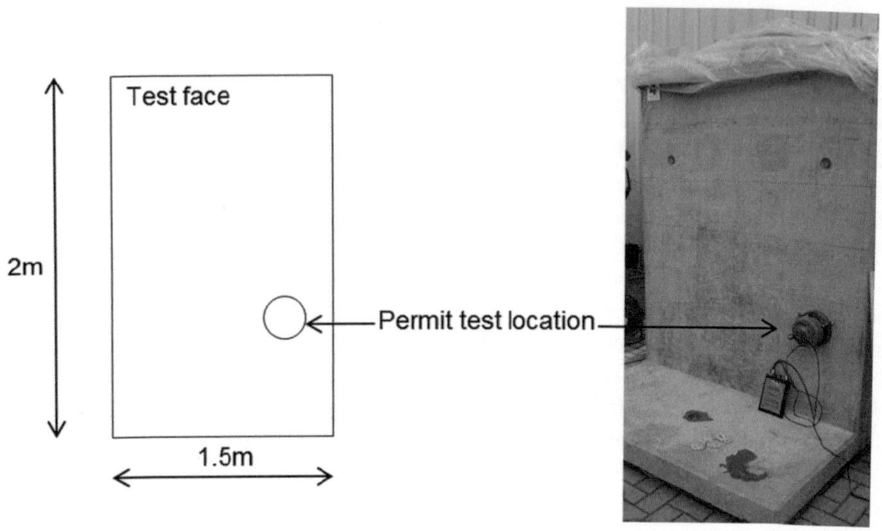

Fig. 11.15 Test location

Table 11.24 Results from Permit test carried out in April 2012

Sample	w/c	CEM	Date of testing	Age at testing (days)	Measured $D_{\text{in situ}}$ (10^{-12} m^2/s)
1	0.44	I	15-Apr-12	18	0.62
2	0.44	I		18	0.14
3	0.54	I		17	0.85
4	0.54	I		17	1.85
5	0.40	II/B/V		16	0.95
6	0.40	II/B/V		16	1.05
7	0.59	II/B/V		14	2.64
8	0.59	II/B/V		14	1.89

As shown in Table 11.24, the test was carried out at a relatively early age of 14-18 days. The microstructure of concrete would not have fully developed at this early age due to insufficient curing of the concrete samples. Therefore, the migration coefficients are likely to be very high. The development of microstructure with curing results in a reduction of diffusivity; this is defined as maturity effect. The rate of reduction of diffusivity or maturity value *m* is understood to be much higher for concretes containing supplementary cementitious materials. *M* values can be determined by quantifying the microstructure development with time. Generally, diffusivity tests are performed at different test ages and a maturity value is determined using the following equation:

$$D_{t\,days} = D_{28\,days} \times \left(\frac{28}{t}\right)^m \qquad (11.3)$$

where $D_{t\ days}$ and $D_{28\ days}$ are the diffusivity at test ages of t and 28 days, respectively, and *m* is the maturity index.

In order to compare the relative performance of all concrete samples, a representative age of 6 months was selected. A maturity index was determined using electrical resistivity measurements performed on controlled concrete samples cured under water. Ideally this should have been either resistivity measured on concrete samples 1-8 in real site exposure or Permit test repeated at selected test intervals such as 28, 56 and 180 days. As the samples are located in The Netherlands, the aforementioned methods were considered unfeasible. *m* values were determined from concrete resistivity and are summarised in Table 11.25. $D_{\text{in situ}}$ values for test ages 106 and 180 days were predicted using Eq. (11.3). These are also presented in Table 11.25. The Permit test was repeated for few samples at a concrete age of 106 days (July 2012) in order to get a second point of reference for comparison and these results are used to study the validity of using a maturity approach for predicting diffusivity. However, there are several limitations in using this approach, mainly the resistivity was determined on samples cured under water, but samples 1-8 were cured differently. Also the second Permit test was performed on different

Table 11.25 Determination of migration coefficient at 6 months and the error between actual and predicted values

Sample	w/b	CEM	Maturity value *m* from resistivity	Predicted *D* value at 180 days ($D_{\text{in situ}}$)	Predicted *D* value at 106 days	Measured *D* value at 106 days ($D_{\text{in situ}}$)	Error %
1	0.44	I	0.15	0.44	0.48	0.34	41
2	0.44	I	0.15	0.10	0.11		
3	0.54	I	0.38	0.34	0.42		
4	0.54	I	0.38	0.75	0.92		
5	0.40	II/B/V	0.56	0.25	0.33		
6	0.40	II/B/V	0.56	0.27	0.36	0.42	13
7	0.59	II/B/V	0.64	0.51	0.72	0.76	5
8	0.59	II/B/V	0.64	0.37	0.52		

test location (left of the test face along the same horizontal axis as previous test). Nevertheless, the magnitude of the *D* values and the error between actual and predicted values will give an indication of the usefulness of this approach.

11.6.3 Predicting the Chloride Transport Through Concrete

Any service life prediction model can be used for predicting the chloride transport. The relationship between the diffusion coefficient (or migration coefficient) used in the model and $D_{\text{in situ}}$ needs to be established first. If a model relies on non-steady state diffusion or migration coefficient, it is also necessary to quantify the binding capacity. In this case, ClinConc service life model [6] was used.

The following assumptions were made:

- D_{nssm} was assumed to be 9.09 $D_{\text{in situ}}$ [7]. If steady state migration models are used D_{ssm} can be taken as 1.81 $D_{\text{in situ}}$.
- Predicted $D_{\text{in situ}}$ for 180 days is taken from Table 11.25.
- Surface chloride concentration for XS3 worst-case scenarios is C_s = 14 g/l.
- Critical chloride concentration for initiation of corrosion is 0.1 % by weight of concrete (assuming 340 kg binder, this is 0.705 % by weight of binder).
- Surface temperature cycle was on average 10 °C (Standard deviation of 5 °C).
- Curing temperature was assumed to be 10 °C; samples exposed to XS3 from 14 days age onwards.

The predicted chloride concentrations obtained from the ClinConc service life model are provided in Figs. 11.16, 11.17, 11.18, 11.19 and 11.20.

As can be observed in Fig. 11.16, concrete used in Sample 1 will not be feasible for an XS3 exposure structure for even 50 years unless the cover is increased dramatically to 70 mm. Figures 11.17 and 11.18 shows that corrosion is only likely

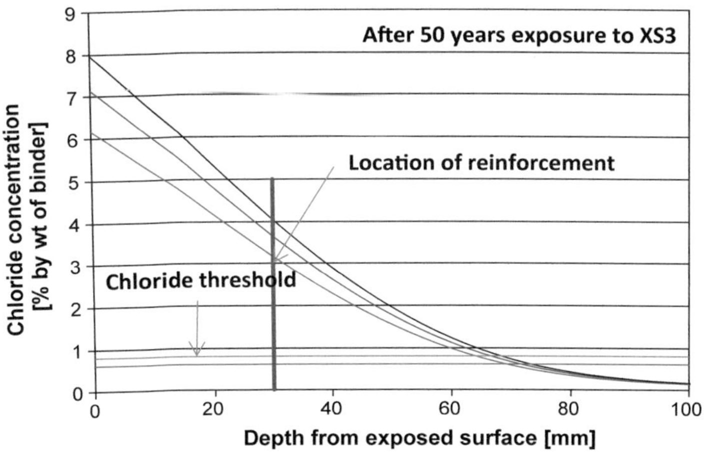

Fig. 11.16 Predicted chloride concentration in Sample 1 after 50 years of exposure to XS3 (Cs 14 g/l), cover 30 mm, $D_{\text{in situ}} = 0.44 \times 10\text{-}12$ m2/s. *Note* the *curve* in the *middle* indicates the average prediction line using absolute values of D whereas the *upper* and *lower curves* indicate the confidence intervals

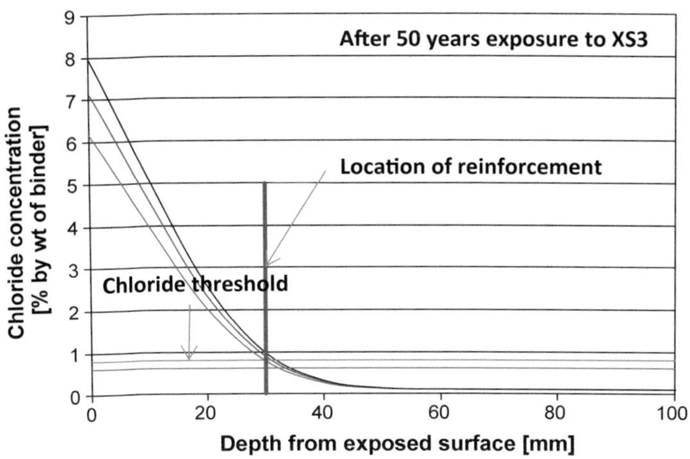

Fig. 11.17 Predicted chloride concentration in Sample 2 after 50 years of exposure to XS3 (C_s 14 g/l), cover 30 mm, $D_{\text{in situ}} = 0.10 \times 10^{-12}$ m^2/s

to begin in Sample 2 by 50 years, so increasing the cover to 40 mm would ensure that such concrete structures will survive for 100 years in XS3 exposure.

Figures 11.19 and 11.20 indicate that corrosion is likely to start in both samples (5 and 7) before 100 years. Increasing the cover to 60 mm or more may result in 100 year service life for Sample 5. This is based on the fact that Sample 5 is made of CEM II/B-V and the assumption that the increased aluminate content may help to

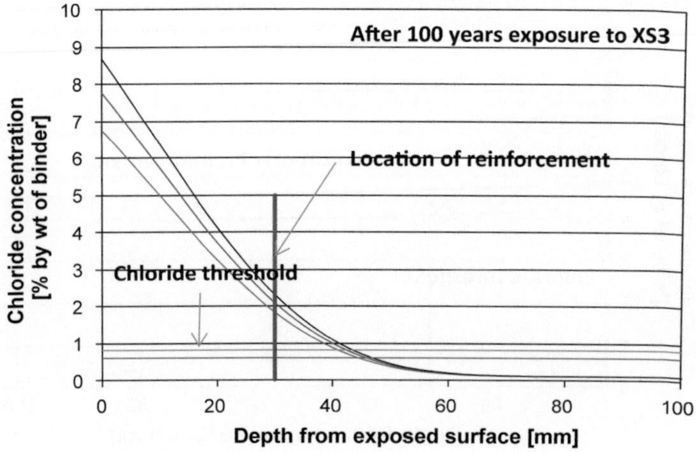

Fig. 11.18 Predicted chloride concentration in Sample 2 after 100 years of exposure to XS3 (C_s 14 g/l), cover 30 mm, $D_{\text{in situ}} = 0.10 \times 10^{-12}$ m^2/s

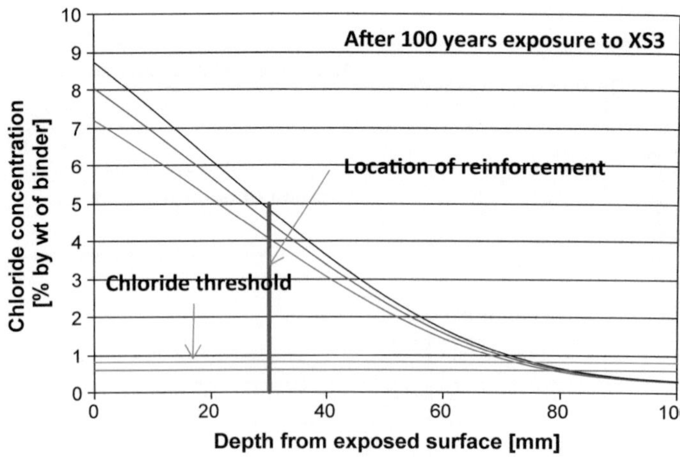

Fig. 11.19 Predicted chloride concentration in Sample 5 after 100 years of exposure to XS3 (C_s 14 g/l), cover 30 mm, $D_{\text{in situ}} = 0.25 \times 10^{-12}$ m^2/s

increase the critical chloride concentration required for corrosion. Increasing the binder content from 340 to 400 kg may also help to achieve the required 100 year service life. Sample 7 however, is not suitable for XS3 exposure.

A summary of the predicted performance of Samples 1–8 in XS3 and XD3 exposure environment for C_s = 14 g/l is provided in Tables 11.26 and 11.27 respectively.

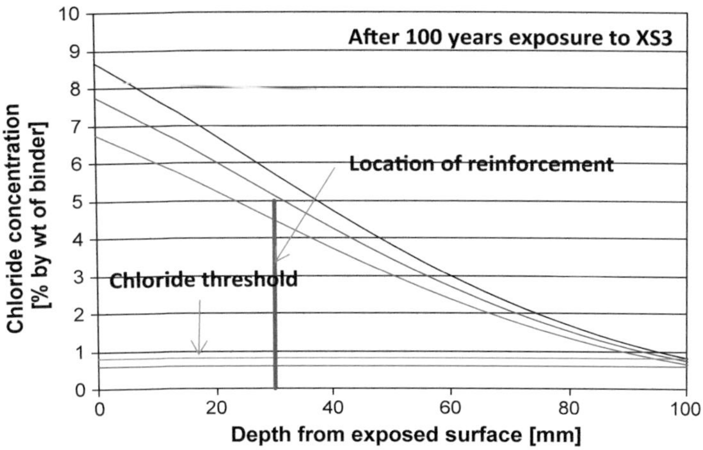

Fig. 11.20 Predicted chloride concentration in Sample 7 after 100 years of exposure to XS3 (C_s 14 g/l), cover 30 mm, $D_{\text{in situ}} = 0.51 \times 10^{-12}$ m²/s

Table 11.26 Summary of the predicted performance of Samples 1–8 in XS3 exposure environment for C_s = 14 g/L

Sample	Predicted D value at 180 days ($D_{\text{in situ}}$)	Actual cover (mm)	Predicted corrosion initiation time	Acceptable for XS3 exposure for 100 year service life	Suggested improvement for 100 year service life
1	0.44	30	<10 years	No	Proper curing, increase binder content and cover ≥ 80 mm
2	0.10	30	~50 years	No	Increase cover to 40 mm
3	0.34	30	~10 years	No	Increase cover to 80 mm
4	0.75	30	<10 years	No	Not suitable
5	0.25	30	~20 years	No	Increase cover to 70 mm
6	0.27	30	~15 years	No	Proper curing, increase binder content and cover ≥ 70 mm
7	0.51	30	≤10 years	No	Not suitable > > 80 mm cover
8	0.37	30	~10 years	No	Increase cover to 80 mm

Table 11.27 Summary of the predicted performance of Samples 1–8 in XD3 exposure environment for C_s = 14 g/L

Sample	Predicted D value at 180 days ($D_{\text{in situ}}$)	Actual cover (mm)	Predicted corrosion initiation time	Acceptable for XD3 exposure for 100 years Service Life	Suggested Improvement for 100 year Service Life
1	0.44	30	~30 years	No	Proper curing, increase cover to ≥ 60 mm
2	0.10	30	≥100 years	Yes	
3	0.34	30	~30 years	No	Proper curing, increase cover to ≥ 55 mm
4	0.75	30	<20 years	No	Proper curing, increase cover to ≥ 70 mm
5	0.25	30	~50 years	No	Increase cover to 40 mm
6	0.27	30	~50 years	No	Proper curing and increase cover to 40 mm
7	0.51	30	~25 years	No	Increase cover to 60 mm
8	0.37	30	~40 years	No	Increase cover to 50 mm or proper curing, increase binder content and cover of 40 mm

11.6.4 Concluding Remarks

These predictions are made on the assumption that critical chloride concentration sufficient to initiate corrosion is 0.705 % by weight of concrete. Therefore revising this value will have a significant effect on the chloride transport and acceptability of the concretes. Performance of samples 1–8 were presented for two exposures in Tables 11.26 and 11.27. A low cover of 30 mm was given in the design brief and this value was used to compare and rate the concrete samples. Suggestions for improving the performance of concrete samples were also given. It is of prime importance that testing is carried out on multiple locations and a holistic approach is used for performance prediction. By combining surface resistivity and Permit based diffusivity test, it is possible to rapidly estimate the overall quality of concrete structure. Such approach will need to be adopted for performance prediction of large-scale structures. An example of such approach can be found in Nanukuttan et al. [8].

11.7 Single Chamber Method

This section is based on information provided by Kei-ichi Imamoto.

11.7.1 Introduction

On the test panels, in situ permeability measurements were performed in April using the Single Chamber Method (SCM [9], compare Sect. 4.2.2.2). In addition, core specimens (150 mm in diameter) were extracted from each panel and tested for accelerated carbonation in a carbonation chamber. Prior to carbonation testing, the cores were at 20 °C and 60 % R.H. for 1 month. The accelerated carbonation exposure started approximately 2 months after the manufacture of the test panels.

11.7.2 Test Results

The permeability measurements and accelerated carbonation results are summarised in Table 11.28. The relationship between the Average Permeability Index (API) and carbonation depth after 26 weeks of exposure (5 % CO_2, 20 °C, 60 % R.H.) is shown in Fig. 11.21. It can be seen that the trend of concrete with CEM I is different from that with CEM II.

11.7.3 Analysis of Permeability Results

The relationship between the Average Permeability Index and expected service life in carbonation environments is discussed in the literature [10, 11]. Based on the

Table 11.28 Average Permeability Index (API) and carbonation depth

Panel	API (kPa)	STDV of API (kPa)	Max. PI[a] (kPa)	Min. PI[a] (kPa)	Carbonation depth[b] (mm)
1	0.28	0.019	0.32	0.23	9.3
2	0.29	0.062	0.44	0.14	7.3
3	0.57	0.065	0.72	0.41	15.9
4	0.59	0.108	0.85	0.33	15.1
5	0.22	0.013	0.25	0.19	9.8
6	0.20	0.016	0.23	0.16	7.8
7	0.54	0.115	0.82	0.27	35.7
8	0.41	0.088	0.62	0.20	30.6

[a]Expected maximum or minimum permeability index with reliance of 95 %
[b]Average carbonation depth after 26 weeks exposure, calculated from 5 individual measurements

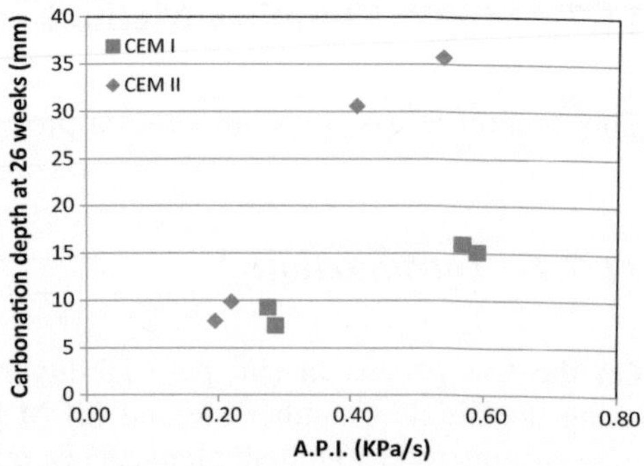

Fig. 11.21 Relationship between A.P.I. and carbonation depth of concrete for concrete with cement CEM I or CEM II/B-V

Table 11.29 Analysis of API results: expected age when carbonation reaches rebars

	Max. age (years)		Min. age (years)		Ranking
Cover	30 mm	50 mm	30 mm	50 mm	
Panel					
1	27	71	23	62	5
2	39	91	19	56	1
3	20	57	16	50	8
4	22	61	15	48	7
5	31	77	26	68	3
6	35	85	27	70	2
7	25	66	15	49	6
8	30	75	16	51	4

analysis, the time when carbonation will reach to the cover depth of reinforcing steel (assumed to be either 30 or 50 mm) and the corresponding durability ranking were calculated, as shown in Table 11.29. It should be noted that the relationship between API and carbonation [10, 11] is based on ordinary Portland cement. The relationship between API and concrete made with blended cements is subject of future research.

11.8 Air Permeability (Packer Test)

This section is based on information provided by Peter Paulini.

11.8.1 Introduction

For the Air Permeability Packer Test, a hole (12 cm deep) was drilled into the panels to fasten the packer and a sealing plate. A peripheral slot (10 mm deep) served as working area for the air pressure. Three to four pressure levels in the range of 3–11 bar were applied on each measuring point. A pressure drop of at least 50 mbar in the vessel had to occur under steady state flow conditions before changing to the next pressure level. The coefficient of permeability was calculated at each pressure level and the mean value is reported.

In addition, a 30 mm diameter drill was used to take drill dust from the outermost 10 mm of cover concrete. The drilling dust was collected in a sealed glass which was then weighed and dried in the laboratory. The samples were dried in two temperature steps (at 50 and 105 °C) until a constant weight was reached. The saturation degree was calculated and used in the interpretation of the in situ permeability results.

11.8.2 Test Results

The measured coefficient of gas-permeability is strongly influenced by the saturation level in the capillary pore system. Therefore, a correction given by Abbas et al. [12] has been applied in order to calculate the coefficient of gas-permeability kT_{dry} for dry concrete.

11.9 Air Permeability (Seal Method)

This section is based on information provided by Okazaki Shinichiro.

11.9.1 Introduction

A team at Ehime University developed the air permeation area clarification method, which is also called the "seal method", and verified the high accuracy of obtained air permeability coefficient k [13]. The relation between the air permeability coefficient k of concrete based on the seal method, moisture content and chloride ion diffusion coefficient is obtained in advance, using reference samples. Then, an arbitrary cover depth of a concrete structure of an arbitrary service life can be evaluated by using the corresponding chloride ion diffusion coefficient calculated from the air permeability coefficient k and moisture content.

To evaluate the chloride diffusion coefficient of a concrete structure indirectly by the Seal Method, the air permeability coefficient k and moisture content are required. The air permeability coefficient k is obtained from permeability measurements and the moisture content is measured by a concrete moisture meter using a conductive probe. Taking into account the variation of air permeability coefficient k and moisture content, the value of k averaged over two points and the value of moisture content averaged over six points are adopted as representative values.

11.9.2 Test Results

The air permeability coefficients were obtained separately for the lower and upper halves of the test panels as presented in Fig. 11.22. When comparing the figures, note the different scale of the y-axis.

11.9.3 Conformity and Service Life Prediction

The in situ air permeability test is proposed as a surveying technique to assess the durability of concrete structures. The moisture content in the concrete must be considered when evaluating the test results (Table 11.30).

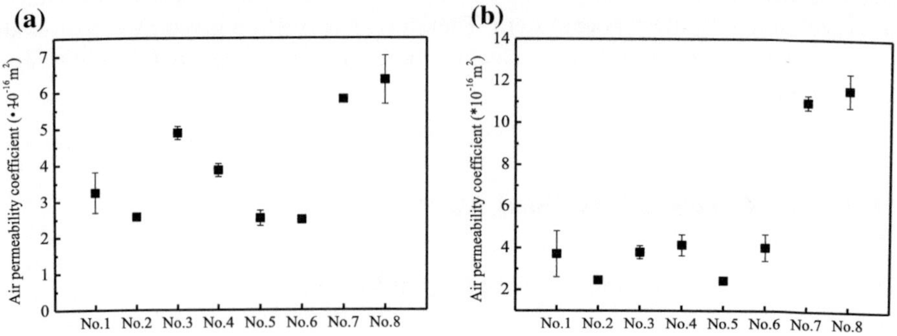

Fig. 11.22 Air permeability coefficient in the **a** top, and **b** bottom halves of the test panels

Table 11.30 Coefficient of air-permeability tested on-site using the Packer Test

Panel	Humidity 105 °C (%-M)	Saturation	In-situ measurements		Correction for saturation	
			kT (m^2)	Rank	kT_{dry} (m^2)	Rank
1	3.94	0.84	4.82 E-28	2	3.89 E-17	2
2	4.48	0.95	8.40 E-18	3	8.54 E-17	6
3	4.40	0.67	1.32 E-17	6	7.82 E-17	5
4	5.30	0.80	1.89 E-17	7	1.44 E-16	7
5	3.64	0.74	3.48 E-18	1	2.35 E-17	1
6	4.23	0.86	2.14 E-17	8	1.81 E-16	8
7	4.03	0.51	1.30 E-17	5	5.88 E-17	4
8	4.36	0.55	1.07 E-17	4	5.19 E-17	3

Figures 11.23 and 11.24 show the calibration air permeability values for moisture content. Based on the calibrated air permeability results, effective and apparent chloride diffusion coefficients are estimated using the method described by Ujike et al. [13], see Table 11.31. The resulting estimated service life is presented in Figs. 11.25 and 11.26.

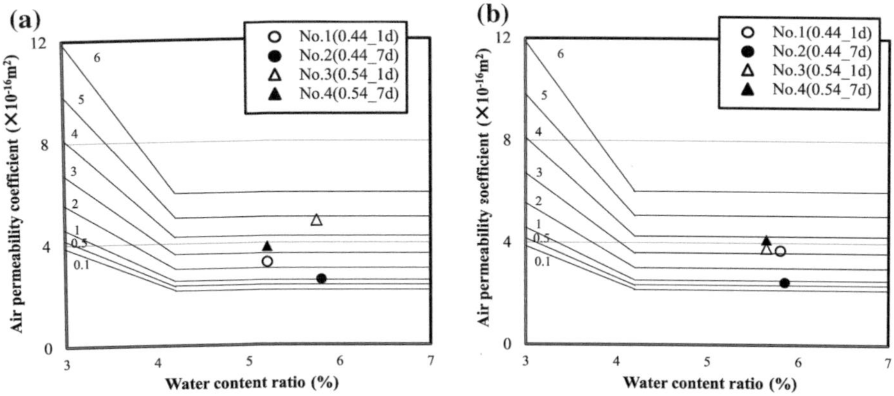

Fig. 11.23 Calibration of air permeability coefficient for concrete water content for the **a** top, and **b** bottom halves of Panels 1–4; the number (0.1–6) and *lines* in the figures correspond to effective chloride diffusion coefficients (cm^2/a)

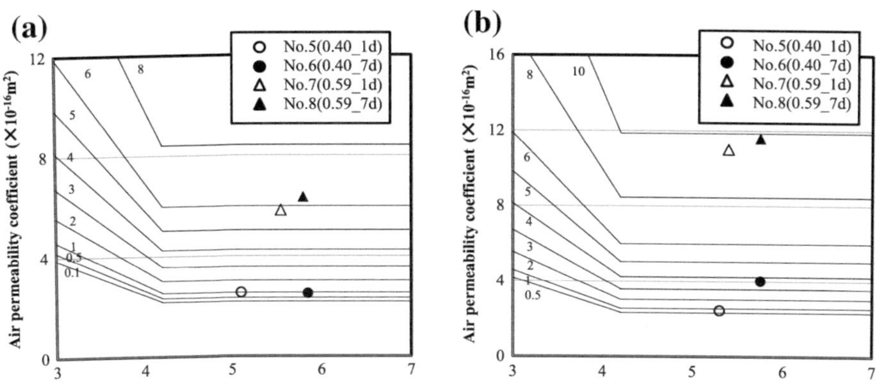

Fig. 11.24 Calibration of air permeability coefficient for concrete water content for the **a** top, and **b** bottom halves of Panels 5–8; the numbers (0.1–10) and *lines* in the figures correspond to effective chloride diffusion coefficients (cm^2/a)

Table 11.31 Predicted chloride diffusion coefficients

Panel	Effective Cl^- diffusion coefficient (cm^2/year)		Apparent Cl^- diffusion coefficient (cm^2/year)	
	Top	Bottom	Top	Bottom
1	2.42	3.19	1.09	1.44
2	1.07	0.82	0.48	0.37
3	4.79	3.32	2.87	1.99
4	3.42	3.83	2.05	2.30
5	0.99	0.80	0.69	0.56
6	0.91	3.67	0.64	2.57
7	5.81	9.47	4.06	6.63
8	6.28	9.81	4.40	6.87

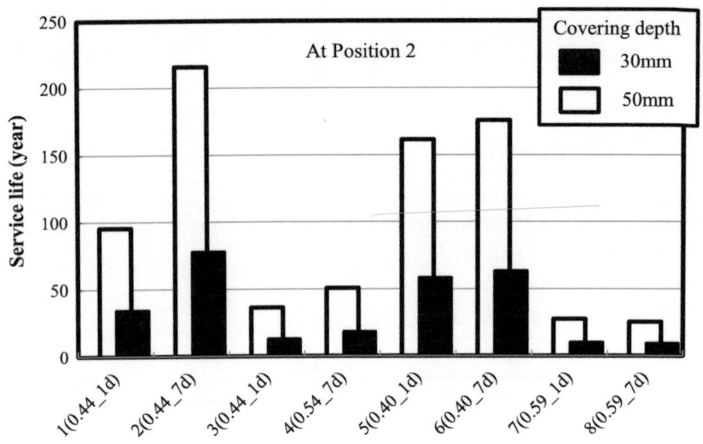

Fig. 11.25 Service life estimate (chloride environment), top halve of the panels

11.10 Water Absorption (SWAT)

This section is based on information provided by K. Hayashi.

11.10.1 Introduction

A non-invasive Surface Water Absorption Test (SWAT) for newly constructed and existing concrete structures was developed by Hayashi and Hosoda [14–18]. The test apparatus can be attached on the surface of concrete structures by a small

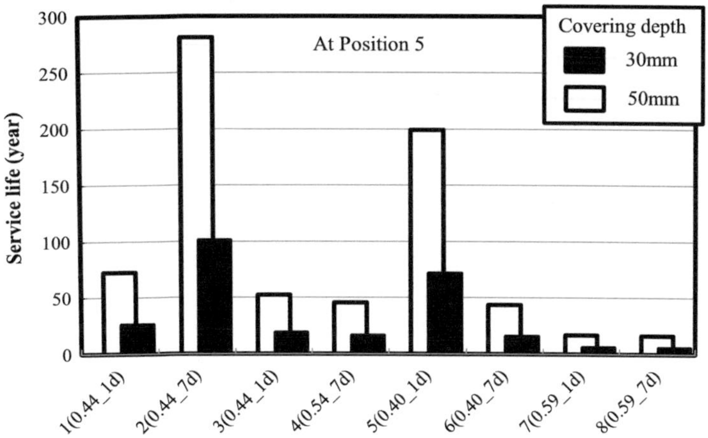

Fig. 11.26 Service life estimate (chloride environment), bottom halve of the panels

vacuum pump. Two or more water cups with rubber gaskets are fixed with a frame and vacuum cells. Water with a small amount of pressure (300 mm head at the beginning) is supplied from the concrete surface and the absorbed water volume is measured automatically by using a pressure sensor and data recording system. From recent research [14–18] it was concluded that SWAT can detect the effects of w/c and curing conditions on covercrete quality of concretes.

In the application tests, four points were tested on each wall with 50 mm cover. The average of four points of water absorption factor at 10 min is shown in Table 11.32. The carbonation rate can be estimated from the test results. However this conversion factor is tentative value. The calculated age at a carbonation progress of 50 mm is presented in Table 11.32.

Table 11.32 Test results

No.	Water absorption factor at 10 min (ml/m^2/s)	Expected age at carbonation progress of 50 mm (years)	Ranking
1	0.185	345	3
2	0.176	368	2
3	0.548	68	7
4	0.477	85	6
5	0.200	313	4
6	0.162	409	1
7	0.974	25	8
8	0.414	108	5

References

1. SIA 262:2013. "Betonbau" (Concrete Construction), Swiss Concrete Code (based on Eurocode 2); 2013.
2. SIA 262/1:2013. "Betonbau—Ergänzende Festlegungen" (Concrete Construction—Supplementary Specifications), Swiss Concrete Standard describing testing methods, specified values and conformity rules. Annex E: "Luftpermeabilität am Bauwerk" (Air-Permeability on Site), 2013 [in German].
3. SN EN 206:2013. "Beton—Teil 1: Festlegung, Eigenschaften, Herstellung und Konformität", Swiss National version of European Standard EN 206; 2013.
4. Autoclam Permeability System Operating Manual. UK:Amphora NDT Limited; 2005.
5. Basheer PAM, Nolan E, Long AE. Near surface moisture gradients and in-situ permeation tests. Constr Build Mater. 2001;**15**(2–3):105-114.
6. Tang L. Service-life prediction based on the rapid migration test and the ClinConc model. In: Proceedings of international RILEM workshop on performance based evaluation and indicators for concrete durability, 19–21 March 2006, Madrid, RILEM PRO 47, pp. 157–164, 2006.
7. Nanukuttan SV, Basheer PAM, Robinson DJ. Determining the chloride diffusivity of concrete in situ using Permit ion migrat*ion test.* In: Proceedings of the concrete platform Conference, April 2007, Belfast, pp. 217–228, 2007.
8. Nanukuttan SV, Green C, Pouryahya P, Magee B, Basheer M, Backus J, Tang L. Combined use of non-destructive testing and factor based service life analysis for assessing the state-of-health of concrete structures. In: International conference on rehabilitation and restoration of structures, 12–16 Feb 2013, Indian Institute of Technology Madras, Chennai, India, 2013.
9. Schönlin K, Hilsdorf H. Evaluation of the effectiveness of curing of concrete structures, ACI SP-100, Concrete Durability, Katharine and Bryant Mather Intern. Conference, **1**, pp. 207–226, Detroit, ACI, 1987.
10. Imamoto K, Schimozawa K, Yamasaki J, Nimura S. Applicability of single-chamber vacuum cell for the evaluation of the air-permeability of concrete walls. In: International conference on concrete Repair, Rehabilitation and Retrofitting (ICCRRR), South Africa, 2005.
11. Imamoto K. Threshold values of air permeability of concrete cover—A case study in Japan. In: International RILEM conference on site assessment of concrete, Masonry and Timber Structures, pp. 169–178, 2008.
12. Abbas A, Cascasses M, Ollivier J-P. Gas permeablitiy of concrete in relation to its degree of saturation. Mater Struct. 1999;32:3–8.
13. Ujike I, Okazaki S, Nakamura T. A study on improvement of in-situ air permeability test for concrete structures. Cem Sci Concrete Technol. 2009;63:189–95.
14. Hayashi K, Hosoda A. Fundamental study on evaluation method of covercrete quality of concrete structures. J Jpn Soc Civil Eng Ser E2 (Mater and Concrete Struct). 2013; 69(1):82–97 (in Japanese).
15. Akmal U, Hosoda A, Hayashi K, Suhara K. Evaluation of covercrete of expansive concrete with external restraint by surface water absorption test. Proc Jpn Concrete Inst. 2012;34 (1):760–5.
16. Akmal U, Hosoda A, Hayashi K. A new surface water absorption test and its applications. In: 14th International Conference of Structural Faults & Repair, Edinburgh, 2012.
17. Akmal U, Hosoda A, Hayashi K, Fujiwara M. Analysis of quality of covercrete subjected to different curing conditions using new surface water absorption test. In: Proceedings of the 13th International Summer Symposium, JSCE, pp. 287–291, 2011.
18. Akmal U, Hosoda A, Hayashi K, Motoshige N. Inspection of cover concrete of an actual structure by surface water absorption test. in: Proceedings of the 9th International Symposium on High Performance Concrete -Design, Verification and Utilization, Rotorua, New Zealand, C9–1, 2011.

Chapter 12
Conclusions

H. Beushausen

Owing to the ageing and often premature deterioration of our infrastructure, the durability of concrete structures has increasingly received attention in the past decades. As a result, a significant amount of test methods for characterizing concrete for its potential durability properties have been developed worldwide. Some of these methods have already been applied in practice for many years and others are still in the research and development phase. The availability of suitable test methods makes it possible to consider performance-based approaches for the design and quality control of concrete. The concept of performance-based design has been around for about a century and the specification and standard testing of compressive strength on cubes or cylinders is probably the best example. However, compressive strength is a clearly defined property and relatively easy to assess, while concrete durability is a very complex issue. The complexity of concrete durability is reflected in the amount and diversity of test methods available, which makes it very difficult to develop a universally accepted approach. The acceptance of the performance based approach for practical applications depends further on the verification that test results can be used as input parameters in service life models that predict the deterioration of concrete structures under real environmental exposure conditions. This requires long-term testing and verification of models against actual performance in the field. Despite these limitations, some of the available performance-based approaches for concrete durability have progressed to a state where they are already applied in practice.

The need for suitable and reliable performance approaches relates to the shortcomings of the traditionally prescriptive design methods for concrete durability. These methods, which are usually based on selecting a limited combination of mix parameters for a range of environmental exposure conditions, largely fail to account for the influence of different binder types as well as mineral and chemical additions on concrete durability. The prescriptive approach therefore often fails to offer a rational basis for the selection of suitable concrete mixes. It also does not allow taking into consideration that concrete durability is largely influenced by construction procedures and workmanship on site. Experience has shown that the

H. Beushausen (✉)
University of Cape Town, Cape Town, South Africa

H. Beushausen and L. Fernandez Luco (eds.), *Performance-Based Specifications and Control of Concrete Durability*,
RILEM State-of-the-Art Reports 18, DOI 10.1007/978-94-017-7309-6_12

durability of concrete structures can only be improved if suitable methods for quality control are in place. Such methods should be based on testing the properties of the in situ structure and linking these to conformity criteria, or ideally, to service life models. The outcome of the quality control testing should enable engineers, contractors and owners to design suitable measures should performance requirements not have been met, in order to avoid costly maintenance and repair measures during the design service life of the structure.

The application of a performance approach for concrete durability shifts a large portion of the responsibility from the design engineer to the concrete supplier and contractor, who have to work as a team to produce a structure that meets the required durability characteristics. The engineer, representing the owner, has to clearly define the requirements and suitable test methods for quality control and conformity assessment. The move from prescriptive towards performance approaches therefore requires all parties involved to gain a good understanding of the factors that influence concrete durability. This in itself can be considered a step into the right direction as specifying authorities engage with the topic and suppliers and contractors are looking at concrete material compositions and construction methods that improve the quality of the structure.

Some of the available performance-based approaches for design and conformity assessment for concrete durability were investigated in the Venlo Application Tests (AT) that were organized and performed by RILEM TC 230-PSC. All of the applied test methods were based on assessing transport properties of the concrete and most of them were relatively simple to use and yielded relevant information for the quality assessment of the test panels. However, the AT also highlighted the challenges associated with the assessment of concrete quality in situ as the prevailing environmental conditions, such as relative humidity of the concrete and the environment, as well as the temperature, may have a significant influence on the test results. As a consequence, many approaches include supplementary test methods that relate to the moisture content in the concrete, which can then be used to calibrate the test results for transport properties.

One of the main outcomes of the Venlo AT was that a range of test methods exist that can be successfully used to rank various concrete members in terms of their durability characteristics. However, there is still a significant need for further work in developing or refining deterioration and service life prediction models for many of the various approaches, in order to develop rational conformity criteria.

For practical application, test methods and related service life models need to be calibrated for locally available materials and the prevailing environmental conditions. The principles for developing fundamental and empirical relationships between performance test results and actual deterioration, mainly relating to carbonation and chloride ingress into concrete, were discussed in detail in this State-of-the-Art Report. The authors hope that this presents a foundation for future research and successful development and implementation of performance-based approaches for concrete durability.

Appendix A
Performance-Based Test Methods and Applications (By Country)

Table A.1 provides and overview on performance test methods used in various countries and reflects the experience of the TC members. An attempt was made to indicate how a particular test is used, for example if a method is still used in research only or if it is already applied in practice for design and conformity assessment. For comprehensive performance specification and design, test methods should be linked to limiting values and, ideally, deterioration models, information on which is also included in the table.

H. Beushausen and L. Fernandez Luco (eds.), *Performance-Based Specifications and Control of Concrete Durability*,
RILEM State-of-the-Art Reports 18, DOI 10.1007/978-94-017-7309-6

Table A.1 Performance-based test method uses and applications (by country)

Country	Test name	Test type	Research	Pre-qualification (Mix design)	Reference test	Specification of limiting values	Durability control of In Situ concrete	Test values linked to deterioration models	Included in national standards	Comments
Australia	BS 1881: Part 122	Water absorption								
	ASTM C 1202	Rapid chloride ions penetration								
	Nord test build 492	Chloride migration resistance								
	Nord test build 443	Chloride diffusion resistance								
	AS 1021.21 (based on ASTM C 642)	Apparent volume of permeable voids								
Austria	Nord test build 492	Chloride permeability	×	×						
	XFA		×							
	Permeability exponent (Paulini-Nasution)	Gas permeability	×							
	Zeiml-Lackner		×							
Canada and USA	ASTM C1202	Chloride permeability	×	×		×			×	Found to be reliable
	ASTM C1556	Chloride diffusion	×	×						Found to be problematic
	ASTM C1585	Water sorptivity	×	×						
	ASTM E96	Water vapour rates	×	×						
	C1202	Five-minute conductivity	×							

(continued)

Table A.1 (continued)

Country	Test name	Test type	Research	Pre-qualification (Mix design)	Reference test	Specification of limiting values	Durability control of In Situ concrete	Test values linked to deterioration models	Included in national standards	Comments
Croatia	HRN EN 12390-8	Water penetration	×	×		×	×		×	Limiting values proposed but not connected to environmental classes
	HRN EN 993-4	Gas permeability	×						×	
	HRN U.M8.300	Capillary absorption	×						×	
	Autoclam	Gas permeability, water permeability and capillary absorption	×							
	ISAT	Initial capillary absorption	×							
	Nord test build 492	Chloride permeability	×	×		×	×			Method recommended if testing is prescribed by the client
	Nord test build 443	Chloride permeability	×							
	ASTM C 1202	Chloride permeability	×							
	Resistivity—end to end (direct method)	Resistivity	×							
	HRN CEN/TR 15177	Freeze-thaw	×	×		×	×		×	Limiting values proposed for XF
	HRN CEN/TS 12390-9	Freeze-thaw-de-icing salts	×	×		×	×		×	Limiting values proposed for XF

(continued)

Table A.1 (continued)

Country	Test name	Test type	Research	Pre-qualification (Mix design)	Reference test	Specification of limiting values	Durability control of In Situ concrete	Test values linked to deterioration models	Included in national standards	Comments
Cyprus	Penetration of chloride ions (ASTM C1543-02 and EN13396:2004)	Chloride ingress	×							
	Rapid chloride permeability (ASTM C1202)	Chloride permeability	×			×				
	Liquid sorptivity test (RILEM TC-116)	Liquid sorptivity	×							
	Liquid permeability (in a closed triaxial cell)	Liquid permeability	×							
France	Torrent	Gas permeability	×				×			
	Gas permeability test (closed to Cembureau)	Gas permeability	×	×				×		
	Water permeability test	Water permeability	×					×		
	PN-EN 12390-8	Water penetration	×	×				×	×	
	NF EN 13369 Annexe G	Water sorptivity		×		×		×	×	
	Mercury intrusion	Pore size distribution	×	×				×		
	Nord test build 492	Chloride migration	×	×				×		
	Nord test build 443	Chloride migration	×	×				×		
	Steady state migration test	Chloride migration	×	×				×		
	Non steady state migration test AASHTO	Chloride migration	×	×				×		
	Diffusion cell test	Chloride diffusion	×					×		
	XP P 18-458	Carbonated depth	×	×				×	×	

(continued)

Table A.1 (continued)

Country	Test name	Test type	Research	Pre-qualification (Mix design)	Reference test	Specification of limiting values	Durability control of In Situ concrete	Test values linked to deterioration models	Included in national standards	Comments
Greece	Penetration of chloride ions (ASTM C1543-02 and EN13396:2004)	Chloride ingress	×							These two methods are mainly used in research. RCP is used for control of in-situ concrete in specific large scale public structures
	Rapid chloride permeability (ASTM C1202)	Chloride permeability	×			×	×			
	EN 12390-8	Water penetration	×							
Japan	Cembureau	Gas permeability	×							
	Figg's invasive method	Gas permeability	×							
	Single chamber (Schönlin and Hilsdorf)	Gas permeability	×							
	Torrent	Gas permeability	×							

(continued)

Table A.1 (continued)

Country	Test name	Test type	Research	Pre-qualification (Mix design)	Reference test	Specification of limiting values	Durability control of In Situ concrete	Test values linked to deterioration models	Included in national standards	Comments
Mexico	Fagerlund method	Water sorptivity	×							Need specimens prepared during concrete fabrication or core extraction if is a bridge in service
	UNE 83988	Electrical resistivity	×	×		×	×	×		Best test of all to estimate durability performance in situ
	ASTM C597	Pulse velocity	×		×		×			Gives homogeneity of concrete.
	ASTM C805	Rebound number	×		×		×			Gives homogeneity of concrete, not as a strength determination
	UNE 112-011	Carbonation	×		×		×	×		Need core extraction
	–	Rebar location by pachometer	×		×	×	×	×		Not a standard, it is only a test procedure to detect location, diameter, and concrete depth. Last value used for durability models
	ASTM C876	Half-cell potential	×		×	×	×			Need to make a drilled hole for electrical connections
	–	Corrosion rate of steel in concrete (Gecor)	×		×	×	×	×		Need to make a drilled hole for electrical connections

(continued)

Table A.1 (continued)

Country	Test name	Test type	Research	Pre-qualification (Mix design)	Reference test	Specification of limiting values	Durability control of In Situ concrete	Test values linked to deterioration models	Included in national standards	Comments
Nether-lands	Nord test build 492	Chloride permeability	×	×				×		XS and XD, limiting values related to cover depth and cement type
	NDT for resistivity	Resistivity	×	Suggested use						Used for production control
New Zealand	AS 1012.21/ASTM C642	Water absorption/porosity								
	ASTM C1585; In-house methods	Water sorptivity						×		
	Nord test build 492	Chloride migration resistance		×	×			×	×	Recommended by national standard
	Nord test build 443	Chloride diffusion resistance						×	×	Recommended by national standard
	ASTM C1260/RILEM AAR-4	Alkali reactivity of aggregate/binder combinations		×						
	AS 4456.9	Abrasion resistance				×			×	
	Wenner (In-house method)	Electrical resistivity						×		
	DIN 1048-5	Water permeability								
	"Taywood In-house method"	Water permeability								

(continued)

Table A.1 (continued)

Country	Test name	Test type	Research	Pre-qualification (Mix design)	Reference test	Specification of limiting values	Durability control of In Situ concrete	Test values linked to deterioration models	Included in national standards	Comments
Northern Ireland/Sweden	Cembureau	Oxygen permeability	×	×		×				
	Figg's invasive Method	Oxygen permeability	×	×						
	Single chamber (Schönlin and Hilsdorf)	Oxygen permeability	×	×						
	Water permeability autoclam	Depth of ingress	×			×				Examples for application presented
	Mercury intrusion	Water porosity	×			×				Examples for application presented
	Tang non steady state	App. chloride diffusion	×			×				Examples for application presented
	Ponding test	Capillary suction								
	RILEM method	Capillary suction								
	ISA test	Capillary suction								
	Figg water absorption	Capillary suction								Found to be probalematic
	Migration	Chloride ingress								
	AASHTO	Chloride ingress								
	Immersion	Chloride ingress								
	Autoclam	Gas permeability								Good performance reported. Soon in Chinese standards
	PERMIT	Chloride ingress								On-site chloride testing

(continued)

Table A.1 (continued)

Country	Test name	Test type	Research	Pre-qualification (Mix design)	Reference test	Specification of limiting values	Durability control of In Situ concrete	Test values linked to deterioration models	Included in national standards	Comments
Poland	PN-EN 12390-8	Water penetration	×	×		×			×	
	Nord test build 492	Chloride permeability	×						–	
	Wenner	Resistivity	×						–	
	Torrent	Gas permeability	×			×	×	×	–	
	Cembureau	Gas permeability	×	×					–	
Portugal	RILEM CPC-18	Carbonation resistance	×	×	×	×		×	×	
	Migration	Chloride ingress	×	×	×	×		×	×	
	AASHTO	Chloride ingress		×						
	Immersion	Chloride ingress	×							
	Diffusion cell	Chloride ingress								
	RILEM CPC11-2	Capillary suction	×				×		×	
	Torrent	Gas permeability	×				×			
	Cembureau	Gas permeability	×	×	×	×		×	×	
	RILEM 25 PEM	Open porosity								
	ISO 7031	Water permeability		×		×				
South Africa	Oxygen permeability index test	Oxygen permeability	×	×		×	×	×	x	
	Chloride conductivity index test	Resistivity/ conductivity	×	×		×	×	×	x	
	Water sorptivity index test	Water sorptivity	×	×		(×)	×			

(continued)

Table A.1 (continued)

Country	Test name	Test type	Research	Pre-qualification (Mix design)	Reference test	Specification of limiting values	Durability control of In Situ concrete	Test values linked to deterioration models	Included in national standards	Comments
Spain	PN-EN 12390-8	Water penetration		×		×			×	Prequalification for exposure classes with chlorides
	UNE 83988-2 wenner	Resistivity	×	×		×	×	×	×	Very good performance
	UNE 83988-1 end to end (direct method)	Resistivity	×	×	×	×		×	×	Reference method
	Torrent	Gas permeability					×			
	UNE 83981 cembureau	Gas permeability	×			×			×	
	UNE 83982 (fagerlund method)	Water sorptivity	×			×			×	
	ASTM D4404 mercury intrusion	Mercury intrusion	×	×						Used as complementary information
	Ponding test (EN 12390-11)	Chloride ingress	×	×	×	×	×	×	×	Reference method for specification
	ASTM G109	Chloride ingress	×						×	Very long
	ASTM C 1202	Migration	×	×		×				
	Nord test build 492	Migration	×							
	Integral test (PNE 83992-2)	Migration	×	×		×		×	×	Very rapid. It can be applied for inhibitors and other protective measures
	Multiregimen (UNE 83987)	Migration	×	×		×		×	×	Good performance reported. Soon in Chinese standards
	Natural exposure (EN 12390-10)	Carbonation	×	×	×	×		×	×	Reference method for specification

(continued)

Table A.1 (continued)

Country	Test name	Test type	Research	Pre-qualification (Mix design)	Reference test	Specification of limiting values	Durability control of In Situ concrete	Test values linked to deterioration models	Included in national standards	Comments
Sweden	Nord test build 492	Chloride permeability	×	×						"This method is on the way to standardization in Europe, US and China"
	NDT for resistivity	Resistivity		Suggested use						
Switzerland	Torrent SIA 262/1	Air permeability	×	×		×	On site		SIA 262/1-E	
	Chloride migration SIA 262/1 (similar to nord test build 492)	Chloride permeability	×	×	×	×			SIA 262/1-B	
	Capillary suction SIA 262/1	Capillary absorption	×	×	×	×			SIA 262/1-A	
	Freeze-thaw-de-icing salts test SIA 262/1	Surface mass loss after 28 cycles	×	×	×	×			SIA 262/1-C	

MIX
Papier aus verantwortungsvollen Quellen
Paper from responsible sources
FSC® C105338

If you have any concerns about our products,
you can contact us on
ProductSafety@springernature.com

In case Publisher is established outside the EU,
the EU authorized representative is:
Springer Nature Customer Service Center GmbH
Europaplatz 3, 69115 Heidelberg, Germany

Printed by Libri Plureos GmbH
in Hamburg, Germany